Modern Mineralogy

Modern Mineralogy

KEITH FRYE

Prentice-Hall, Inc., Englewood Cliffs, New Jersey

Library of Congress Cataloging in Publication Data

Frye, Keith.
Modern mineralogy.

Bibliography:
1. Mineralogy. I. Title.
QE363.2.F79 1974 549 73-4962
ISBN 0-13-595686-2

10 9 8 7 6 5 4 3 2 1

Printed in the United States of America

PRENTICE-HALL INTERNATIONAL, INC., *London*
PRENTICE-HALL OF AUSTRALIA, PTY. LTD., *Sydney*
PRENTICE-HALL OF CANADA, LTD., *Toronto*
PRENTICE-HALL OF INDIA PRIVATE LIMITED, *New Delhi*
PRENTICE-HALL OF JAPAN, INC., *Tokyo*

Contents

2 Crystal Structures 52

3 Crystal Symmetry 93

4 Physical Properties of Minerals 153

5 Radiant Energy and Crystalline Matter 171

Preface

Mineralogy has grown sporadically, each spurt being initiated by a development in instrumentation or by an idea borrowed from another field. (A brief history of mineralogical advancement is to be found at the beginning of Chapter 3.) Thus, geometrical crystallography was a completed subject before the advent of X-ray diffraction, which permitted the elucidation of the ionic structure of crystals. And most of the common crystal structures were determined before chemical bonding was at all understood. But no reason exists, today, for presenting the subject of mineralogy in the order in which it developed historically. My purpose, in *Modern Mineralogy*, is to introduce the reader first to topics with which he already has some cognizance, and then to use them as a logical base for the remainder of the discipline.

My first indebtedness in the preparation of this book is to the many professors who tried, at different times, to teach the various parts of mineralogy to me: Drs. Fred Foreman, S. S. Goldich, J. W. Gruner, R. H. Jahns, Barclay Kamb, E. F. Osborn, Rustum Roy, and O. F. Tuttle. I am especially indebted to Nancy Ferguson Frye, my wife, for editing the manuscript at the several stages of preparation. Mrs. Gail Revels typed the manuscript. This book was written while the author was associate professor of geophysical sciences at Old Dominion University.

I will welcome comments and criticisms from the readers of this book.

KEITH FRYE
Norfolk, Virginia

1

Principles of Crystal Architecture

Mineralogy is the study of minerals. Minerals are crystalline materials. This means that the atoms or ions which comprise them are arranged in an ordered and repeated pattern in three dimensions. In a vapor, a liquid, or a glass, the constituents—ions, atoms, or molecules—are distributed randomly throughout the space that the gas, liquid, or glass occupies. In any mineral—any crystalline substance—the ionic, atomic, or molecular positions are ordered and predictable. This characteristic is important, for the crystal structure, more than any other single factor, determines those physical properties by which each mineral species is described and by which each mineral sample is identified. Physical properties include everything from hardness and cleavage to the way in which the mineral behaves in a beam of visible light or X rays.

A mineral has a definite chemical composition, which can be expressed by a chemical formula. There is no restriction, however, that limits the complexity or precision of this formula. A wide range of substitutions of one ion for another is possible, and is allowable by the definition as stated. But it must be possible, after analysis, to write a precise formula for a given mineral sample, and this precise formula must fit into the general formula for the mineral species.

The subject of mineralogy covers not only mineral specimens found in nature, but indeed all crystalline matter. Most introductory geology textbooks define the word *mineral* in some variation of that form given by Edward Salisbury Dana:

> A mineral is a body produced by the processes of inorganic nature, having usually a definite chemical composition and, if formed under favorable conditions, a certain characteristic atomic structure which is expressed in its crystalline form and other physical properties.*

As happens with many other formal definitions in science, *mineral* is defined, as above, at the beginning of the study of the subject, but the word actually is used in a somewhat different sense most of the time.

One such variance derives from the concept of genesis as it applies in a scientific definition. If two substances cannot be distinguished by measurements on the substances themselves, if pedigree is required to differentiate them, then their distinction cannot be a truly scientific one. Today, many mineralogists employed in academic and industrial research laboratories synthesize solid, crystalline compounds. They then proceed to call these substances minerals and give them mineral-like names regardless of whether or not their synthetic *mineral* has a natural counterpart. Thus, the requirement that a mineral be produced by the processes of nature is artificial and at variance with usage among mineralogists themselves.

Since the principles of chemistry and structure and the techniques of analysis are the same, regardless of genesis, the origin of a substance need not be taken into consideration in a study of the nature of minerals. The minerals discussed in this book, however, are mostly naturally occurring ones, and, in general, the ones commonly found in the rocks of the crust of the earth.

1-1. Bohr Model of the Atom

The constituents of minerals must be studied in order to understand mineral properties, these constituents being atoms or ions. The study of atoms or ions requires some knowledge about the still lesser constituents of which they are made, that is, electrons, protons, and neutrons. The multitude of elementary particles that have been discovered in high-energy physics studies may be ignored, and attention turned to the *Bohr model* of the atom.

The Bohr model of the atom has three components—electrons, protons, and neutrons. An *electron* has a single negative charge. A *proton* has a single positive charge and a mass 1836.1 times that of the electron. A *neutron* has no electronic charge and a mass 1838.6 times that of the electron. Neutrons and protons make up the nucleus of the atom and, taken together, are called *nucleons*. Electrons occur outside the nucleus, and they make up the bulk of the space occupied by an atom. The diameter of atoms is on the order of 1–2 Å, but the diameter of an atomic nucleus is on the order of 0.0001 Å.

Bohr first described atoms by analogy with the solar system, the sun rep-

*W. E. Ford, *Dana's Textbook of Mineralogy*, 4th ed. (New York: John Wiley & Sons, Inc., 1932).

resenting the nucleus and the planets representing the orbiting electrons. But it was noted, from the manner in which atoms absorb or emit energy in discrete quanta, that the energies absorbed or emitted are unique for each element (see Section 5-2). In contrast to the solar system, therefore, an electron cannot take up just any orbit about the nucleus, but only certain orbitals, which are arranged in concentric shells and are separated by discrete energy gaps (Fig. 1-1). This fact is described in quantum theory, wherein each electron about an atom is assigned four principal quantum numbers.

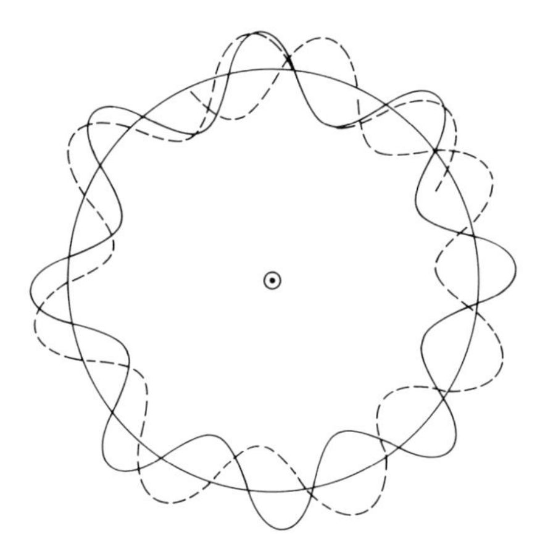

Figure 1-1 Schematic drawing of an electron wave in orbit about an atomic nucleus (solid wave). If the wavelength were slightly different (dashed wave), the wave would be out of phase with itself and destroyed by wave interference. The next highest energy state occurs when exactly that quantum of energy necessary to shorten the wavelength sufficiently to result in one more complete cycle in the electron wave is added to the electron.

Quantum numbers are used to label and characterize orbital electrons. They are always integers and are defined in the following manner: the *principal quantum number n* refers to the fact that the angular momentum of an electron can only take on certain integral values, 1, 2, 3, . . . ; the *azimuthal quantum number l* indicates that certain of the electron orbitals can be described as nonspherical, and has values of 0, 1, 2, . . . , $n-1$; the *magnetic quantum number m* refers to the orientation of the electron orbital in space, and has allowed values of 0, ± 1, ± 2, . . . , $\pm l$; and the *spin quantum number s* has to do with the sense of rotation of the electron itself, and has values of $\pm\frac{1}{2}$. The *Pauli exclusion principle* states that in an atom no two electrons can have the same four quantum numbers.

Another variance from the idea that atoms are solar systems in miniature comes from orbital shapes. The *s* orbitals ($l = 0$) are spherical, but all other

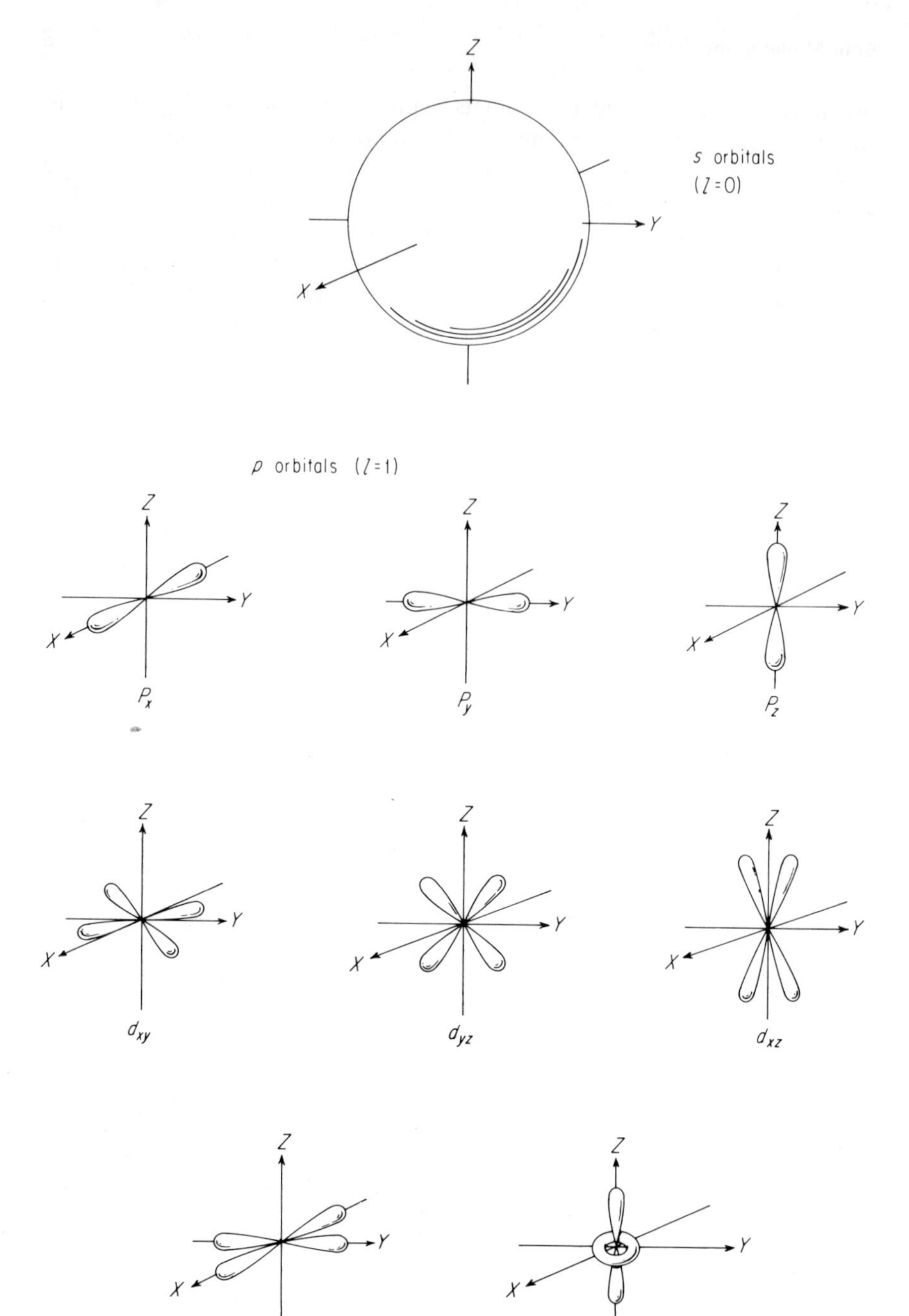

Figure 1-2 Schematic drawings of the shapes of electron orbitals. When nonspherical orbitals are involved in chemical bonding, the bonds tend to be directional.

electron orbitals ($l > 0$) have strong directional components (Fig. 1-2). (Note that the letter s is used to designate both the spin quantum number and the orbitals for which $l = 0$. Which of the two meanings of this letter is intended should be clear from the context.) Orbital shape has a strong influence on the kind of chemical bond formed (Section 1-4).

1-2. Periodic Chart of the Elements

The periodic chart of the elements is a classification of the chemical elements on the basis of their chemical properties. These chemical and, therefore, geochemical properties are in turn based on the configuration of occupied electron orbitals in the atom or ion. The three particles, electrons, protons, and neutrons; the four electron quantum numbers, n, l, m, and s; and the Pauli exclusion principle may be used to illustrate how the periodic chart of the elements, which is based on the electronic structure of the atom, is built up.

A neutral atom is one that has the same number of electrons in orbitals about it as it has protons in its nucleus. The number of neutrons in the nucleus is variable, but is the same as or somewhat greater than the number of protons.

Element number 1, H (hydrogen), the simplest of all atoms, consists of one proton with one electron in orbit about it. The quantum numbers of the electron are $n = 1$, $l = 0$, $m = 0$, and $s = +\frac{1}{2}$.

It is also possible to have an atom made up of one electron, one proton, and one neutron. Since the identity of an element depends on its atomic number, the number of protons in the nucleus, such an atom is chemically identical to H, but has an atomic mass of 2. Changing the number of neutrons produces an *isotope* of an element, and in the case of H, the isotopes are given separate names. The H isotope with one neutron in the nucleus is deuterium (D), and the isotope with two neutrons in the nucleus is tritium (T).

Proceeding to build up the periodic chart of the elements, the addition of one proton and one electron to the tritium isotope of H results in element number 2, He (helium). The quantum numbers of the electrons are $n = 1$, $l = 0$, $m = 0$, $s = +\frac{1}{2}$; and $n = 1$, $l = 0$, $m = 0$, $s = -\frac{1}{2}$. No more electrons are possible with $n = 1$. Therefore, the first electron shell is full with two electrons; no others are permitted by the Pauli exclusion principle.

The next distant shell is the one for which $n = 2$. Element number 3, Li (lithium), has, in addition to the two s electrons in the first shell, a third s electron in the next shell for which the quantum numbers are $n = 2$, $l = 0$, $m = 0$, and $s = +\frac{1}{2}$. Lithium has an atomic number of 3, corresponding to the three protons in the nucleus. The most common isotope of Li has four neutrons in the core, giving it an atomic mass of 7. However, natural Li also contains a small percentage of an isotope with three neutrons and with an atomic mass of 6.

It should be obvious that descriptions of the nuclear and electronic structure will very quickly become unwieldy in their repetition if some simplifiying notation and explanation are not introduced. The periods of the periodic chart of the elements are determined by the principal quantum number n. For some purposes (see Chapter 5) the electron shells and the electrons themselves are indicated by capital letters: K for $n = 1$, L for $n = 2$, M for $n = 3$, N for $n = 4$, O for $n = 5$, P for $n = 6$, and Q for $n = 7$. Within the periods, the letter s refers to electrons with $l = 0$, of which there are two in each period. The letter p refers to electrons with $l = 1$, of which there are six in each period. The letters d and f refer to electrons having l values of 2 and 3, respectively. This notation uses the principal quantum number to indicate the electron shell of the particular electron, and the letter s, p, d, or f to refer to the kind of subshell. Thus, Li has two electrons in the $1s$ shell and one electron in the $2s$ shell, and this configuration is indicated in the following manner: $1s^2 2s^1$. Quantum notation is summarized in Table 1-1.

To return to the progressive construction of the periodic chart of the elements, number 4, Be (beryllium), has a nucleus consisting of four protons

Table 1-1

Electron Distribution and Quantum Notation

Shell	Subshell	Number of orbitals	Maximum electrons	
K ($n = 1$)	$1s$ ($l = 0$)	1	2	2
L ($n = 2$)	$2s$ ($l = 0$)	1	2	8
	$2p$ ($l = 1$)	3	6	
M ($n = 3$)	$3s$ ($l = 0$)	1	2	18
	$3p$ ($l = 1$)	3	6	
	$3d$ ($l = 2$)	5	10	
N ($n = 4$)	$4s$ ($l = 0$)	1	2	32
	$4p$ ($l = 1$)	3	6	
	$4d$ ($l = 2$)	5	10	
	$4f$ ($l = 3$)	7	14	
O ($n = 5$)	$5s$ ($l = 0$)	1	2	50*
	$5p$ ($l = 1$)	3	6	
	$5d$ ($l = 2$)	5	10	
	$5f$ ($l = 3$)	7	14	
P ($n = 6$)	$6s$ ($l = 0$)	1	2	72*
	$6p$ ($l = 1$)	3	6	
	$6d$ ($l = 2$)	5	10	
Q ($n = 7$)	$7s$ ($l = 0$)	1	2	98*

*Maximum possible electrons for each shell. This number is not realized in naturally occurring atoms.

and five neutrons, and has four orbiting electrons. The fourth electron is an s electron with spin opposite to that of the third. The electronic configuration can be written $1s^2 2s^2$.

The fifth element, B (boron), has five nuclear protons and either five or six neutrons. It has five electrons. Since, for s electrons, l and, therefore, m equal 0 and $s = \pm\frac{1}{2}$, no other s electron can exist in the L shell because it would have quantum numbers identical to one of the other two. However, l may take on values up to $n-1$. Since $n = 2$ for the L shell, l may have a value of 1. If $l = 1$, then $m = 0, +1, -1$, and $s = \pm\frac{1}{2}$ for each, giving six possible p electrons. The electron configuration for B can be written $1s^2 2s^2 2p^1$ (Table 1-2).

Elements 6 through 10—C (carbon), N (nitrogen), O (oxygen), F (fluorine), and Ne (neon)—each are formed by adding successively one proton to the nucleus and one p electron to the L shell. The element Ne completes the L shell with an electronic configuration $1s^2 2s^2 2p^6$ (Table 1-2). To continue to build the elements periodically by adding protons to the atomic nucleus, a new electron shell must be started to accommodate the additional electrons. The new shell, M, has the principal quantum number $n = 3$. The first two electrons, those for element 11, Na (sodium), and for element 12, Mg (magnesium), are $3s$ electrons with quantum numbers $n = 3$, $l = 0$, $m = 0$, and $s = \pm\frac{1}{2}$. Elements 13 through 18—Al (aluminum), Si (silicon), P (phosphorous), S (sulfur), Cl (chlorine), and Ar (argon)—are formed by adding six $3p$ electrons with $l = 1$; $m = 0, +1$, or -1; and with $s = \pm\frac{1}{2}$. Argon has an electronic configuration of $1s^2 2s^2 2p^6 3s^2 3p^6$.

This does not complete the M shell, since $3d$ electrons with $l = 2$ are possible. However, these orbitals are not filled next. The relative energies of the orbitals change with atomic number (Fig. 1-3).

With the exception of the d electrons, all possible M shell quantum numbers have been used up in the electronic configuration of Ar. The fourth period in the periodic chart has atoms with electrons in the N shell ($n = 4$). Element 19 is K (potassium), and the single electron beyond the Ar core ($1s^2\ 2s^2\ 2p^6\ 3s^2\ 3p^6$) is the $4s^1$ with quantum numbers $n = 4$, $l = 0$, $m = 0$, and $s = \frac{1}{2}$. As with the preceding period, the next electron that can be added is also an s electron, and the $4s$ shell is completed with Ca (calcium), element 20.

At this point in the development of the periodic chart, the order of addition of electrons to the atom changes. Before the $4p$ electron orbitals are filled, electrons begin filling the still vacant $3d$ orbitals. Scandium (Sc), element 21, is the first element in the first transition series. Although the $3d$ orbitals are filled after the $4s$, the $3d$ electrons actually end up at a lower energy level than the $4s$ electrons (Fig. 1-3). As stated earlier, the d electrons are those for which $l = 2$. The $3d$ electrons may have values of $m = 0, +1, -1, +2$, or -2, and $s = \pm\frac{1}{2}$. There are, therefore, 10 $3d$ electrons, and 10

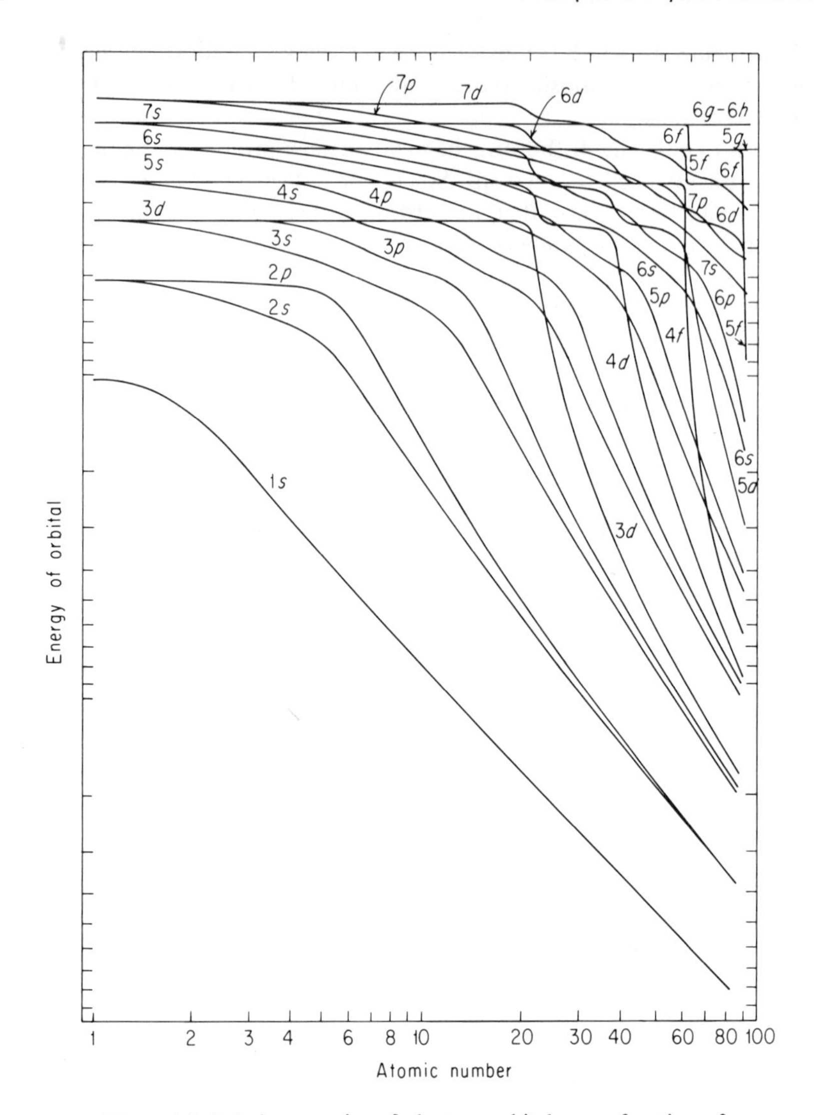

Figure 1-3 Relative energies of electron orbitals as a function of atomic number. The energy levels change their relative positions as they become occupied by electrons. Thus, the $3d$ orbitals, when unoccupied, have a higher energy than the $4s$ orbitals, but a lower one when occupied. (After W. J. Moore, *Physical Chemistry*, 3rd ed. Englewood Cliffs, N.J.: Prentice-Hall, Inc., 1962, p. 508.)

transition elements. The elements are those from numbers 21 to 30: Sc, Ti (titanium), V (vanadium), Cr (chromium), Mn (manganese), Fe (iron), Co (cobalt), Ni (nickel), Cu (copper), and Zn (zinc). The electronic configuration of Zn is $1s^2 2s^2 2p^6 3s^2 3p^6 3d^{10} 4s^2$ (Table 1-2).

The next electrons to be added are the $4p$ electrons, of which there are six. The elements thus created are those with atomic numbers 31 through 36: Ga (gallium), Ge (germanium), As (astatine), Se (selenium), Br (bromine), and Kr (krypton). Krypton like Ne and Ar before it, has a completed s and p electronic configuration, and is a noble gas. The electronic configuration of Kr is given in Table 1-2. The $4d$ ($l = 2$) and $4f$ ($l = 3$) orbitals are as yet unfilled.

Two s electrons are added outside the Kr core, the $5s$ electrons of Rb (rubidium) and Sr (strontium). The $4d$ electrons are then added. And, thus, the next 10 elements are the second transition series ($n = 3$, $l = 2$). These are elements 39 through 48. Their names and symbols may be found in the periodic chart of the elements (Fig. 1-4) and their electronic structure in Table 1-2. The six $5p$ electrons complete the fifth period of the periodic chart with elements 49 through 54. The last, Xe (xenon), has the electronic configuration given in Table 1-2. Possible shells that have not yet been filled are the $4f$ ($l = 3$), the $5d$ ($l = 2$), and the $5f$ ($l = 3$).

Elements 55, Cs (cesium), and 56, Ba (barium), are formed by adding the two $6s$ electrons. As with the fifth period, the sixth is here interrupted with the filling of a $5d$ orbital. Element number 57, La (lanthanum), is technically the first element in the third transition series. However, a second interruption in the sequence occurs after La, and the $4f$ shell is filled next (Fig. 1-3). The quantum numbers of these elements are $n = 4$; $l = 3$; $m = 0, \pm 1, \pm 2$, or ± 3; and $s = \pm \frac{1}{2}$. There are a total of 14. These elements are known as the first rare-earth series, or the lanthanide series. Lanthanum is generally counted among them, owing to its chemical similarity to the elements of the series.

With element 72 the filling of the $5d$ orbitals resumes and is complete with element 80, Hg (mercury). The sixth period is then completed with the filling of the six $6p$ orbitals of the element Rn (radon), a noble gas. Period 7 follows the pattern established in period 6. Elements 87 and 88 have $7s$ electrons. The fourth transition series with the filling of the $6d$ orbitals is started with element 89, Ac (actinium). This is interrupted by the second rare-earth series, the actinide series, and the filling of the 14 $5f$ electron orbitals. The naturally occurring elements stop with element 92, U (uranium), although all the actinide transuranium elements have been synthesized. Table 1-2 summarizes the electronic configuration of the atoms.

From the preceding discussion it should be clear that the periodic chart of the chemical elements (Fig. 1-4) is more than merely a tabulation. Their basic properties depend upon the nature of their outer electrons, the *valence electrons*, which are available for forming chemical bonds. Outer electronic configurations are repeated in a regular manner throughout the periodic chart. Certain elements are chemically similar, and, therefore, occur in similar mineralogical conditions, because their outer electron configurations are the same. From knowing nothing more than the order in which electrons

Table 1-2

Period, element, and atomic number			K	L		M			N				O				P			Q
			1s	2s	2p	3s	3p	3d	4s	4p	4d	4f	5s	5p	5d	5f	6s	6p	6d	7s
1	H	1	1																	
	He	2	2																	
2	Li	3	2	1																
	Be	4	2	2																
	B	5	2	2	1															
	C	6	2	2	2															
	N	7	2	2	3															
	O	8	2	2	4															
	F	9	2	2	5															
	Ne	10	2	2	6															
3	Na	11	2	2	6	1														
	Mg	12	2	2	6	2														
	Al	13	2	2	6	2	1													
	Si	14	2	2	6	2	2													
	P	15	2	2	6	2	3													
	S	16	2	2	6	2	4													
	Cl	17	2	2	6	2	5													
	Ar	18	2	2	6	2	6													
4	K	19	2	2	6	2	6		1											
	Ca	20	2	2	6	2	6		2											
	Sc	22	2	2	6	2	6	1	2											
	Ti	21	2	2	6	2	6	2	2											
	V	23	2	2	6	2	6	3	2											
	Cr	24	2	2	6	2	6	5	1											
	Mn	25	2	2	6	2	6	5	2											
	Fe	26	2	2	6	2	6	6	2											
	Co	27	2	2	6	2	6	7	2											
	Ni	28	2	2	6	2	6	8	2											
	Cu	29	2	2	6	2	6	10	1											
	Zn	30	2	2	6	2	6	10	2											
	Ga	31	2	2	6	2	6	10	2	1										
	Ge	32	2	2	6	2	6	10	2	2										
	As	33	2	2	6	2	6	10	2	3										
	Se	34	2	2	6	2	6	10	2	4										
	Br	35	2	2	6	2	6	10	2	5										
	Kr	36	2	2	6	2	6	10	2	6										
5	Rb	37	2	2	6	2	6	10	2	6			1							
	Sr	38	2	2	6	2	6	10	2	6			2							
	Y	39	2	2	6	2	6	10	2	6	1		2							
	Zr	40	2	2	6	2	6	10	2	6	2		2							
	Nb	41	2	2	6	2	6	10	2	6	4		1							
	Mo	42	2	2	6	2	6	10	2	6	5		1							
	Tc	43	2	2	6	2	6	10	2	6	6		1							
	Ru	44	2	2	6	2	6	10	2	6	7		1							
	Rh	45	2	2	6	2	6	10	2	6	8		1							
	Pd	46	2	2	6	2	6	10	2	6	10									
	Ag	47	2	2	6	2	6	10	2	6	10		1							
	Cd	48	2	2	6	2	6	10	2	6	10		2							
	In	49	2	2	6	2	6	10	2	6	10		2	1						
	Sn	50	2	2	6	2	6	10	2	6	10		2	2						

Table 1-2 (Cont.)

Period, element, and atomic number			K	L		M			N				O				P			Q
			1s	2s	2p	3s	3p	3d	4s	4p	4d	4f	5s	5p	5d	5f	6s	6p	6d	7s
	Sb	51	2	2	6	2	6	10	2	6	10		2	3						
	Te	52	2	2	6	2	6	10	2	6	10		2	4						
	I	53	2	2	6	2	6	10	2	6	10		2	5						
	Xe	54	2	2	6	2	6	10	2	6	10		2	6						
6	Cs	55	2	2	6	2	6	10	2	6	10		2	6			1			
	Ba	56	2	2	6	2	6	10	2	6	10		2	6			2			
	La	57	2	2	6	2	6	10	2	6	10		2	6	1		2			
	Ce	58	2	2	6	2	6	10	2	6	10	2	2	6			2			
	Pr	59	2	2	6	2	6	10	2	6	10	3	2	6			2			
	Nd	60	2	2	6	2	6	10	2	6	10	4	2	6			2			
	Pm	61	2	2	6	2	6	10	2	6	10	5	2	6			2			
	Sm	62	2	2	6	2	6	10	2	6	10	6	2	6			2			
	Eu	63	2	2	6	2	6	10	2	6	10	7	2	6			2			
	Gd	64	2	2	6	2	6	10	2	6	10	7	2	6	1		2			
	Tb	65	2	2	6	2	6	10	2	6	10	9	2	6			2			
	Dy	66	2	2	6	2	6	10	2	6	10	10	2	6			2			
	Ho	67	2	2	6	2	6	10	2	6	10	11	2	6			2			
	Er	68	2	2	6	2	6	10	2	6	10	12	2	6			2			
	Tm	69	2	2	6	2	6	10	2	6	10	13	2	6			2			
	Yb	70	2	2	6	2	6	10	2	6	10	14	2	6			2			
	Lu	71	2	2	6	2	6	10	2	6	10	14	2	6	1		2			
	Hf	72	2	2	6	2	6	10	2	6	10	14	2	6	2		2			
	Ta	73	2	2	6	2	6	10	2	6	10	14	2	6	3		2			
	W	74	2	2	6	2	6	10	2	6	10	14	2	6	4		2			
	Re	75	2	2	6	2	6	10	2	6	10	14	2	6	5		2			
	Os	76	2	2	6	2	6	10	2	6	10	14	2	6	6		2			
	Ir	77	2	2	6	2	6	10	2	6	10	14	2	6	9					
	Pt	78	2	2	6	2	6	10	2	6	10	14	2	6	9		1			
	Au	79	2	2	6	2	6	10	2	6	10	14	2	6	10		1			
	Hg	80	2	2	6	2	6	10	2	6	10	14	2	6	10		2			
	Tl	81	2	2	6	2	6	10	2	6	10	14	2	6	10		2	1		
	Pb	82	2	2	6	2	6	10	2	6	10	14	2	6	10		2	2		
	Bi	83	2	2	6	2	6	10	2	6	10	14	2	6	10		2	3		
	Po	84	2	2	6	2	6	10	2	6	10	14	2	6	10		2	4		
	At	85	2	2	6	2	6	10	2	6	10	14	2	6	10		2	5		
	Rn	86	2	2	6	2	6	10	2	6	10	14	2	6	10		2	6		
7	Fr	87	2	2	6	2	6	10	2	6	10	14	2	6	10		2	6		1
	Ra	88	2	2	6	2	6	10	2	6	10	14	2	6	10		2	6		2
	Ac	89	2	2	6	2	6	10	2	6	10	14	2	6	10		2	6	1	2
	Th	90	2	2	6	2	6	10	2	6	10	14	2	6	10	1	2	6	1	2
	Pa	91	2	2	6	2	6	10	2	6	10	14	2	6	10	2	2	6	1	2
	U	92	2	2	6	2	6	10	2	6	10	14	2	6	10	3	2	6	1	2
	Np	93	2	2	6	2	6	10	2	6	10	14	2	6	10	4	2	6	1	2
	Pu	94	2	2	6	2	6	10	2	6	10	14	2	6	10	5	2	6	1	2
	Am	95	2	2	6	2	6	10	2	6	10	14	2	6	10	7	2	6		2
	Cm	96	2	2	6	2	6	10	2	6	10	14	2	6	10	7	2	6	1	2
	Bk	97	2	2	6	2	6	10	2	6	10	14	2	6	10	8	2	6	1	2
	Cf	98	2	2	6	2	6	10	2	6	10	14	2	6	10	10	2	6		2
	Es	99	2	2	6	2	6	10	2	6	10	14	2	6	10	11	2	6		2
	Fm	100	2	2	6	2	6	10	2	6	10	14	2	6	10	12	2	6		2
	Md	101	2	2	6	2	6	10	2	6	10	14	2	6	10	13	2	6		2
	No	102	2	2	6	2	6	10	2	6	10	14	2	6	10	14	2	6		2
	Lr	103	2	2	6	2	6	10	2	6	10	14	2	6	10	14	2	6	1	2

Figure 1-4 Periodic chart of the elements. Elements on either side of the inert gases form ions having the electronic structure of the adjacent inert gas. (Courtesy of Instruments for Research and Industry, 1968.)

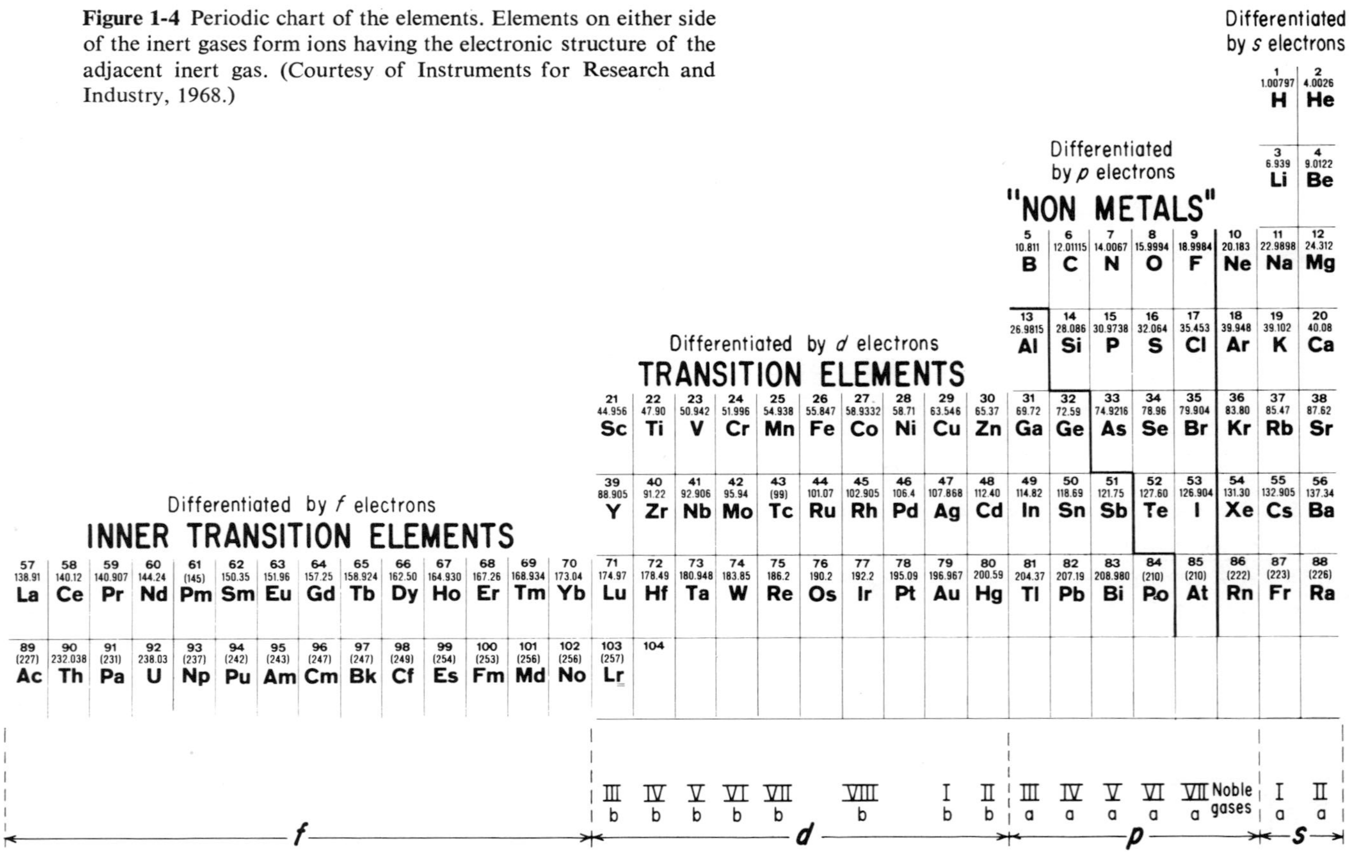

Differentiated by *s* electrons

Differentiated by *p* electrons

"NON METALS"

III a	IV a	V a	VI a	VII a	Noble gases	I a	II a
						1 1.00797 H	2 4.0026 He
						3 6.939 Li	4 9.0122 Be
5 10.811 B	6 12.01115 C	7 14.0067 N	8 15.9994 O	9 18.9984 F	10 20.183 Ne	11 22.9898 Na	12 24.312 Mg
13 26.9815 Al	14 28.086 Si	15 30.9738 P	16 32.064 S	17 35.453 Cl	18 39.948 Ar	19 39.102 K	20 40.08 Ca
31 69.72 Ga	32 72.59 Ge	33 74.9216 As	34 78.96 Se	35 79.904 Br	36 83.80 Kr	37 85.47 Rb	38 87.62 Sr
49 114.82 In	50 118.69 Sn	51 121.75 Sb	52 127.60 Te	53 126.904 I	54 131.30 Xe	55 132.905 Cs	56 137.34 Ba
81 204.37 Tl	82 207.19 Pb	83 208.980 Bi	84 (210) Po	85 (210) At	86 (222) Rn	87 (223) Fr	88 (226) Ra

Differentiated by *d* electrons

TRANSITION ELEMENTS

III b	IV b	V b	VI b	VII b	VIII b			I b	II b
21 44.956 Sc	22 47.90 Ti	23 50.942 V	24 51.996 Cr	25 54.938 Mn	26 55.847 Fe	27 58.9332 Co	28 58.71 Ni	29 63.546 Cu	30 65.37 Zn
39 88.905 Y	40 91.22 Zr	41 92.906 Nb	42 95.94 Mo	43 (99) Tc	44 101.07 Ru	45 102.905 Rh	46 106.4 Pd	47 107.868 Ag	48 112.40 Cd
71 174.97 Lu	72 178.49 Hf	73 180.948 Ta	74 183.85 W	75 186.2 Re	76 190.2 Os	77 192.2 Ir	78 195.09 Pt	79 196.967 Au	80 200.59 Hg
103 (257) Lr	104								

Differentiated by *f* electrons

INNER TRANSITION ELEMENTS

57 138.91 La	58 140.12 Ce	59 140.907 Pr	60 144.24 Nd	61 (145) Pm	62 150.35 Sm	63 151.96 Eu	64 157.25 Gd	65 158.924 Tb	66 162.50 Dy	67 164.930 Ho	68 167.26 Er	69 168.934 Tm	70 173.04 Yb
89 (227) Ac	90 232.038 Th	91 (231) Pa	92 238.03 U	93 (237) Np	94 (242) Pu	95 (243) Am	96 (247) Cm	97 (247) Bk	98 (249) Cf	99 (254) Es	100 (253) Fm	101 (256) Md	102 (256) No

← *f* → ← *d* → ← *p* → ← *s* →

are added to the atoms and the atomic number of the atom, the reader should be able to predict much of the chemical behavior of an element.

1-3. Classification of the Chemical Elements

Although each element is individual and unique, certain elements may be grouped together because they have mineralogic and chemical properties in common. Elements may be classified on the basis of the outermost electrons of their atoms, upon which atomic and ionic properties depend. The first type is comprised of atoms that have a complete octet of *s* and *p* electrons. It is a very stable electronic configuration, so these elements, the noble gases, do not tend either to lose or to gain electrons in chemical reactions. Crystals of fluorides of some of these elements have been synthesized, XeF_4 for instance, but are not found in nature.

Atoms in which the differentiating electrons occupy the shells of highest energy comprise the second type. They have incomplete *s* or *p* orbitals. The ones with unfilled *s* orbitals are metallic, but they grade to nonmetallic elements with an increased number of *p* electrons. This group of elements tends to gain or lose sufficient electrons in chemical reactions to give the ion of the element the preceding or the following noble gas electronic configuration. The ions O^{2-}, Cl^-, K^+, and Ca^{2+} all have the electronic configuration of the noble gas atom Ar ($1s^2 2s^2 2p^6 3s^2 3p^6$), which, as noted above, is a very stable configuration.

The third type of atom includes the elements of the transition series. The electrons which differentiate these elements lie in the second highest energy level, the *d* level. The outermost electrons, the *s* electrons, are the same for all the elements of each series. Because of this similarity and the fact that the energies of the *d* orbitals are not greatly different, many of the atoms have more than one oxidation state. The mineralogically important element Fe has an electronic configuration of $1s^2 2s^2 2p^6 3s^2 3p^6 3d^6 4s^2$. Under the moderately oxidizing conditions that exist within the crust of the earth, Fe loses its two 4*s* electrons and becomes Fe^{2+}. But under the extremely oxidizing conditions of most of the surface of the earth, one of the 3*d* electrons is also lost to form the ion Fe^{3+}. Because, in a crystalline enviroment, *d* orbitals develop energy differences, many of them equal to a particular wavelength of visible light, these elements tend to form colored compounds. The subject of color is taken up in some detail in Section 5-3.

The rare-earth elements of both the actinide and the lanthanide series have electronic configurations that differ in the third highest energy level. They comprise the fourth type of element. The highest *s* orbital is full, and the next lower *p* level is also full. Electrons of the *f* shell, which differentiate the rare-earth elements, are energetically well removed from chemical effects.

For this reason, the elements of each rare-earth series are quite similar chemically. Also, because electrons are being added beneath the outer two shells of the atom, the outermost electrons are subject to progressively greater nuclear attraction with increase in atomic number. For this reason, the lanthanide elements become progressively smaller with increasing atomic number.

1-4. Chemical Bonds

With a few notable exceptions, elements do not commonly occur alone, but rather are compounded with other elements. The kinds of bonds that occur between elements depend upon the electronic configurations of the atoms which make up the compounds. The principles that govern the aggregation of elements into compounds have been the object of a very large part of chemical research.

Compounds are formed when two or more different elements acquire a lower energy by combining, one with the other. The evolution of energy in the form of heat shows that the energy of a system is lowered when spontaneous reactions occur. The reverse of these reactions, the separation of a compound into its constituent elements, generally requires that energy, either in the form of heat or of electricity, be put into the system. One result of the study of the principles of compound formation is a conceptualization of chemical bonds into ideal types. A particular bond can be described in terms of one or more of these types. Four ideal types of bonds are the ionic, the covalent, the metallic, and the van der Waals.

In the ideal *ionic bond* an electron is completely transferred from the cation to the anion, in many cases giving both ions a noble gas electronic configuration. The resulting electrostatic attraction between the positively charged cation and the negatively charged anion holds the ions together in a crystal. Very detailed X-ray diffraction studies have shown that in crystals such as NaCl the electron is completely transferred from the Na^{+} to the Cl^{-} ion (Fig. 1-5).

Ionic bonding is nondirectional in the sense that, because the attraction between ions is essentially electrostatic, each ion tends to surround itself with as many oppositely charged ions as their respective radii permit (Section 1-10). This type of bond is generally formed between elements from opposite sides of the noble elements, and never between similar elements. No molecules are formed by ionic bonding. Minerals with ionic bonds are mechanically strong and hard, have high melting points, and conduct electricity only by ion movement at temperatures near their melting points (Table 1-3).

In a *covalent bond* the two atoms involved attain a more stable electronic configuration, again commonly that of a noble gas, by sharing one or more

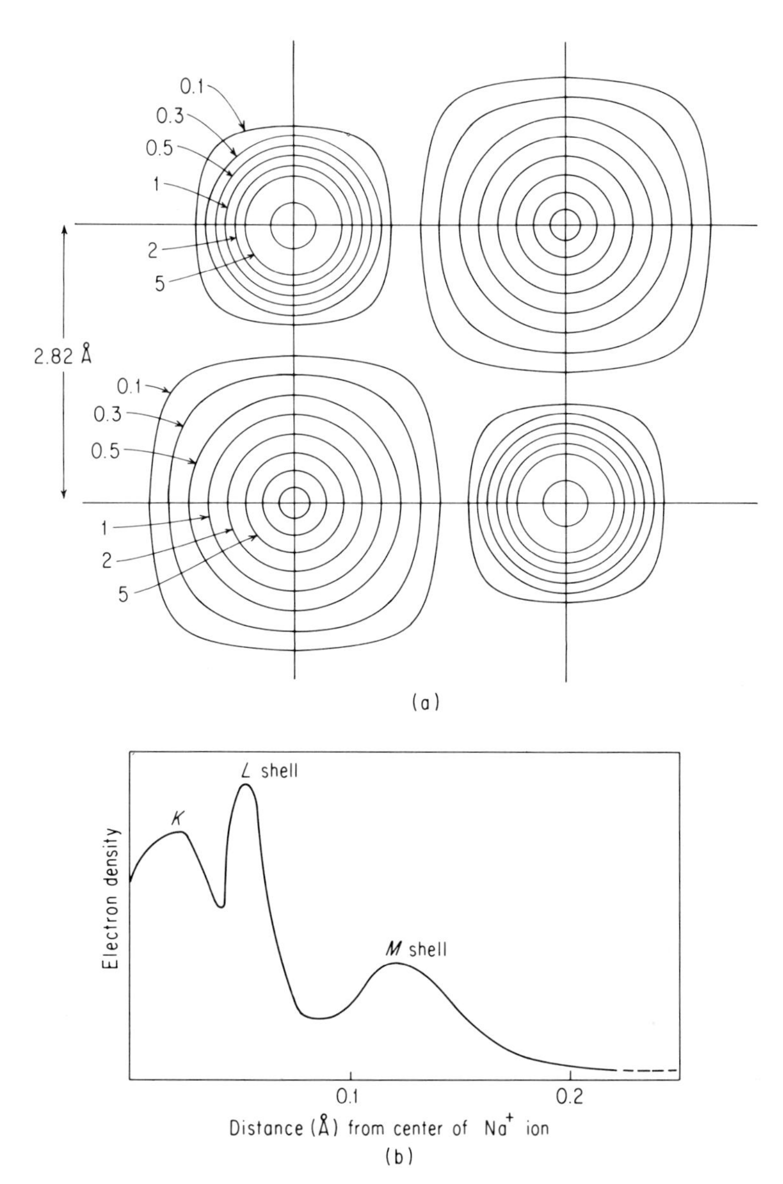

Figure 1-5 Electron densities in ions plotted as a function of distance from the nucleus. (a) Contour map of electron density on the cube face of NaCl. The contours indicate the number of electrons per cubic angstrom of volume. (b) Electron density in a Na^+ ion. Although the ionic radius of Na^+ in octahedral coordination is 1.10 Å, most of the electrons are concentrated at less than 0.2 Å from the nucleus. Atoms and ions typically have a very low electron density in their outer regions. (After H. Witte and E. Wolfel, *Revs. Modern Phys. 30,* No. 1, 1958.)

Table 1-3

Properties Associated with Bond Types

Names	*Nature of bond*			*Minimum electron density between atoms*	*Cation coordination*	*Gives these properties to compounds*
Ionic (heteropolar) (electrovalent)	Electrostatic attraction between charged particles (ions) formed by loss or gain of electrons by atoms (+), (−)	Nondirectional	Never between similar units	No discrete groups or molecules	Usually ⋝ VI, efficient packing	Mechanically strong; hard; high m.p.; colorless; high thermal expansion; ionic conductor
Covalent (homopolar)	Sharing of electron pairs by two atoms: :Cl:Cl: (electron-dot diagram) Zn:S	Directional	Often between similar units	Often causes discrete molecules; e.g., organic solids	$\leq$ IV, poor packing due to directional forces	Fairly strong; fairly hard (except diamond); lower m.p.; *short liquid range;* colorless; low expansion; insulators
Metallic	Positive nuclei in electron "gas." Electrons move in discrete, levels, zones, bands	Nondirectional		No discrete units	> VI, often XII, perfect packing	Variable strength and hardness; long liquid range; opaque; moderate expansion; electronic conductor
Van der Waals	Weak forces caused by induced dipoles resulting from juxtaposition of molecules	Nondirectional	XeF_4	Does not *cause* discrete units	XII	Weak; very soft; very low m.p.; very high expansion; insulators

electrons between them. In this case no ion is formed. Because of the specific location of the shared electrons, covalent bonds are directional, and bonding occurs in some directions and not in others. Covalent bonds may occur between two atoms of the same element, as well as between different kinds of atoms. Molecules are often formed. Mechanically, crystals with covalent bonding are electrical insulators and generally have a short liquid range between melting and boiling.

In ionic bonds the valence electron is transferred completely from the cation to the anion, and in covalent bonds the electron is localized between two atoms. But in metals the valence electrons are free to move from ion to ion as far as the piece of matter extends. *Metals* may be thought of as positive ions bathed in an electron gas. No directional forces are involved and no molecules are formed. Because electrons are free to move from ion to ion throughout a metallic substance, metals are good conductors of electricity and are opaque to light (Section 5-3). Metallically bonded substances have variable strength and hardness and have a long liquid range (Table 1-3).

When the juxtaposition of atoms or molecules results in an induced dipole in an atom or ion, it becomes *polarized.* These weak forces of electrostatic attraction are called *van der Waals bonds.* Although van der Waals bonding does not occur in pure form in minerals, its forces do influence the structure of a mineral.

These four bond types are models, or ideals. Real chemical bonds in minerals may partake of the character of two or more of these ideals. Figure 1-6 shows the different types of ideal chemical bonds as corners of a tetra-

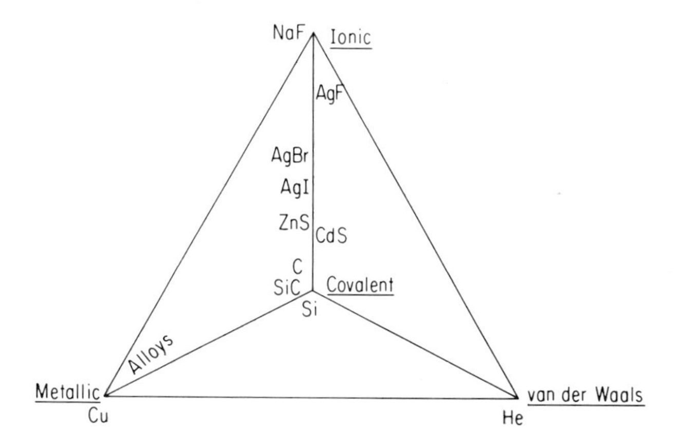

Figure 1-6 Types of chemical bonds with examples. Each apex of the tetrahedron represents an idealized bond type. Crystals having bonds of intermediate character are plotted along the edges of the tetrahedron or in its interior. In minerals having more than one kind of chemical bond, each bond must be plotted separately.

hedron and real chemical bonds as points between. Thus, the bonding in Cu is almost purely metallic in character, the bonding in NaF is ionic, the bonding in C (diamond) is covalent, and the bonding in noble element crystals is van der Waals. Some alloys of metals, however, have a certain amount of covalent character in their metallic bonds. Larger halides of Ag and transition metal sulfides have considerable covalent character in their ionic bonds. The very important Si–O bond in silicates is approximately half ionic and half covalent in character. *Resonance* is the term for this sort of partial bond character. This does not mean that the Si–O bond alternates between being ionic and being covalent; it is partially both at all times.

Most of the minerals to be dealt with have more than one kind of chemical bond. They are heterodesmic as opposed to homodesmic, in which there is but one type of chemical bond.

In the minerals that have more than one kind of chemical bond, the structural pattern is governed by the strongest bonds present, but the physical properties of the minerals are determined by the weakest bonds present (Section 4-4).

1-5. Nature of Ions

To deal with minerals, which are the important constituents of the earth's crust, is to deal with chemical bonds that are predominantly ionic in character. For this reason, ions and ionic bonds will be treated in some detail. Keep in mind, however, that most bonds in minerals are not completely ionic, and that many bonds exist which are best described as covalent or metallic.

Particles that do not have the same number of electrons as protons are called ions: *cations* if they have lost electrons and have acquired a positive charge, and *anions* if they have gained electrons and have acquired a negative charge. The electronic charge, or valence, depends upon the number of electrons removed or added in forming the ion. Because the size of an atom is determined, in part, by the electrostatic attraction between the positive nucleus of the atom and the negative electron cloud about the nucleus, a change in the number of electrons results in a change in the size of the particle. Since cations have lost electrons, they are considerably smaller than their atoms. Anions are correspondingly larger.

The charge on an ion is dependent on its chemical environment and on its electronic structure. At present, however, the concern is with the effect of electronic structure on the degree of ionization that can take place, rather than with chemical environment. *Ionization potentials*, the amounts of energy required to remove electrons from atoms or ions, are given for each element in Table 1-4, along with its outer electron configuration. The lower the ionization potential, the easier the electron is to remove by a chemical reaction. The discussion of ionic character, which follows, should be studied with reference to the data in Table 1-4 and Fig. 1-3.

The noble gas elements are those in which the outer electron shells are complete, having two *s* electrons and six *p* electrons, a total of eight. That this is a very stable electronic configuration is demonstrated by the difficulty in involving the noble gases in chemical reactions.

In atoms, as discussed in Section 1-3, the noble gas structure is one of the most stable electronic configurations. In minerals, the ions F^- and Cl^- always occur with a -1 charge. Neutral halogen atoms have an electronic configuration of two *s* electrons and five *p* electrons. In acquiring an additional electron in a chemical reaction, the halogen elements acquire the very stable noble gas electronic structure of eight outer electrons.

Similarly, Na and the rest of the alkali metals attain a noble gas configuration upon losing one valence electron. (Metallic Na does not exist in nature and is hard to keep in the laboratory.)

The transition metals and the rare-earth elements aside, almost all ions found in minerals have a noble gas electronic structure, which is formed either by the addition or the removal of electrons from the neutral atom. They are ions of the second type.

The neutral transition elements have two incomplete outer electronic shells. They have two *s* electrons, but no *p* electrons, and their *d* shell is incomplete. The first three or four transition elements in each period form ions from which all the *d* and both the *s* electrons have been removed. These elements, like the alkali and alkali-earth elements, have ions with a noble gas electronic configuration. The remainder retain some of their *d* electrons. The fact that the outer shell, or the outer two shells, of the ions remains incomplete has important ramifications in the properties of the chemical compounds into which these elements enter. These electrons interact with the electrons of the surrounding anions, giving the resultant chemical bond nonionic characteristics.

The neutral rare-earth elements have three incomplete outer electronic shells. In naturally occurring minerals, these elements almost uniformly lose their two *s* electrons, plus one 5*f* electron. In the lanthanide rare earths, this outer electron similarity results in the *lanthanide contraction.* With increasing atomic number, the ions of the lanthanide elements become progressively smaller because of the combined effect of the regular increase in nuclear charge and of the constancy of the outer electronic configuration, which is $5s^2 5p^6$. The only two actinide rare-earth elements with half-lives long enough to be mineralogically significant, Th and U, generally acquire a noble gas electronic configuration in their ionic form, although other valence states of U do occur.

In many discussions, ions are treated as though they were hard spheres. This treatment is valid to a first approximation. However, the size of an ion

Table 1-4

IONIZATION POTENTIALS OF THE ELEMENTS[a]

Element	Outer electrons	*Ionization potential (eV)*		
		+1[b]	+2	+3
H	$1s^1$	13.595		
He	$1s^2$	24.580	54.40	
Li	$2s^1$	5.390	75.6193	122.420
Be	$2s^2$	9.320	18.206	153.850
B	$2s^2 2p^1$	8.296	25.149	37.920
C	$2s^2 2p^2$	11.264	24.376	47.864
N	$2s^2 2p^3$	14.54	29.605	47.426
O	$2s^2 2p^4$	13.614	35.146	54.934
F	$2s^2 2p^5$	17.42	34.98	62.646
Ne	$2s^2 2p^6$	21.559	41.07	64
Na	$3s^1$	5.138	47.29	71.65
Mg	$3s^2$	7.644	15.03	80.12
Al	$3s^2 3p^1$	5.984	18.823	28.44
Si	$3s^2 3p^2$	8.149	16.34	33.46
P	$3s^2 3p^3$	11.0	19.65	30.156
S	$3s^2 3p^4$	10.357	23.4	35.0
Cl	$3s^2 3p^5$	13.01	23.80	39.90
Ar	$3s^2 3p^6$	15.755	27.62	40.90
K	$4s^1$	4.339	31.81	46
Ca	$4s^2$	6.111	11.87	51.21
Sc	$3d^1 4s^2$	6.56	12.89	24.75
Ti	$3d^2 4s^2$	6.83	13.63	28.14
V	$3d^3 4s^2$	6.74	14.2	29.7
Cr	$3d^5 4s^1$	6.76	16.6	(31)
Mn	$3d^5 4s^2$	7.432	15.70	(32)
Fe	$3d^6 4s^2$	7.896	16.16	43.43
Co	$3d^7 4s^2$	7.86	17.3	
Ni	$3d^8 4s^2$	7.633	18.2	
Cu	$3d^{10} 4s^1$	7.723	20.34	29.5
Zn	$3d^{10} 4s^2$	9.391	17.89	40.0
Ga	$4s^2 4p^1$	6.00	20.43	30.6
Ge	$4s^2 4p^2$	8.13	15.86	34.07
As	$4s^2 4p^3$	9.84	20.1	28.0
Se	$4s^2 4p^4$	9.750	21.3	33.9
Br	$4s^2 4p^5$	11.84	19.1	25.7
Kr	$4s^2 4p^6$	13.996	26.4	36.8
Rb	$5s^1$	4.176	27.36	(47)
Sr	$5s^2$	5.692	10.98	
Y	$4d^1 5s^2$	6.6	12.3	20.4
Zr	$4d^2 5s^2$	6.95	13.97	24.00
Nb	$4d^4 5s^1$	6.77		24.2
Mo	$4d^5 5s^1$	7.18		
Ru	$4d^7 5s^1$	7.5		
Rh	$4d^8 5s^1$	7.7		

Table 1-4 (Cont.)

Element	*Outer electrons*	*Ionization potential (eV)*		
		+1	+2	+3
Pd	$4d^{10}$	8.33	19.8	
Ag	$4d^{10}5s^1$	7.574	21.4	35.9
Cd	$4d^{10}5s^2$	8.991	16.84	38.0
In	$5s^25p^1$	5.785	18.79	27.9
Sn	$5s^25p^2$	7.332	14.5	30.5
Sb	$5s^25p^3$	8.64	(18)	24.7
Te	$5s^25p^4$	9.01		30.5
I	$5s^25p^5$	10.44	19.4	
Xe	$5s^25p^6$	12.127	(21.1)	32.0
Cs	$6s^1$	3.893	23.4	(35)
Ba	$6s^2$	5.210	9.95	
La	$5d^16s^2$	5.61	11.4	(20.4)
Ce	$4f^26s^2$	(6.91)	14.8	
Pr	$4f^36s^2$	(5.76)		
Nd	$4f^46s^2$	(6.31)		
Sm	$4f^66s^2$	5.6	11.4	
Eu	$4f^76s^2$	5.67	11.4	
Gd	$4f^75d^16s^2$	6.16		
Tb	$4f^96s^2$	(6.74)		
Dy	$4f^{10}6s^2$	(6.82)		
Yb	$4f^{14}6s^2$	6.2		
Lu	$4f^{14}5d^16s^2$	5.0		
Hf	$5d^26s^2$	5.5±	(14.8)	
Ta	$5d^36s^2$	6±		
W	$5d^46s^2$	7.98		
Re	$5d^56s^2$	7.87		
Os	$5d^66s^2$	8.7		
Ir	$5d^9$	9.2		
Pt	$5d^96s^1$	8.96		
Au	$5d^{10}6s^1$	9.223	19.95	
Hg	$5d^{10}6s^2$	10.434	18.65	34.3
Tl	$6s^26p^1$	6.106	20.32	29.7
Pb	$6s^26p^2$	7.415	14.96	(31.9)
Bi	$6s^26p^3$	8±	16.6	25.42
Rn	$6s^26p^6$	10.745		
Ra	$7s^2$	5.277	10.099	
U	$5f^36d^17s^2$	4±		

[a]After W. S. Fyfe.

[b]*Note:* The first ionization potential represents the energy of the process

$$X \longrightarrow X^+ + e,$$

the second the process

$$X^+ \longrightarrow X^{2+} + e,$$

etc.

is influenced by its physical environment, for instance when being squeezed by pressure (see also Section 1-7). The distribution of electrons in a Na^+ ion is shown in Fig. 1-5. In general, the density of electrons is quite low throughout most of the volume of any ion. For this reason the apparent size of an ion varies depending upon its environment.

1-6. Determination of Ionic Properties

The *size* of an ion is its most important property in determining the type of crystalline compound into which it will fit. Its electronic *charge* is of secondary importance. Of tertiary importance is a property called *polarizability* (which may be thought of as deformability). This generality warrants repetition. The ionic properties that determine the kind of crystalline compound into which an element will go are, in order of importance, size, charge, and polarizability.

Although the relationship of ionic charge to electronic structure of the atom has been discussed, and references have been made to sizes of ions, it is necessary now to show how these things about atoms and ions are known. That is, what laboratory experiments have been performed to substantiate such statements?

(In any scientific discussion, only one type of definition is permissible, and that is the operational definition. A distinction that cannot be related to an observation of a property or a reaction, and which cannot be universally verified, is not a valid scientific distinction and is of no worth whatsoever in a scientific discussion.)

Although the discussion of an atomic theory of matter goes back, at least, to Greek philosophers of the fifth century B.C., no solid experimental basis for it was provided until the eighteenth century. In 1785, Lavoisier showed that there was no change in mass in a chemical reaction; that is, the mass of the products equals exactly the mass of the reagents. In 1799, Proust observed that different samples of a substance contain its elementary constituents in the same proportions. Other contemporary experiments showed that every time 1 g of hydrogen was reacted with 8 g of oxygen, 9 g of water resulted. By 1805, Dalton had sufficient data to state the hypothesis that elements consist of atoms, that all the atoms of one element are identical, and that com- pounds result from the combination of a certain number of atoms of o element with a fixed number of atoms of another element. This law of *co stant proportions* described what was observed in many chemical reactio But the idea remained hypothetical until the atom itself could be studi a period of about a century.

In 1895, Röntgen observed a new kind of radiation, which, like light, traveled in straight lines and exposed photographic emulsions. But, unlike light, it was invisible. Because the nature of this radiation was unknown, he called it X radiation. Its nature remained unknown until 1912, when von Laue reasoned that if X rays were electromagnetic radiation and that if crystals were made up of regularly spaced atoms that might act as scattering centers, then crystals should defract X rays. Note the two *if*'s in the previous statement. There was no direct evidence that X rays were indeed a kind of electromagnetic radiation and there was no direct evidence that crystals were composed of atoms. The lag of 17 years between the discovery of X rays and the question asked by von Laue indicates just how speculative atomic theory and radiation theory were at that time. (It is generally true in major scientific advances that the discovery is 90 percent asking the right question and 10 percent answering it.) Von Laue's associates placed a crystal of $CuSO_4 \cdot 5H_2O$ between an X-ray source and a photographic plate. On their second try, diffraction spots appeared on the plate. Those spots, the results of coherent scattering of X rays, were the first direct evidence for the existence of atoms.

For direct evidence of the mass and electronic charge on ions, it is possible to skip the experimentation of half a century and go directly to the results of experiments that can be done with a mass spectrometer. This instrument depends upon the physical fact that an electric current traveling in a magnetic field exerts a force which is at right angles both to the direction of current flow and to the direction of magnetic flux. A moving ion constitutes an electric current flow.

Without going into the details of the design and construction of mass spectrometers, it is possible to see what kind of direct evidence confirms the existence of ions. A vapor consisting of ions can be produced by heating a substance in a vacuum. Since the ions are charged, they can be accelerated by an electric field. If high-velocity ions are passed through a magnetic field, their rectilinear paths are bent in a fashion that is mathematically related to their mass and their charge. By considering the geometry of the instrument, the velocity of the ions, the strength of the magnetic field, and the amount of deflection from a straight path that an ion suffers, it is possible to compute the exact charge-to-mass ratio for a given particle. This information, when combined with the results of the previously mentioned experiments, reveals not only the existence of atoms and ions, but their individual charges and masses as well.

1-7. Ionic Sizes

From the preceding discussion of X-ray diffraction, it would seem that the size of an ion in a crystal would be one of the easier properties to measure accurately. This is true for values that are accurate in a general sense; but when great precision is required, the values for ionic sizes become elusive.

The size of an ion depends upon its electronic structure and upon its environment (see Fig. 1-5). As will be made apparent in subsequent discussion, there are different kinds of crystalline environments that influence the apparent sizes of ions.

Measurements of diffraction angles give interplanar distances in crystalline compounds (Fig. 5-23). For the moment, only consider the fact that in crystals the ions or atoms are arranged in a three-dimensional periodic array. As can be seen from the structure of NaCl (Fig. 2-2), or from any structure model of a crystal, the ions appear as though they were in layers. The distance between these layers is measurable by X-ray diffraction. Although this does not measure the radius of either ion, it does yield the sum of the two radii. However, the same measurement can be made on a crystal of KCl to arrive at the sum of the radii of K^+ and Cl^- ions. The same measurements can be made for the fluorides of Na^+ and K^+ and for other alkali halides of the same structure. If the radius of one of the ions can be measured by some independent means, all the rest can be calculated from it. Such means do exist, but they are beyond the scope of this book.

Alternatively, a reasonable size for one of the ions may be assumed and the others calculated from that, with the reservation that the results must turn out to be compatible. That is, when finished, the calculated ionic radii should always add up to the measured interplanar distance. By making a sufficient number of different measurements on different crystalline compounds with the halite structure, an internally consistent set of ionic radii should be developed.

An ion does not have, however, a rigid outer surface (Fig. 1-5). The distance between ions in a crystal is an exact balance of attractive forces between ions of opposite charge and repulsive forces between electron clouds of individual ions. In the halite structure, each ion is surrounded by six ions with an opposite charge. If, instead, it were surrounded by four or eight ions of the opposite charge, the attractive forces between the ions of opposite charge would be somewhat different, the interionic distances would be affected, and the ionic sizes would not be the same. It is an observed fact that when a cation changes from an environment of six neighbors to an environment of eight neighbors, its effective size increases. Similarly, in changing to an environment of four neighbors, its effective size decreases (Table 1-5).

So far, ions have been visualized as perfect spheres. This is relatively accurate for ions with a noble gas electronic configuration and an electronic charge of 2 or less. However, with highly charged cations or cations with incomplete outer shells, there is always a directional component to the chemical bond (Fig. 1-2). The bond is no longer purely ionic, but has some characteristics of covalency. This, too, will affect the measured sizes of ions. Thus, it should be clear that measuring the size of an ion in one crystalline environment will not always predict its size in another.

The radii given in Table 1-5 are based on the work of Whittaker and Muntus. Other sets of ionic radii, including especially those of V. M. Goldschmidt, Linus Pauling, and L. H. Ahrens, have been published and continue to be used in crystal chemical computations. Each set gives slightly different values for the radii, because, since their measurement is indirect, different kinds of experimental data may be used to calculate them. An important point to remember when using ionic radii is that for any single problem all radii should be taken from the same source.

Certain trends and relations among ionic radii are obvious. Within a single period, ions with the same electronic configuration become progressively smaller with increasing atomic number and electronic charge. The decrease in size, for instance, from Na^{+} (1.10 Å) to S^{6+} (0.37 Å) is quite regular, as should be expected from the preceding description of atomic structure. The number of electrons remains constant while the nuclear charge increases. Therefore, the attraction for the electrons increases also. Anions that have exactly the same electronic structure, on the other hand, show an increase in size with increase in valency. The increase in size from F^{-} (1.25 Å) to O^{2-} (1.32 Å) can be explained by the fact that O^{2-} has one less proton in the nucleus and, therefore, there is less nuclear attraction on the electrons.

Within a single column of the periodic chart, ions with the same kind of electronic configuration and the same charge increase in size with increasing atomic number. The increase in ionic radii from Na^{+} (1.10 Å) to Cs^{+} (1.78 Å) is to be expected, considering the number of electron shells added to the atom with each higher period.

Mention has already been made of an interesting reversal of what might be expected at first consideration, the lanthanide contraction. Because the added electrons in the lanthanide series go into the $4f$ shell, which is in the interior of the ion as defined by the location of the $5s$ and $5p$ electrons, the added nuclear charge exerts progressively greater attraction on the outer electrons, and each succeeding rare-earth ion is smaller than the previous one. An important result is that the transition metal ions of the third transition series are not substantially larger than the corresponding ions of the second transition series in spite of there being an additional electron shell in ions of the third transition series. Thus, there are no Hf minerals, since all the Hf of the earth's crust is found in Zr minerals.

It should also be noted that there are few cations substantially larger than 1 Å in radius, and that the common anions are considerably larger than 1 Å. It follows that, in most minerals, more volume is taken up by the anions than by the cations. Crystal structure may be thought of as a periodic arrangement of the anions, with the cations generally occupying the intervening spaces. Since O^{2-} constitutes over 90 percent of the volume of the crust of the earth, crustal minerals may be regarded as masses of closely packed O^{2-} ions with the smaller cations residing in the interstices.

Table 1-5

IONIC RADII (Å) OF IONS ACCORDING TO COORDINATION NUMBER[a]

Cation	*III*	*IV*	*VI*	*VIII*	*X*	*XII*
Ag^{+}	—	—	1.23	1.38	—	—
Al^{3+}	—	0.47	0.61	—	—	—
As^{5+}	—	0.42	0.58	—	—	—
B^{3+}	0.10	0.20	—	—	—	—
Ba^{2+}	—	—	1.44	1.50	1.60	1.68
Be^{2+}	0.25	0.35	—	—	—	—
C^{4+}	—	—	—	—	—	—
Ca^{2+}	—	—	1.08	1.20	1.36	1.43
Cd^{2+}	—	0.88	1.03	1.15	—	—
Ce^{3+}	—	—	1.09	1.22	—	—
Ce^{4+}	—	—	0.88	1.05	—	—
Cl^{-}	—	1.67	1.72	—	—	—
Co^{2+}	—	0.65	0.83[b]	—	—	—
Co^{3+}	—	—	0.69[b]	—	—	—
Cr^{3+}	—	—	0.70	—	—	—
Cs^{+}	—	—	1.78	1.82	1.89	1.96
Cu^{+}	—	—	—	—	—	—
Cu^{2+}	—	—	0.81	—	—	—
F^{-}	1.22	1.23	1.25	—	—	—
Fe^{2+}	—	0.71[b]	0.86[b]	—	—	—
Fe^{3+}	—	0.57[b]	0.73[b]	—	—	—
Hg^{2+}	—	1.04	1.10	1.22	—	—
I^{-}	—	—	2.13	1.97	—	—
K^{+}	—	—	1.46	1.59	1.67	1.68
La^{3+}	—	—	1.13	1.26	1.36	1.40
Li^{+}	—	0.68	0.82	—	—	—
Mg^{2+}	—	0.66	0.80	0.97	—	—
Mn^{2+}	—	—	0.91[b]	1.01	—	—
Mn^{4+}	—	—	0.62	—	—	—
Mn^{6+}	—	0.35	—	—	—	—
Mo^{4+}	—	—	0.73	—	—	—
Mo^{6+}	—	0.50	0.68	—	—	—
Na^{+}	—	1.07	1.10	1.24	—	—
Ni^{2+}	—	—	0.77	—	—	—
O^{2-}	1.28	1.30	1.32	1.34	—	—
P^{5+}	—	0.25	—	—	—	—
Pb^{2+}	—	—	1.26	1.37	—	1.57
Ra^{2+}	—	—	—	1.56	—	1.72
Rb^{+}	—	—	1.57	1.68	1.74	1.81
S^{6+}	—	0.37	—	—	—	—
S^{2-}	—	1.56	1.72	1.78	—	—
Sb^{3+}	—	0.85	—	—	—	—
Sb^{5+}	—	—	0.69	—	—	—
Sc^{3+}	—	—	0.83	0.95	—	—
Se^{6+}	—	0.37	—	—	—	—
Se^{2-}	—	1.88	—	1.90	—	—

Table 1-5 (Cont.)

Cation	*III*	*IV*	*VI*	*VIII*	*X*	*XII*
Si^{4+}	—	0.34	0.48	—	—	—
Sn^{4+}	—	—	0.77	—	—	—
Sr^{2+}	—	—	1.21	1.33	1.40	1.48
Ta^{5+}	—	—	0.72	0.77	—	—
Th^{4+}	—	—	1.08	1.12	—	—
Ti^{4+}	—	—	0.69	—	—	—
U^{4+}	—	—	—	1.08	—	—
U^{6+}	—	0.56	0.81	—	—	—
V^{3+}	—	—	0.72	—	—	—
V^{5+}	—	0.44	0.62	—	—	—
W^{4+}	—	—	0.73	—	—	—
W^{6+}	—	0.50	0.68	—	—	—
Y^{3+}	—	—	0.98	1.10	—	—
Zn^{2+}	—	0.68	0.83	0.98	—	—
Zr^{4+}	—	—	0.80	0.92	—	—

[a]Based on the data of E. J. W. Whittaker and R. Muntus, "Ionic Radii for Use in Geochemistry," *Geochim. Cosmochim. Acta 34*, 952–953 (1970).
[b]Radius for all *d* electrons unpaired (high-spin state) (Section 5-3).

1-8. Polarizability of Ions

The third factor which determines the way that a particular ion behaves in a crystal structure is its polarizability. Unfortunately, the polarizability of an ion is not so well defined operationally as are its size and charge. If the apparent size or shape of an ion is strongly dependent upon its crystalline environment, it is said to have a high polarizability. If it tends to behave more rigidly as a sphere of constant size in all environments, it is said to have low polarizability. In general, large ions, monovalent ions, and ions that have a non-noble gas electronic structure tend to be polarizable. Small ions with noble gas structures have low polarizability.

Put crudely, any behavior of an ion that cannot be explained by its size and its charge is too often said to result from its polarizability, and the effects of resonance are too often included under this heading.

1-9. Close Packing

The preceding discussion on ionic properties leads to the observation that ions in crystalline structures may be treated, to a first approximation, as small hard spheres. The approximation is accurate enough for those ions with a noble gas electronic configuration, and only slight deviations occur for those ions with incomplete outer electron shells. Since the way in which different ions fit together in minerals depends, in large measure, upon their sizes, the ways in which space can be filled with spheres of different sizes are of next importance.

The simplest case to consider is that of filling space with ions of one kind, that is, all of one size. As is familiar to all pool players, there is one most efficient packing arrangement for racking billiard balls on the table. This is a hexagonal, two-dimensional array, as shown in Fig. 1-7.

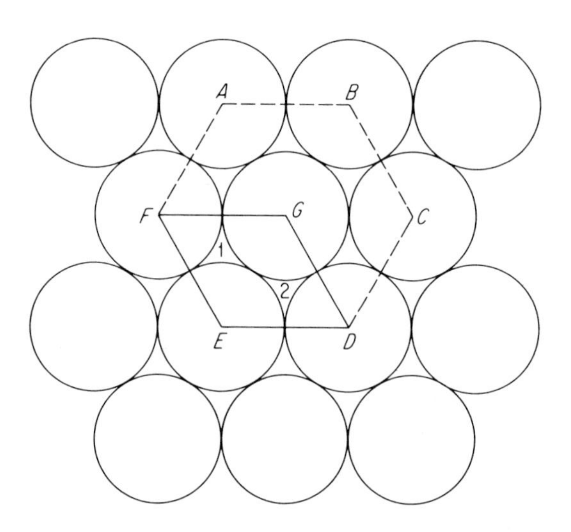

Figure 1-7 Closest-packed layer. Each circle (*G*) is surrounded by six other circles (*ABCDEF*). Each circle is also surrounded by six spaces. Each space (*1*) is surrounded by three circles (*EFG*). The closest-packed layer may be considered as a pattern of parallelograms (*DEFG*), each containing one circle ($\frac{1}{6}D + \frac{1}{3}E + \frac{1}{6}F + \frac{1}{3}G$) and two spaces (*1* and *2*). Thus, there are twice as many spaces as circles in the pattern. If these circles are considered to represent spheres, then an additional sphere may be placed over space *1*, resting on spheres *EFG*, or over space *2*, resting on spheres *DEG*, but not over both simultaneously.

This array should be so familiar, even to those who have never racked up billiard balls, that a detailed discussion of the properties of a closest-packed layer is warranted. (People habitually ignore the familiar to the extent that the results of an analysis of something familiar generally involves surprises.) Each sphere is surrounded by six other spheres. Each space, or void, has three spheres about it. If the void is imagined to be divisible, one third of each void may be assigned to an adjacent sphere. In addition to the six spheres about each sphere, there are six voids about each sphere. Since there are six voids about each sphere and three spheres about each void, there are twice as many void spaces in this closest-packed layer as there are spheres. The radius of the largest sphere that could fill one of those voids, without separating the original spheres from mutual contact, is 0.155 times the radius of the larger spheres.

To fill three-dimensional space, a succession of closest-packed layers stacked one upon the other may be considered. When two stacked layers are examined closely, it is clear that there are two kinds of three-dimensional voids among the spheres (Fig. 1-8). The first type is formed in those places

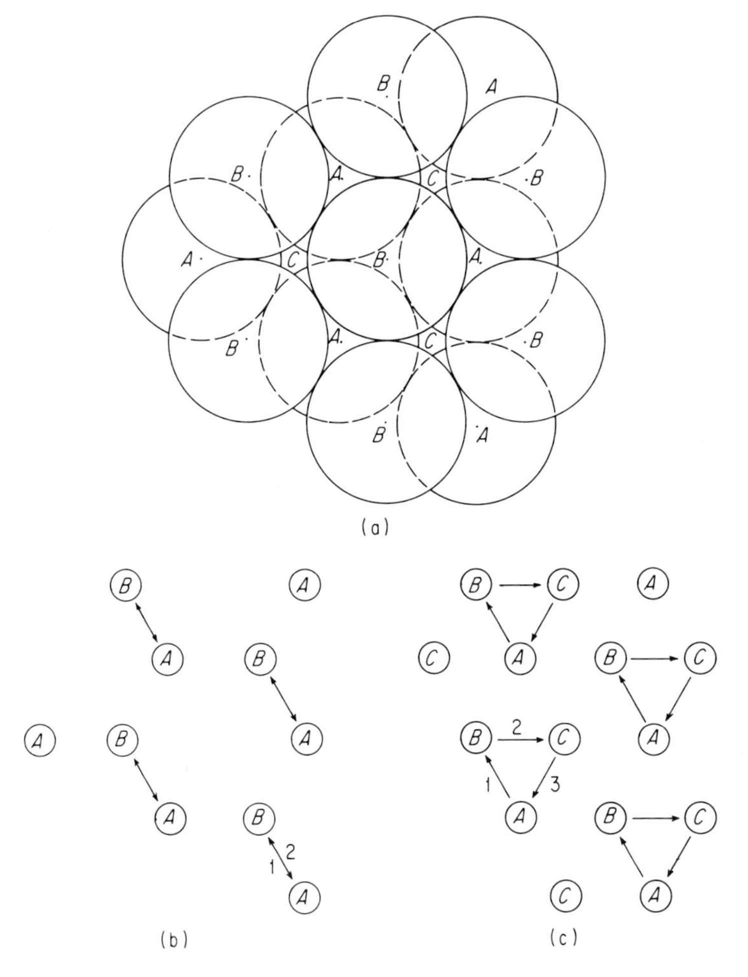

Figure 1-8 Hexagonal (HCP) and cubic (CCP) close packing. (a) Two layers of spheres with closest packing. Layer *B* is on top of layer *A*. A third layer may be added such that its spheres are directly over those in layer *A*, or such that its spheres are at *C*, which is over neither *A* nor *B*. (b) Hexagonal close packing. Layers alternate between position *A* and position *B* with a repeat unit of two layers. (c) Cubic close packing. Layers alternate from position *A* to position *B* to position *C* with a repeat unit of three layers.

where one sphere of the second layer rests on top of a void space in the first layer. In this arrangement, a three-dimensional void is surrounded by four spheres.

By imagining lines connecting the four sphere centers about one of the voids, the outline of a tetrahedron is seen (Fig. 1-9a). Thus, this sort of void is called a *tetrahedral void.* The number of ions about one of these voids is called its *coordination number* (CN). Therefore, a tetrahedral void has a CN of 4. If two complete layers are taken, it is obvious that there are also larger voids between the layers, which are surrounded by three spheres from each layer.

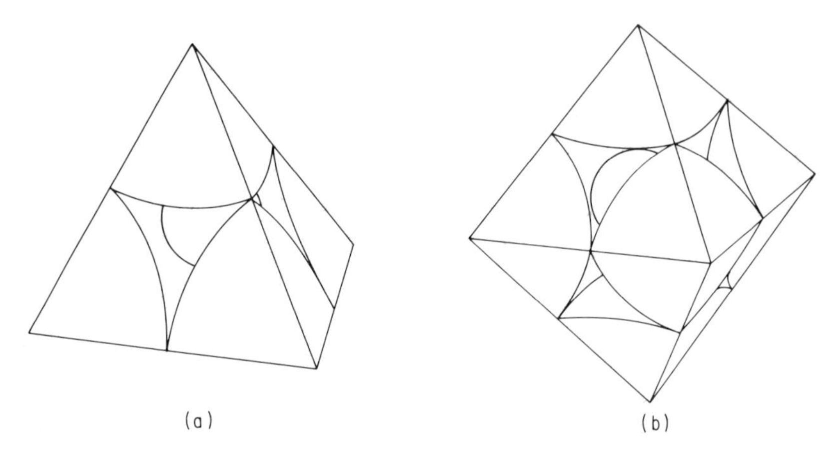

Figure 1-9 Shapes of void spaces. There are two kinds of void spaces in both HCP and CCP. (a) Tetrahedral void formed among four touching spheres. (b) Octahedral void formed among six touching spheres. The number of tetrahedral and octahedral voids is the same in HCP and CCP, but their distribution is different.

The CN of the larger of the two voids between two closest-packed layers is 6. Again, if lines connecting the six sphere centers about the larger void are imagined, the outline of an octahedron can be seen (Fig. 1-9b). (Note carefully, octahedral void means that the void has the geometry of an octahedron, not that there are eight surrounding ions. An octahedron has six corners.)

To continue to build up closest-packed layers into a full three-dimensional array, a third layer is added to the first two. There is now an option (Fig. 1-8). It is possible to place the third layer so that its spheres are directly over the spheres of the first layer. It is also possible to place it in a staggered position so that its spheres are not directly over those of the first layer. If the first case is chosen, a fourth layer is added directly over the second, a fifth over the first and third, and so forth. This structure is built of closest-packed layers that repeat in units of two. In the second case, the fourth layer is over the first, the fifth over the second, the sixth over the third, and so forth, making a repeat unit of three layers.

The two kinds of close packing are distinguished on the basis of their symmetry, a subject dealt with in detail in Chapter 3. If the repeat unit has two members, the symmetry of the three-dimensional close packing is the same as that of the closest-packed layer. It is hexagonal. Each sphere in a layer is surrounded by six others, and a rotation of the pattern by one sixth of a full circle gives an identical pattern. If the repeat unit has three members, its symmetry is more completely described by reference to a cube, as can be seen in Fig. 1-10. The two kinds of close packing are called *hexagonal close packing* (HCP) and *cubic close packing* (CCP), respectively.

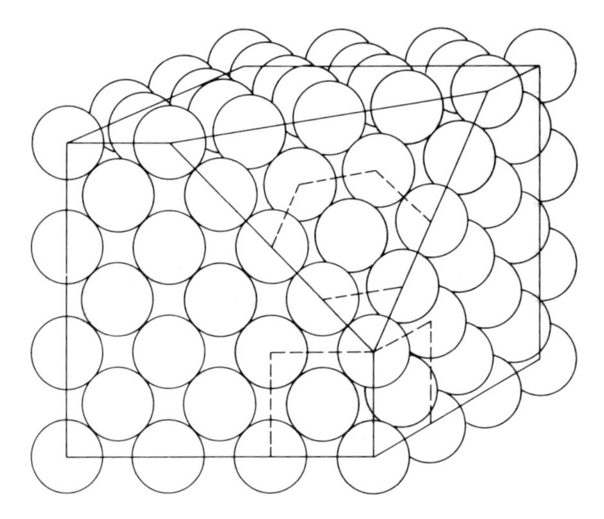

Figure 1-10 One of the closest-packed layers in face-centered cubic (FCC) lattice. FCC and CCP are identical structures. There are four different closest-packed layers in CCP, one of which is shown. The closest-packed layers truncate each of the eight corners of the cube, in the form of an octahedron.

The pattern in which a sphere's neighbors are arranged about it depends upon whether the sphere is in HCP or CCP array. In both arrays, each sphere is surrounded by six octahedral voids, which means that the number of octahedral voids in either array is equal to the number of spheres present. There are eight tetrahedral voids about each sphere, or twice as many such voids as spheres in either array. In HCP, however, the octahedral voids are in pairs aligned at right angles to the planes of closest-packed layers (Fig. 1-11a).

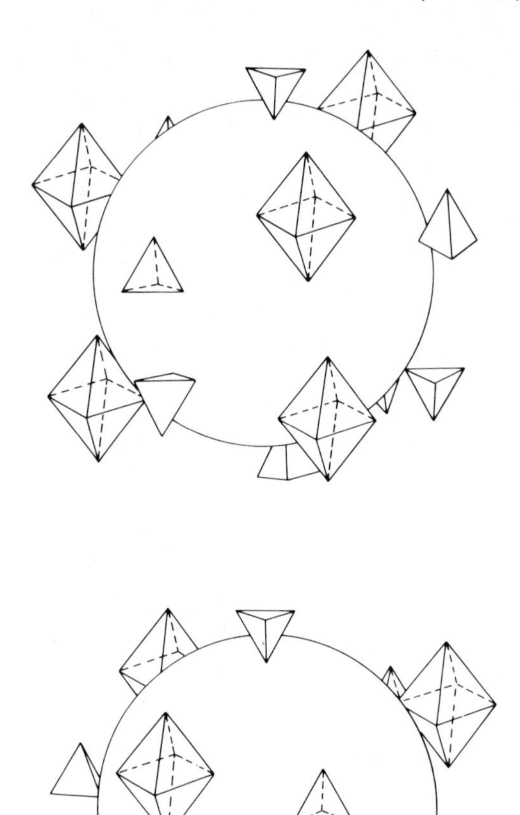

Figure 1-11 Distribution of voids (schematic polyhedra) about a sphere in close packing. (a) Voids in HCP. Note that both octahedral and tetrahedral voids are paired and aligned. (b) Voids in CCP. The voids are distributed such that there is maximum separation between each kind of void.

The tetrahedral voids are in pairs between the octahedral voids, plus an additional one above and one below the sphere. In CCP, no such pairing exists, and octahedral and tetrahedral voids alternate throughout the structure (Fig. 1-11b).

Many pure metals crystallize as HCP or CCP, and many ionic minerals have anions arranged in a rough approximation of HCP or CCP with their cations occupying some of their tetrahedral and octahedral voids. However, not all cations are just the right size to fill these octahedral and tetrahedral voids, and other packing arrays must be considered.

1-10. Radius Ratios

The packing of uniform spheres was analyzed in the preceding section. Since most minerals are made of different kinds of ions, an analysis of the geometry of packing different-sized spheres in space must be undertaken. The number of anions that can surround a cation depends both on the size of the anion and the size of the cation. This relationship is best expressed as a ratio. *Radius ratio* is the numerical result of dividing the radius of the cation by the radius of the anion (r^+/r^-).

Theoretical coordination numbers, based purely on radius ratio and on geometry, are given in Fig. 1-12 and Table 1-6. Since most common minerals are compounds of various cations with oxygen, Table 1-7 compares theoretic-

Table 1-6

COORDINATION NUMBERS (CN) ASSOCIATED WITH RADIUS RATIOS (r^+/r^-)[a]

r^+/r^-	*CN*	*Configuration*	*Example*
0.155	II	Linear	$(HF_2)^-$
0.155–0.225	III	Trigonal planar	CO_3^{2-}
0.225–0.414	IV	Tetrahedral	SiO_2
0.414–0.732	IV	Square planar	$Ni(CN)_4^{2-}$
0.414–0.732	VI	Octahedral	NaCl
0.732–1.000	VIII	Cubic	CsCl

[a]For cases when the cation is larger than the anion, the ratio is reversed.

Table 1-7

RELATIONSHIP BETWEEN OBSERVED AND PREDICTED CATION COORDINATION

Cation	*Radius* (Å)	r^+/r^- (O^{2-})	*Predicted CN*	*Observed CN*
Ca^{2+}	1.08	0.82	VIII	VI
Ca^{2+}	1.20	0.89	VIII	VIII
Na^+	1.10	0.84	VIII	VI
Na^+	1.24	0.93	VIII	VIII
Fe^{2+}	0.86	0.65	VI	VI
Mg^{2+}	0.80	0.60	VI	VI
Fe^{3+}	0.73	0.55	VI	VI
Ti^{4+}	0.69	0.52	VI	VI
Al^{3+}	0.61	0.46	VI	VI
Al^{3+}	0.47	0.36	IV	IV
Be^{2+}	0.35	0.27	IV	IV
Si^{4+}	0.34	0.26	IV	IV
S^{6+}	0.20	0.15	III	IV

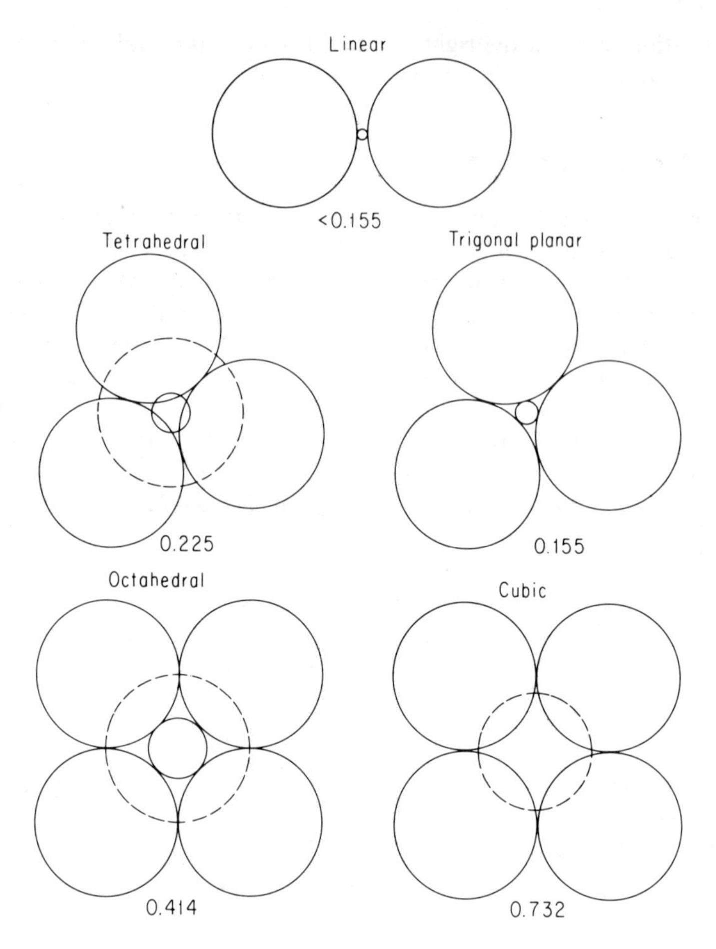

Figure 1-12 Radius ratios for coordination polyhedra. The radius ratio is the radius of the cation (smaller) divided by the radius of the anion (larger). The radius ratio given is that for an exact fit, the anions being in mutual contact and in contact with the cation.

ally predicted coordination numbers with those actually found in minerals. It can be observed that those ions which deviate from the predicted coordination are those whose radius ratio with O^{2-} falls very close to the boundary between two different configurations.

So far, ions have been treated as though they were rigid spheres. And as can be seen from Table 1-7, this is a good approximation. Many of the apparent discrepancies in the table disappear if the polarizability of ions and a tendency to form partially covalent bonds are taken into account. In general, the larger ions are the more polarizable. In the case of common minerals, the most abundant large ion is O^{2-}. Simple pressure squeezes ions together in a crystal and makes them effectively smaller; O^{2-} is squeezed more than the cations. With increasing pressure, the number of O^{2-} that can be coordinated by a cation tends to increase. Commonly, those ions with coordination numbers greater than the number predicted on the basis of geometry alone are to be found in minerals formed at a very high pressure.

The minimum radius ratio represents the case in which the surrounding anions touch one another and the cation exactly fills the void. In other words, the perfect geometric fit occurs (Fig. 1-12). If the cation is too small for a perfect fit, the next smaller coordination can be expected. This fact has been referred to as the *no-rattling-around rule*. On the other hand, if the cation is somewhat oversized for a perfect fit, the anions are merely somewhat spread apart.

1-11. Coordination Polyhedra

In addition to eliciting the principles that underlie crystal architecture, one of the goals of mineralogy is the description of specific crystal structures (Chapter 2). One way of describing crystal structures is to state the location, in the unit cell, of all ions in the crystal. This is relatively simple for structures such as halite or fluorite, which are simple and have a high degree of symmetry. It becomes quite complex and confusing for structures with large unit cells and many ions.

A second way of describing crystal structures is to state the kind of array in which the anions are found, such as close packed, and to state the kinds of voids in which the cations are found. As shall be seen, this kind of description is quite adequate for a structure like corundum, in which the O^{2-} ions are approximately HCP.

A third way of describing crystal structures makes use of the coordination polyhedra of anions about a particular cation, and of showing how these polyhedra are linked together (Fig. 1-13).

The simplest polyhedron is planar trigonal, such as the CO_3^{2-} or the BO_3^{3-} radical. If one anion is common to two different planar trigonal polyhedra, that anion is shared by the two polyhedra. In geometric terms, the polyhedra are sharing corners.

The smallest three-dimensional polyhedron is the tetrahedron, for example, the SiO_4^{4-} group. If one anion is common to two tetrahedra, corners are shared. By sharing corners, it is possible to build up rings, chains, sheets, and three-dimensional networks from the basic silica tetrahedron (Figs. 2-15 and 2-16). If two anions are common to adjacent tetrahedra, edges are shared. It would be geometrically possible to share faces of tetrahedra by three anions being common to two adjacent tetrahedra, but mineralogically this never happens because of mutual repulsion between close cations.

The central cation may be surrounded by six anions at the corners of an

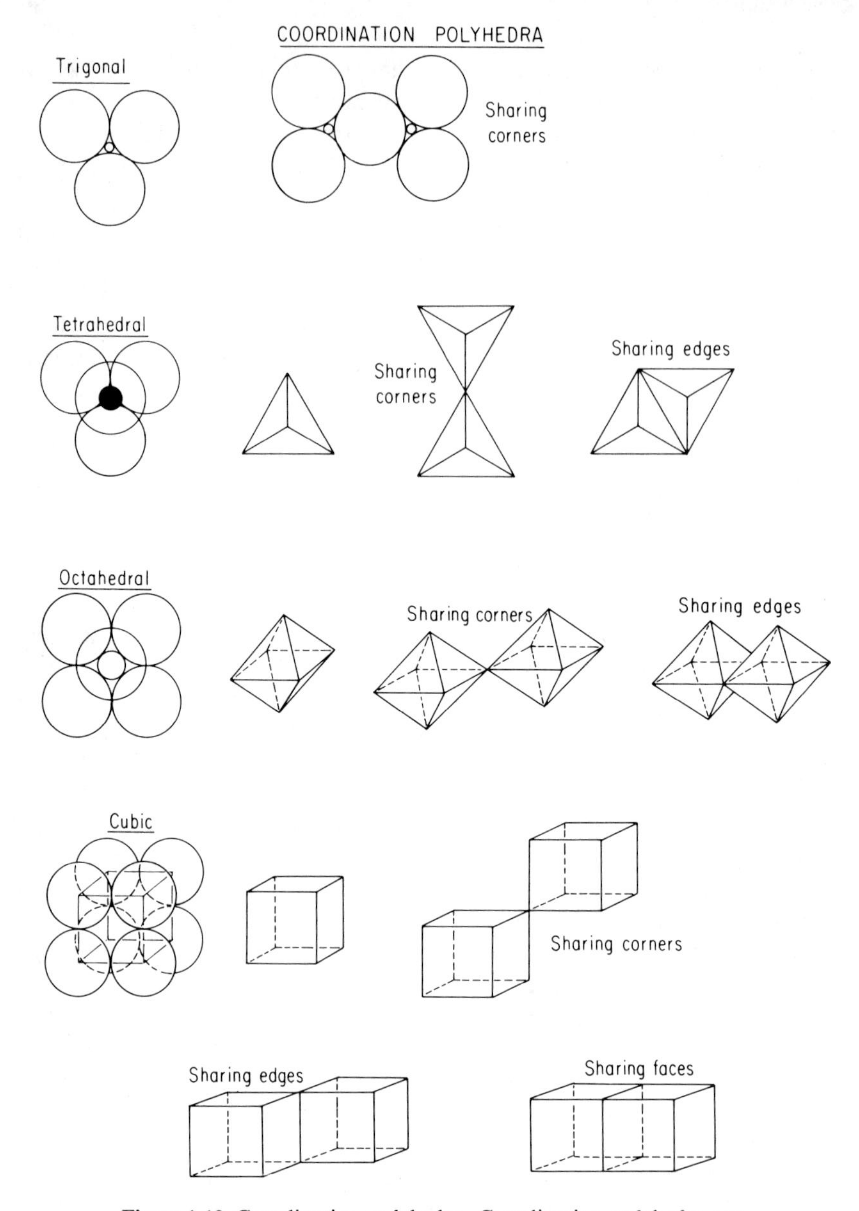

Figure 1-13 Coordination polyhedra. Coordination polyhedra are groups of anions about a central cation. They may be visualized in terms of the geometric figure formed by imaginary lines that connect the nuclei of the anions. If one anion is common to two coordination polyhedra, the polyhedra share corners. If two anions are common, edges are shared. If three or four anions are common, faces are shared. Coordination polyhedra in real crystal structures tend to share as few anions as possible.

octahedron. This unit can be indicated merely by drawing the outline of an octahedron. Again, depending upon the number of anions shared between two adjacent octahedra, there may be corner sharing, edge sharing, or face sharing, the last being somewhat rare in minerals.

If eight anions surround a central cation, they take the geometric form of a cube. Again, cubes may share corners, edges, or faces.

As described in Section 1-9, the CCP array of spheres is exactly equivalent to the FCC (face-centered cubic) array. This array may be regarded as octahedral coordination polyhedra linked up in three dimensions by tetrahedra, the octahedra sharing corners or edges. All ionic arrays, other than close packed, result in less efficient space filling. Two very common arrays are the primitive cubic (PC) (Fig. 1-14a) and the body-centered cubic (BCC) array of spheres (Fig. 1-14b).

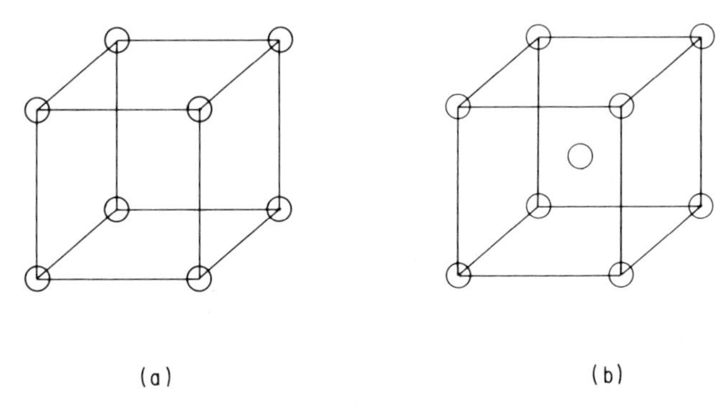

Figure 1-14 Primitive (PC) and body-centered cubic (BCC) structures. (a) The voids in a PC structure have a coordination number of 8. (b) The voids in a BCC structure have a coordination number of 6, but not all at equal distance.

Crystal structures are described in Chapter 2 by whichever method makes the best sense of the way in which the ions go together.

1-12. Pauling's Rules

So far, the effects of size, charge, and polarizability on ionic mineral structures have been covered. The other requirement for a real mineral structure is that electrostatic neutrality must be maintained. That is, the total charge on the cations must exactly equal that on the anions.

The study of crystals and crystal structures has led to a series of observations known as Pauling's rules. Real minerals display these characteristics, and hypothetically conceived structures that do not are usually found to be nonexistent and impossible to create.

1. A coordination polyhedron of anions is formed about each cation. The cation–anion distance is determined by the sum of the respective radii, and the coordination number is determined by the radius ratio.

The ionic bond is formed as a result of the transfer of an electron from cation to anion, and ions may be regarded as mutually attracting, charged spheres. Each cation tries to surround itself with as many anions as possible. This number depends upon the relative sizes of the ions.

2. In a stable ionic structure, the valence of each anion, with changed sign, is exactly or nearly equal to the sum of the strengths of the electrostatic bonds to it from the adjacent cations.

Electrostatic neutrality must be maintained to have a stable structure, and electrostatic neutrality is maintained over very short distances. In the NaCl structure (Fig. 2-2), each Na^+ is surrounded by six Cl^- ions and each Cl^- by six Na^+ ions. Therefore, one sixth of the positive charge of the cation may be assigned to each anion, and likewise one sixth of the negative charge of the anion to each cation.

3. The presence of shared edges and especially of shared faces in a coordinated structure decreases its stability; this effect is large for cations with large valence and small coordination number.

This empirical fact can be understood with reference to the strong repulsive force between two cations. They tend to be as far apart as possible in a crystal structure, and where elements of two polyhedra are shared, the polyhedra are distorted to get the cations as far apart as possible.

4. In a crystal containing different cations those with large valence and small coordination number tend not to share polyhedron elements with each other.

This rule is also understood with reference to the repulsive forces between cations. This may be thought of as cations trying to shield themselves as completely as possible with anions.

If crystal structures are thought of as anions in some approximation of a close-packed array, there are a limited number of different kinds of void spaces to fill with cations. As is demonstrated in Section 1-13, substitution of one ion for another is common, but substituting ions are generally quite similar to the ones replaced. This rule is most applicable to predominately ionic minerals. Because in silicates the Si—O bond is partially covalent and strongly directional, the structures of silicates tend to be determined by the manner in which the silica tetrahedra are linked up. This leaves room for a wider variety of cation sites and makes silicate chemistry much more complicated than the crystal chemistry of purely ionic minerals.

1-13. Ionic Substitution and Crystalline Solutions

Idealized crystal structures were invented as a conceptual means of simplifying a vast array of mineralogic data. Certain deviations from these idealized structures occur in such a regular and systemmatic fashion that they, too, have been idealized. For instance, there are many examples of one ion proxying for another without radically altering the basic structure of the mineral. The two most important examples of such substitution in practical mineralogy are Fe^{2+} for Mg^{2+} and Al^{3+} for Si^{4+}. The most important factor governing whether or not one ion will proxy successfully for another in a structure is the relative size of the two ions.

It has been observed by metallurgists that substitution occurs if the substituting ion is less than 15 percent different in size than the replaced ion. This may serve as a general guide to ionic substitution in many minerals, but it needs further qualification. Of greater importance than the 15 percent difference is the cation-to-anion radius ratio, which determines the coordination polyhedron in the structure. With the O^{2-} ion, Al^{3+} lies close to the border between octahedral coordination and tetrahedral coordination (Table 1-7). The radius of Si^{4+}, on the other hand, is well below the range of allowable sizes for octahedral coordination with oxygen. For this reason Al^{3+} commonly proxies for Si^{4+} in silicate structures, but Si^{4+} never proxies for Al^{3+} in the octahedral sites.

In considering ionic similarity in aqueous solution, the identity of electronic charge is the most important factor in determining chemical characteristics. In a mineral, where coupled substitution and stuffing are possible, deficiences in electronic charge balance may be made up elsewhere in the crystal (Fig. 1-17). Thus, size permitting, there may be charge differences of one in ionic substitution.

Table 1-8 lists some of the more common substitutions found in minerals, along with their coordination numbers and ionic radii. It is similarity in size that permits Mg^{2+} to substitute so readily for Fe^{2+} and difference in size that prevents Mg^{2+} from substituting readily for Ca^{2+}. The proximity in size between Ge^{4+} and Si^{4+} and between Ga^{3+} and Al^{3+} accounts for the fact that there are no Ge or Ga minerals, even though these elements are more abundant in the continental crust than more common mineral-forming elements, such as W, Hg, and Ag.

In any detailed discussion of ionic substitution, especially on the extent of

Table 1-8

COMMON IONIC SUBSTITUTION PAIRS

Ion	*CN*	*Radius* (Å)	*Ion*	*CN*	*Radius* (Å)
Mg	VI	0.80	Fe^{2+}	VI	0.86
Al	IV	0.47	Si	IV	0.34
F	VI	1.25	O	VI	1.32
Ba	IX	1.55	K	IX	1.63
Ge	IV	0.48	Si	IV	0.34
Ga	VI	0.70	Al	VI	0.61
Hf	VIII	0.91	Zr	VIII	0.92
Mn^{4+}	VI	0.62	Fe^{3+}	VI	0.73
Mn^{2+}	VI	0.91	Fe^{2+}	VI	0.86
La–Lu	VIII	1.26–1.05	Ca	VIII	1.20
Na	VIII	1.24	Ca	VIII	1.20

percentage of substitution, it is also necessary to specify the temperature and pressure under which the mineral was formed. Higher temperature will increase the tolerance of a crystal structure for foreign ions. Pressure may work either way, depending upon the relative polarizabilities of all the ions involved.

When one ion may substitute for another in a crystal structure, the resulting minerals exemplify a *crystalline solution,* which is analogous to the more familiar liquid solutions in chemistry in that each end member of the series is to a greater or lesser degree soluble in the other. *Solid solution* refers to the same situation, but this term fails to emphasize the fact that crystalline matter is involved, and not, for example, glass or some other noncrystalline solid.

(Some people mistakenly use the term *isomorphous series* for crystalline solutions. Since it is not necessary for the end members or for any of the intermediate compounds of a crystalline solution series to have the same crystal structure, isomorphous series is a misnomer for many crystalline solutions. Another term has been used, which is a faulty translation of a German term. The German term for alloy is *Mischmetall* and for crystalline solution is *Mischkristall.* Quite erroneously, these have been translated literally, causing no end of confusion.)

There are three major types of crystalline solution: substitutional, interstitial, and omission. In *substitutional crystalline solution,* one ion quite simply proxies for another. The degree of substitution may be partial or complete. An example of complete miscibility is the olivine series, where all compositions between pure forsterite [Mg_2SiO_4] and pure fayalite [Fe_2SiO_4] exist. An example of partial miscibility is the solubility of Fe^{2+} into the sphalerite [ZnS] structure. Sphalerite may incorporate up to 50 mole percent of "FeS," although no isostructural iron sulfide end member exists.

Natural *interstitial crystalline solutions* include variable amounts of H^+ in quartz and many ions in zeolites.

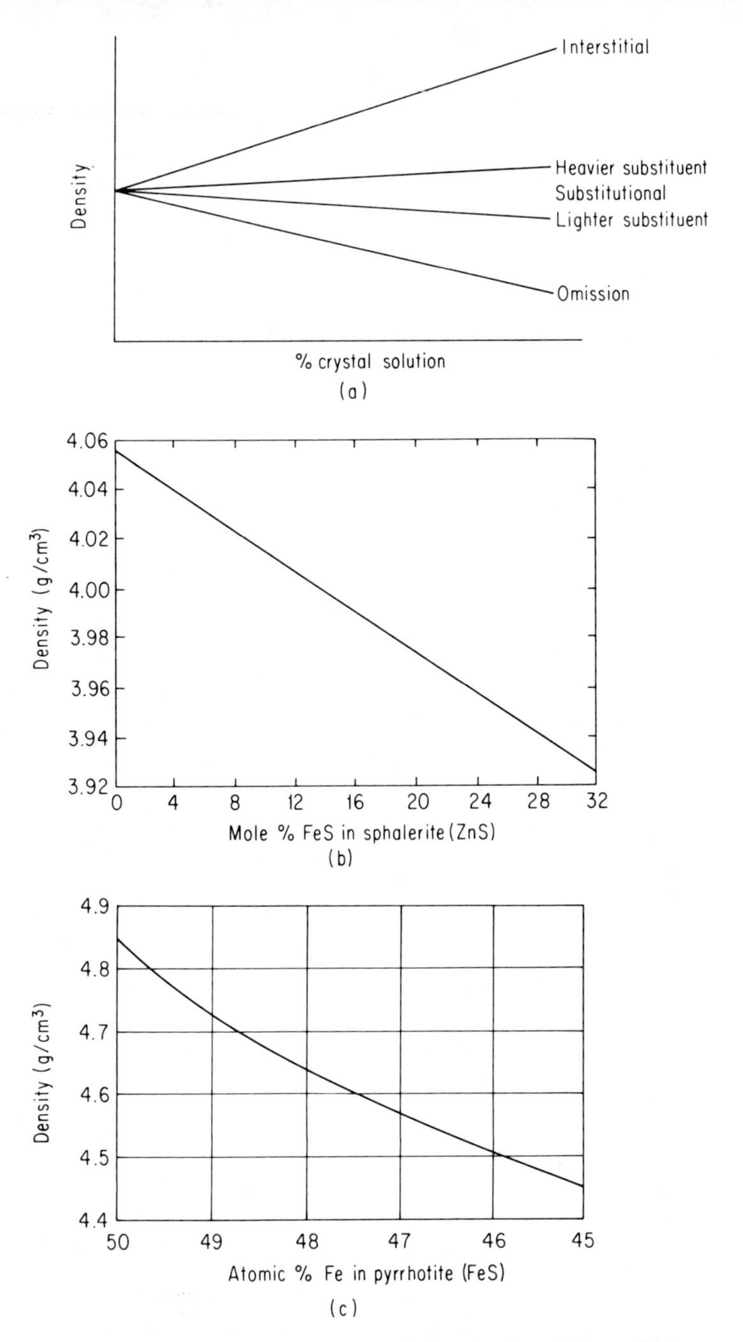

Figure 1-15 Influence of crystal solution on density. (a) Interstitial crystalline solution increases the density of a crystal. Omission crystalline solution decreases the density of a crystal. Substitutional crystalline solution either increases or decreases the density, depending upon whether the substituent ion is heavier or lighter than the substituted ion. (b) Density of sphalerite as a function of Fe^{2+} (atomic weight 55.85) substituting for Zn^{2+} (atomic weight 65.37). (c) Density of pyrrhotite as a function of omission of Fe^{2+} from the crystal lattice.

Omission crystalline solution involves the systematic omission of one ion coupled with a valence change in another. The mineral pyrrhotite almost never exists as pure FeS; there are always more S ions than Fe ions in the structure (Fig. 1-15). Electrostatic neutrality is maintained, since some of the Fe ions are ferric.

The density relations in a crystalline solution series are shown in Fig. 1-15. Interstitial crystalline solutions increase in density as the degree of solution is increased, because more matter is being crammed into the same space. Omission crystalline solutions tend to decrease markedly in density with increased omission. In substitutional crystalline solutions, the relative atomic mass and ionic size must be taken into account to determine whether the density increases or decreases with increasing solution. The amount of density change, however, is generally less than with interstitial or omission crystalline solution.

In point of fact, geometrically perfect crystals exist only in the mineralogist's mind. Actual crystalline solution has a little of everything in anything. Many reagent-grade crystalline compounds on the chemical laboratory stockroom shelf are of less than 99 percent purity for the substance listed on the label. Certain crystals are available from special sources that are *five nines pure*. That is, they are 99.999 percent of the stated material. This means that in some of the purest crystals obtainable from modern chemical technology 0.001 percent of the ions are foreign, that they contain more than 10^{18} foreign ions per mole of substance.

1-14. Structural Defects

Ionic substitution may be thought of as one kind of *defect*—a chemical defect—in an ideal crystal. In addition to this, a vast array of physical defects are found in real crystals. In general, a *physical defect* describes an ion that is out of place in the crystal lattice, such that certain chemical bonds are weakened or are missing altogether. Some of these defects depend upon the past history of the crystalline substance, and, therefore, are susceptible to control. Others, however, are temperature dependent. Complete absence of physical defects, therefore, can occur thermodynamically only at absolute zero, which is, of course, thermodynamically unattainable.

Various types of physical defects, or *dislocations*, that occur in crystals are illustrated schematically in Fig. 1-16 and their characteristics are summarized in Table 1-9. *Interstitial defects* (Fig. 1-16a) are those in which a small, foreign atom or ion intrudes into the structure, causing a slight displacement of ions in the regular array. Atoms of C in steel are interstitials. The *Frenkel defect* (Fig. 1-16b), named after the man who first described it, is one in which an ion is displaced from its regular lattice site to another nearby site where it

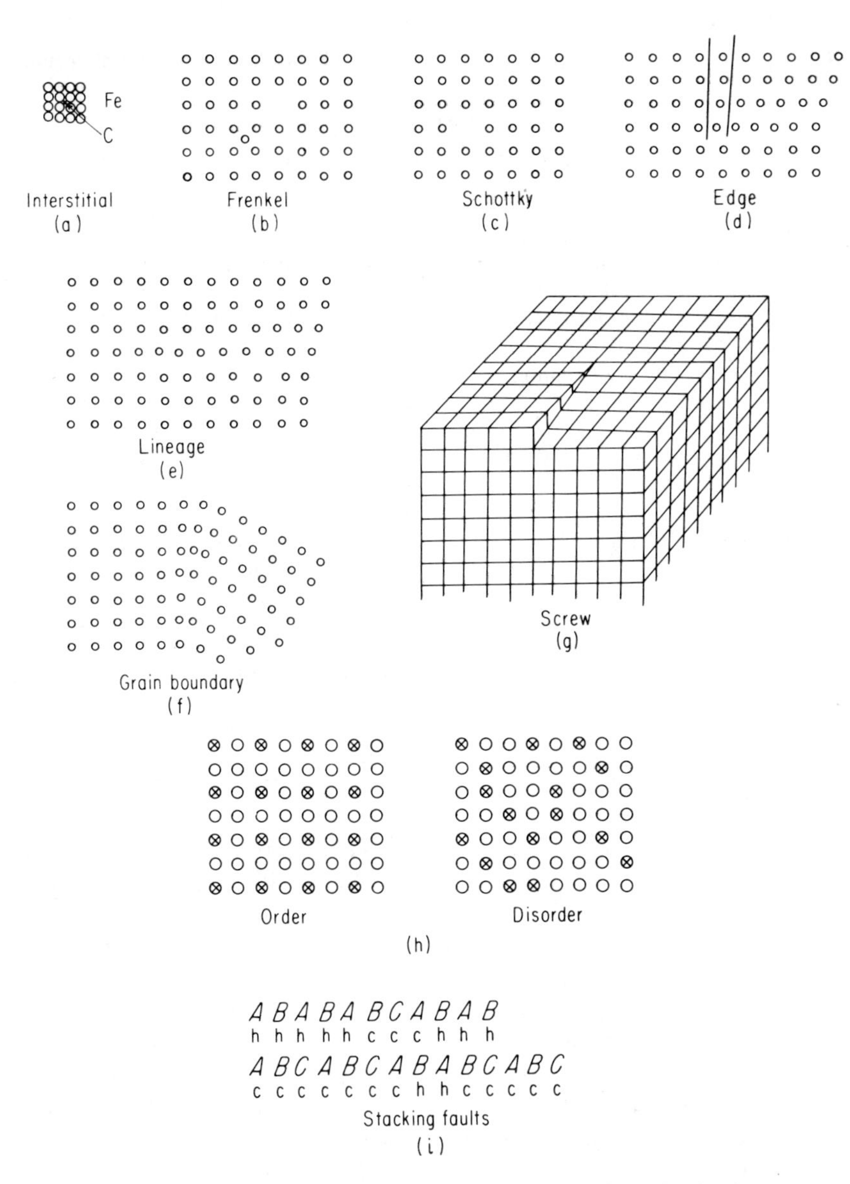

Figure 1-16 Defects in crystal structures. (a) Interstitial of C in Fe as in steel. (b) A Frenkel defect results if an ion is displaced from a lattice site. (c) A Schottky defect results if an ion is missing from the lattice. (d) An edge defect is a plane of ions extending partway through the structure. (e) A lineage results from edge defects coming in at regular intervals. (f) A grain boundary is where two lattices meet. (g) Screw dislocation, in which one or more layers wind about an axis in the form of a helix. (h) Two ions in an ordered array and two ions with the same ratio, but in a disordered array. (i) HCP stacking with one layer in a CCP position, and CCP with one layer missing causes an HCP unit.

Table 1-9
POSSIBLE IMPERFECTIONS IN CRYSTALS

Type of imperfection		*Description of imperfection*
Point defects:	Interstitial	Extra atom in an interstitial site
	Schottky defect	Atom missing from correct site
	Frenkel defect	Atom displaced to interstitial site creating nearby vacancy
Line defects:	Edge dislocation	Row of atoms marking edge of a crystallographic plane extending only partway in crystal
	Screw dislocation	Row of atoms about which a normal crystallographic plane appears to spiral
Plane defects:	Lineage	Boundary between two adjacent perfect regions in the same crystal that are slightly tilted with respect to each other
	Grain boundary	Boundary between two crystals in a polycrystalline solid
	Stacking fault	Boundary between two parts of a closest packing having alternate stacking sequences

appears as an interstitial. The *Schottky defect* (Fig. 1-16c) is, quite simply, an ion missing from its structural position. The electronic charge imbalance so caused may be neutralized by a second Schottky defect of ions with opposite charge, by change of oxidation state of another ion in a regular structural position, or by some other structural adjustment. These defects are known, categorically, as *point defects*.

A second group of defects is known as *line defects*. A point defect is a disturbance of the crystal structure that drops off concentrically from the point, whereas the line defect involves an entire row of ions with some kind of bond irregularity. Where a plane of ions extends only partway into a crystal lattice and terminates there, the line of termination is an *edge defect* (Fig. 1-16d). Ions just beyond the end of the incomplete plane have their bonds weakened by stretching.

A second type of line defect is the *screw dislocation* (Fig. 1-16g). Here a plane of ions does not close upon itself, but continues as a spiral. All ions in such a spiral are in normal crystalline environments except those at the axis of the spiral, which is the line of defect.

A third group of defects is known as *plane defects*, in which an entire sheet of ions has an irregular crystalline environment. A *lineage* (Fig. 1-16e) is created by a series of edge dislocations at regular intervals, and, therefore, a series of defect lines in a plane. The structure on one side of the plane is slightly tilted with respect to that on the other side. Additional tilting of the two sides results in a true grain boundary, wherein all the ions at the interface are in irregular environments (Fig. 1-16f).

Another planar defect is the *stacking fault* (Fig. 1-16i), in which an HCP stacking sequence is interrupted by a layer in CCP position, or a CCP array is missing one of the layers in a three-layer sequence.

Three dimensionally, there is the possibility of *order* and *disorder* (Fig. 1-16h) in a structure. Two different ions may be similar enough to be distributed randomly in a particular coordination polyhedron in a crystal lattice with but a small addition of energy. This is disorder, which is typical of minerals of the appropriate chemistry at high temperatures. Slow cooling, however, may permit the formerly randomized ions to take up regular lattice positions, or become ordered. Order–disorder effects are important in the Al^{3+} and Si^{4+} distributions in the feldspar minerals.

All minerals deviate from ideality in some respect or other. The amount and kind of physical defects which may exist in a particular mineral, like ionic substitution, may be typical of that mineral, and, for that reason, aid in its characterization and identification.

1-15. Isomorphism and Model Structures

Some general statements concerning the crystal chemistry of minerals have been made, some of the principles set forth and applied, and some terms used without formal definition. Certain rigorous clarification now is called for.

First, it has been noted that many different minerals have the same order and structure in their constituent ions. These resemblances range from exactly analogous to generally similar structures. Minerals in which structure is exactly analogous, and in which each and every ion in the one mineral has a counterpart in an identical environment in the other, are called *isotypes.* Minerals in which structure is basically the same, but not all ions in the one mineral have counterparts in the other, are termed *isostructural.* Examples include many half-breed and stuffed derivatives (Fig. 1-17 and Table 1-10).

Table 1-10

EXAMPLES OF COUPLED SUBSTITUTION

	Si	Si	O_4	Quartz
	Al	P	O_4	Berlinite
	Al	Al	O_3	Corundum
	Fe	Ti	O_3	Ilmenite
	Si	Si	Si_2O_8	Coesite
K	Al	Si	Si_2O_8	Sanidine
Na	Al	Si	Si_2O_8	Albite
Ca	Al	Al	Si_2O_8	Anorthite

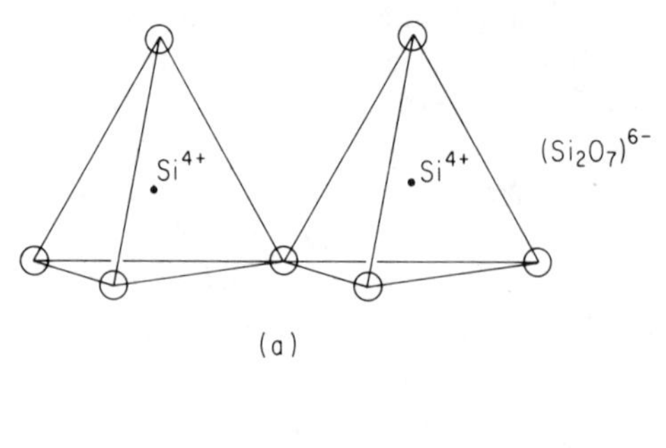

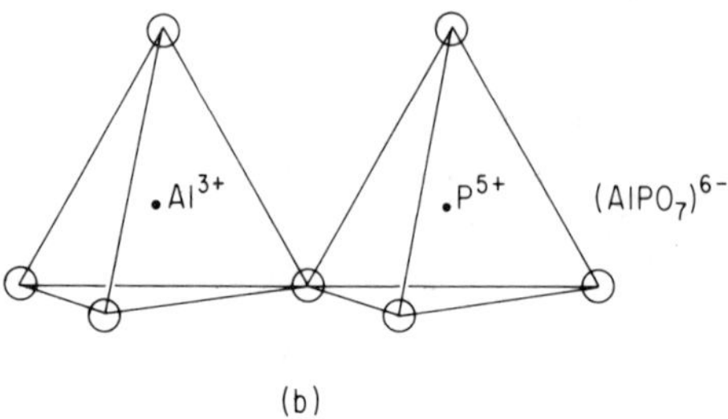

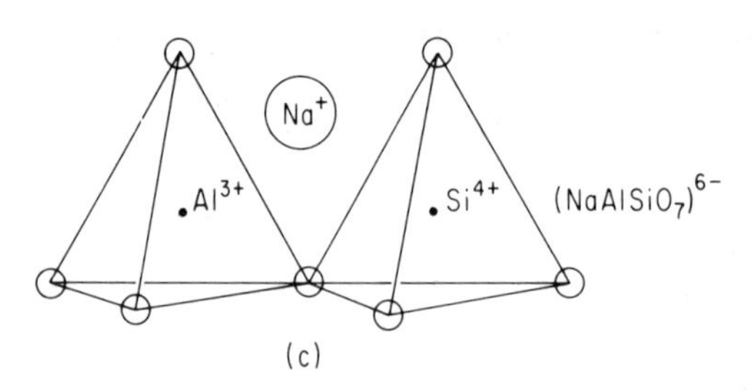

Figure 1-17 Derivative structures. (a) Parent structure of $Si_2O_7^{6-}$ group. (b) Half-breed derivative with a coupled substitution of Al^{3+} and P^{5+} for Si^{4+}. (c) Stuffed derivative with Al^{3+} and Na^+ replacing Si^{4+}.

If two minerals have the same general internal structure, closely similar external morphology, and similar X-ray diffraction patterns, they are said to be *isomorphous.*

In addition to demonstrating similarities among the structures of several apparently different minerals, studies of isostructural matter have importance in experimental mineralogy. Silicate minerals are rather difficult to work with in the laboratory, and some relationships among silicate minerals can be studied by proxy, by using experiments on isostructural matter. Such studies involve what are called *model structures.*

There are two kinds of model structures in current, general use in experimental mineralogy. A *size model* is one in which the ratio of the radius of the cation to the anion is the same. Thus, it is possible to predict much about the crystal chemistry of the compound BaSe by studying that of periclase [MgO]. A *charge model* is one in which the electronic charges of the ions are in a simple ratio. Thus, LiF is a charge model of periclase, and BeF_2 is a charge model of silica [SiO_2].

Derivative structures are ones in which the anions are in the same configuration as in the parent structure, but in which the cations are different. A *half-breed derivative* has two cations of different valences replacing a single ionic species in the parent structure, the average charge of the half-breed cations being the same as the charge on the parent-structure cations. An example of this is the alternating substitution of Al^{3+} and P^{5+} for Si^{4+} in berlinite [$AlPO_4$] (Fig. 1-17b and Table 1-10). A *stuffed derivative* has a cation of a lower ionic charge replacing a cation of higher ionic charge in the parent structure, with electrostatic neutrality being maintained by stuffing another cation into the structure. An example of this is the substitution of Al^{3+} for Si^{4+}, with the addition to the structure of one Na^+ ion for each Al^{3+} being substituted (Fig. 1-17c and Table 1-10), as in albite [$NaAlSi_3O_8$].

1-16. Polymorphism

It is possible for a compound to exist in more than one crystalline structure. A familiar example is C, which has two common structural forms, graphite and diamond. In elementary chemistry courses, this phenomenon, called *polymorphism*, is studied in the several crystalline forms of elemental S. In mineralogy, it is the rule, rather than the exception, that several polymorphs exist for each chemical compound, and each polymorph is given its own mineral name. As is proved in Section 6-2, there are limits to the numbers of different polymorphs that may coexist stably. In general, at an arbitrarily selected pressure and temperature, one polymorph is *stable* and all others are *metastable*. *Phase* is the general term given each different polymorph as well as the liquid and gaseous form of a given element or compound. The change from one phase to another is called a *phase change* or a *phase transformation*. A diagram that shows the stability relationships among phases as a function of temperature and pressure is a *phase diagram* (Fig. 1-18).

Phase transformations may be classified according to the ease or difficulty with which they are made—that is, according to the rate at which they occur. A phase transformation is rapid when the high-temperature or high-pressure form inverts to the form that is stable under laboratory conditions too quickly to be studied except by dynamic methods. If the high-pressure or high-temperature phase can be preserved under laboratory conditions, the transformation is sluggish or quenchable. Clearly, the latter is more important mineralogically, although the former may be significant in deep-crustal and upper-mantle studies. However, phases that invert rapidly are not found at the surface of the earth as minerals.

There are three main types of transformation in the rapid class: dilational, displacive, and nonstructural. An example of a *dilational phase transformation* occurs when NaCl, which has the CsCl structure under very high pressures, inverts back to the NaCl structure on release of pressure, changing the cation coordination number from 8 to 6. A *displacive transformation* occurs when the primary coordination about the cation remains constant, but the number of second-nearest neighbors, that is, the next most distant anions after the coordination polyhedron, changes. An example of a displacive transformation is the alpha–beta change in quartz, where the silicon ion remains in tetrahedral coordination but the angle between tetrahedra changes. Although this transformation is nonquenchable, certain crystal forms can occur only in the beta (high-temperature) type. Crystals with these forms occur in some volcanic rocks. The bonding in the crystal has reverted to the alpha type, but the crystal faces preserve the beta form.

An example of a *nonstructural transformation* is the Curie point of a ferrimagnetic mineral. In studies of remnant paleomagnetism, the magnetic character of the minerals is impressed only after they have cooled below the Curie point, and is erased again on reheating.

There are three main types of transformation in the sluggish class: reconstructive, polytypic, and order–disorder. In a *reconstructive transformation* some of the main chemical bonds are broken and remade. This class is further divided as to whether primary or secondary coordination about the cation is changed. The calcite–aragonite inversion is a reconstructive transformation involving primary coordination, in which the coordination of O^{2-} about the Ca^{2+} changes from VI to IX. Another is the inversion from quartz (CN = 4) to stishovite (CN = 6), which has the rutile structure.

Reconstructive transformations involving secondary coordination changes are common for the silica minerals. The differences among quartz, tridymite, and cristobalite are in the way in which the SiO_4 tetrahedra are interconnected.

Polytypism, also called *one-dimensional polymorphism*, occurs when two polymorphs differ only in the stacking of identical two-dimensional sheets. The only difference between the sphalerite and the wurtzite structure is that the S^{2-} ions in sphalerite are in CCP array and in wurtzite they are in HCP array (Fig. 2-4). As shall be seen in the discussion of the silicate minerals, polytypism is common among the layered silicates, especially among the clay and mica minerals.

Order–disorder relationships occur when two different ions can occupy the same structural site. This is most important in the mineralogy of the silicates, where Al^{3+} substitutes for Si^{4+} in tetrahedral coordination. In the high-temperature, disordered structure, the location of the Al^{3+} ions that replace the Si^{4+} ions is completely random, and statistically all sites are identical. In the low-temperature, ordered form, there is more than one kind of tetrahedral site, and these occur in an ordered distribution in the structure. One of these tetrahedral sites is occupied only by Al^{3+} ions and the others only by Si^{4+} ions. In minerals showing order–disorder transformation, all grades exist from completely ordered to completely disordered.

Several phase diagrams are given in Fig. 1-18. The different regions in pressure–temperature space where a particular phase is stable are outlined for different chemical compositions. At the pressures and temperatures indicated by the boundaries of the regions, two phases are equally stable.

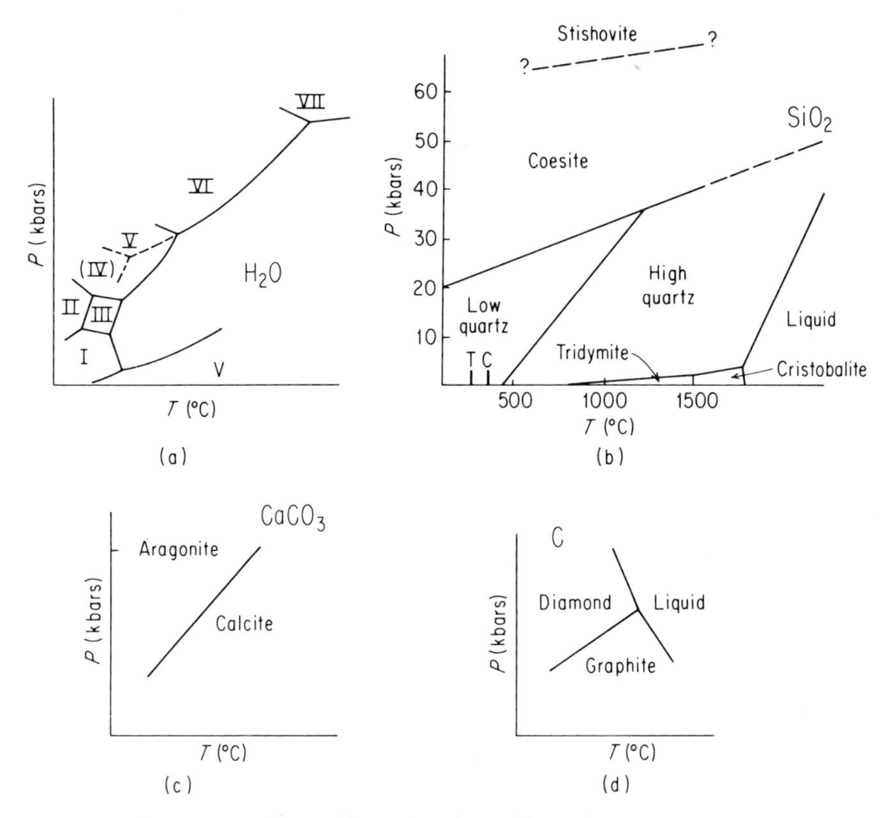

Figure 1-18 Polymorphism. (a) Six stable and one metastable crystal structures of ice. (b) Crystal structures of SiO_2. (c) Crystal structures of $CaCO_3$. (d) Crystal structures of C. Other examples of polymorphism among common minerals include kyanite, sillimanite, and andalusite [Al_2SiO_5] (Fig. 4-1); diaspore and boehmite [AlO(OH)]; sphalerite and wurtzite [ZnS] (Fig. 2-4); and cinnabar and metacinnabar [HgS].

At the triple point where three of these regions meet at a definitely prescribed temperature and pressure, all three phases are equally stable and may coexist in equilibrium (see Section 6-2).

Some generalizations about polymorphs may be derived from LeChâtelier's rule that internal changes occur to ameliorate external changes in environment. High-pressure polymorphs generally have a higher density than low-pressure ones. High-temperature polymorphs generally have a lower density and a higher symmetry than do low-temperature ones.

1-17. Pseudomorphism

In natural processes at or near the surface of the earth, it is not uncommon for one mineral to replace another while maintaining the external form of the first mineral. Such processes lead to *pseudomorphs* of one mineral after another.

A special type of pseudomorphism, called *paramorphism*, occurs when a polymorphic transformation takes place in a well-formed crystal. The existence of alpha quartz crystals with beta quartz forms in some volcanic rocks is one example of paramorphism. Calcite crystals having crystal faces that, for reasons of symmetry, could only have formed on the mineral aragonite are another example.

Pseudomorphism may involve gain or loss of material, or may involve partial or total replacement. Crystals of native Cu that have the crystal form of cuprite [Cu_2O] involve a loss of O^{2-} ions. Crystals of gypsum [$CaSO_4 \cdot 2H_2O$] after anhydrite [$CaSO_4$] involve a gain of H_2O. Goethite [$FeO(OH)$] with the form of pyrite [FeS_2] involves a partial exchange of O^{2-} and OH^- for S^{2-} ions. Quartz with the forms of calcite or petrified wood involves complete replacement of SiO_2 for $CaCO_3$ or for cellulose.

1-18. Noncrystalline Matter

Certain noncrystalline materials merit discussion because they bear on the source of certain minerals or are the natural result of others. Volcanic glass, a noncrystalline material, is the source of some minerals. The structure of a glass has a certain short-range order but no long-range order. That is, the Si^{4+} ions in a glass are always in tetrahedral coordination and the Al^{3+} ions are either in tetrahedral or octahedral coordination, just as they are in minerals. In contrast to minerals in which the coordination polyhedra occur in a regular repeated array, in a glass the coordination polyhedra do not repeat their pattern over distances greater than a few ionic radii. The ions in a glass are not completely at random, but long-range order is lacking. Simple devitrification, or leaching and hydration, soon converts volcanic glass into crystalline matter.

Precipitation of dissolved silica yields a colloidal material rather than distinct crystals. Modern silica gels form in an alkaline environment when seasonal variations in the alkalinity of natural waters range from a pH greater than 9, where silica is somewhat soluble, to a pH less than 9, where it is quite insoluble. As the pH of the water drops below 9, colloidal silica accumulates as a gel. Dehydration of a silica gel yields opal, which in turn crystallizes to chalcedony, a microcrystalline variety of silica.

A *metamict* mineral is one in which the crystal structure has been broken down by nuclear radiation. Figure 4-3 shows the radioactive decay scheme for U, along with the valence and ionic radii for the daughter ions. As can be seen from Fig. 4-3, the daughter ions change size and charge rather drastically with each successive radioactive decay. The changes tend to break up the crystal structure in the vicinity of a decaying element. If there is sufficient radioactive element in a crystal, long-range order is broken up and the material becomes metamict. Most thorite [$ThSiO_4$] is metamict, and old zircons [$ZrSiO_4$], which have appreciable Th^{4+} substitution for Zr^{4+}, may be metamict. The crystal structure can be restored, in many cases, by heating the mineral.

A mineral also may become partially metamict when it contains an oxidizable ion such as Fe^{2+}. Alpha particles, either from radioactive decay in the ferrous mineral or from an adjacent mineral, will oxidize the Fe^{2+} ion to the Fe^{3+} ion, bringing about a change both in charge and in ionic size. A radioactive mineral enclosed in biotite develops a halo of discolored, partially metamict material about it.

FURTHER READINGS

EVANS, R. C., *An Introduction to Crystal Chemistry*, 2nd ed. New York: Cambridge University Press, 1964.

FYFE, W. S., *Geochemistry of Solids*. New York: McGraw-Hill Book Company, 1964.

PAULING, L., *The Nature of the Chemical Bond*, 3rd ed. Ithaca, N.Y.: Cornell University Press, 1960.

SMITH, F. G., *Physical Geochemistry*. Reading, Mass.: Addison-Wesley Publishing Company, Inc., 1963.

WELLS, A. F., *Structural Inorganic Chemistry*, 3rd ed. New York: Oxford University Press, Inc., 1962.

2

Crystal Structures

Since the discovery that crystalline materials diffract X rays, one of the main thrusts of mineralogy has been toward the elucidation of particular crystal structures by means of X-ray diffraction. Within months after the von Laue experiment was conducted in Germany (Section 1-6), the Braggs, father and son, in England, were able to work out the structure of halite and of other alkali halide crystals.

The method by which an individual crystal structure is determined involves interpretation of its diffraction pattern (Section 5-8) in terms of the symmetrical properties of crystals. Crystal symmetry, however, was worked out mathematically on the basis of crystal morphology during the nineteenth century, that is, prior to the advent of X-ray diffraction (see Chapter 3). Although historically backward, it is logical and more illuminating to proceed from the actual distribution of ions in space to the geometric treatment of the distribution pattern.

Crystal structures are described in terms of their unit cells. The structural unit cell is analogous to the chemical formula. The chemical formula gives the combining ratios of the elements in a compound, reduced to the lowest common denominator. The unit cell is the smallest parallelepiped, within the structure, that contains all the ions and all the geometrical relationships among the ions of the crystal. Thus, by repeating the unit cell indefinitely in all three directions, the entire structure can be built from identical parts. There may be one formula per unit cell, or several, but the number of chemical formulas per unit cell is always an integer.

A useful way of classifying ionic structures is to do so first on the basis of their cation-to-anion ratio, and second on the basis of the cation coordination number (CN). The geologically important silicate minerals, because of the partial covalency of the Si–O bond, must be treated differently. Depending upon the Si : O ratio, the silica tetrahedra may share one, two, three, or all four of their corners to build up rings, chains, sheets, and three-dimensional networks (Figs. 2-15 and 2-16). Because of the directional nature of the Si–O bond, which dominates the structure, the coordination of other cations may be highly irregular. Therefore, the silicates are classified on the basis of the Si : O ratio, and, thus, on the number of shared corners of the silica tetrahedra, with a secondary division based on structural symmetry.

This chapter treats specific crystal structures of minerals, that is, the spacial relationships among ions in minerals. There are several methods for showing this graphically. A simple perspective sketch is quite adequate to show spacial relationships among ions in simple ionic structures. More complex structures may be made clear from perspective sketches of the outlines of the coordination polyhedra (Section 1-11), showing how polyhedral elements are arranged and shared. Many minerals, especially oxides, have structures in which the O^{2-} ions are in an approximate HCP or CCP array (Section 1-9). Such structures may be described by stating which type of close packing exists and which of the voids (coordination polyhedra) are occupied by specific cations.

There are two methods of describing the spacial relationships among ions in a crystal structure that are always complete and precise, but somewhat difficult for the beginner to visualize. First, it is possible to state the three-dimensional geometric point coordinates of each and every ion by giving the unit-cell edge the value of 1 in each direction and the ion coordinates as fractions of it. Second, it is possible to project the contents of the unit cell, either in terms of individual ions or in terms of coordination polyhedra, onto one face of the unit cell and to designate heights from the base in terms of fractions of the unit-cell length above the opposite face. Crystal structures are described best by the method which makes them most obvious.

The crystal structures described in this chapter include simple ionic structures and common silicate structures. As is shown in later chapters, geometric symmetry (Chapter 3), physical properties (Chapter 4), electromagnetic properties (Chapter 5), and genesis (Chapter 7) relate directly back to the crystal structure of the mineral species.

2-1. Structures of Naturally Occurring Elements

Of the 90 chemical elements that occur in the continental crust of the earth, only a few are to be found as native minerals. Most of these are transition metals, which are usually alloyed with some other metal. The most common crystal structure found in native transition metals is FCC (Section 1-9, Fig.

1-10). These metals include Cu, which may be alloyed with minor amounts of Fe, Ag, Bi, Sn, Pb, or Sb; Ag, which may contain up to 10 percent Au or lesser amounts of Cu; Au, which may contain as much as 35 percent Ag; Pt, with 10 to 20 percent Fe, up to 3 percent Rh and Ir, and lesser amounts of Cu and Os; and Pb with traces of Ag. Native Fe comes from both terrestrial and extraterrestrial (meteoritic) sources. The metal from both sources is in a BCC array (Fig. 1-14b), and is commonly alloyed with Ni—2 percent Ni in Fe from terrestrial sources and up to 20 percent from meteoritic sources.

Native S is found in crystals made up of molecules consisting of rings of eight S atoms. Native C occurs in two polymorphs. The low-temperature form, graphite, consists of C in sheets of six-member rings that have no chemical bonds in the direction at right angles to the sheets (Fig. 3-6). The high pressure polymorph, diamond, consists of C ions in a three-dimensional tetrahedral array (Fig. 2-1). This structure may be visualized as consisting of

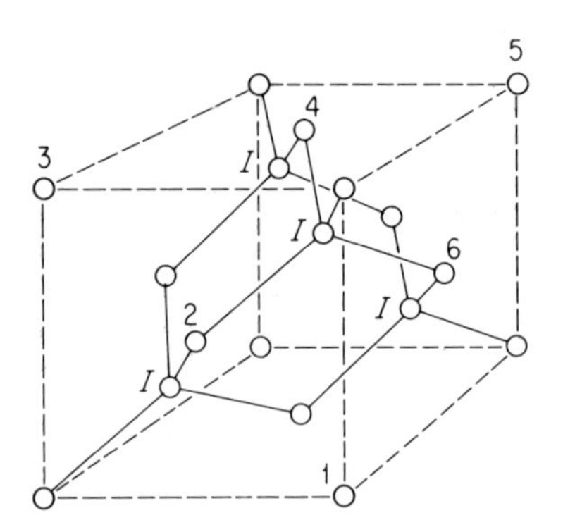

Figure 2-1 Diamond structure. This structure is based on FCC with four additional ions in the interior of the unit cell. The numbered ions are a closest-packed layer. If the interior ions (I) were replaced by Zn^{2+} and the FCC ions by S^{2-}, this structure would be that of sphalerite (Fig. 2-4a). Dashed lines indicate unit-cell outline and solid lines indicate bond directions. Each C ion is bonded to four others.

C ions in a FCC pattern with additional C ions in the interior of four of the eight cube corners. (Actually, these should not be referred to as C ions, since the bonding is covalent.)

2-2. *AX* Structures

The simplest mineral structures occur in compounds in which there is a 1 : 1 cation (A) to anion (X) ratio, and in which the charge is necessarily the same for both. These are referred to as AX structures. There are four major kinds.

The halite structure (Fig. 2-2) is FCC. The Cl^- ions occur at the corners and at the face centers of the unit cell. The Cl^- ions at the face centers clearly are shared between two unit cells and, therefore, count as half an ion in each cell. Since there are six faces on a cube, there are a total of three Cl^- ions per

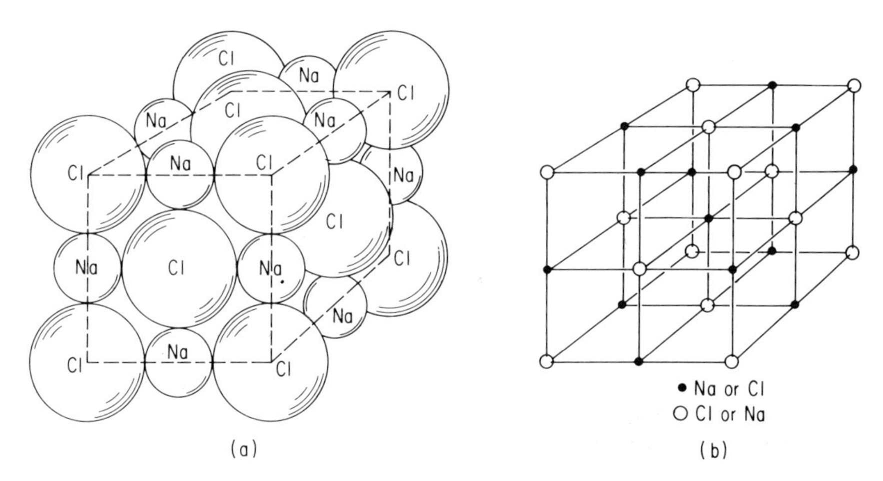

Figure 2-2 Halite structure. (a) Scale drawing of unit cell with relative sizes of ions shown. (b) Outline of unit cell showing the positions of all contained ions. Both drawings are to the same scale for the unit cell. The unit cell of halite may be shown with either the Na^+ or the Cl^- ions in the FCC positions. All ions are in octahedral coordination.

unit cell at the face centers. The Cl^- at the corner is shared among eight unit cells and, therefore, only one eighth of it belongs to each unit cell. Since there are eight corners on a cube, this adds one Cl^- ion to the previously counted three, for a total of four Cl^- ions per unit cell.

The Na^+ ions lie along the centers of the edges of the unit cell, plus one in the center of the cube. Each ion on an edge is shared among four different unit cells, so that each edge ion contributes one fourth of a Na^+ ion to the unit cell. A cube has 12 edges, making a total of three edge Na^+ ions. These, added to the one in the center, yield four per unit cell, which is exactly equal to the number of Cl^- ions already counted. Note that it is immaterial whether the unit cell is defined with Cl^- ions at the corners, or with Na^+ ions at the corners. (The coincidence of unit-cell corners with ions is a convention, not a necessity.)

Each Na^+ ion in the halite structure is surrounded by six Cl^- ions. That is, it has octahedral coordination. And each Cl^- ion is surrounded by six Na^+ ions.

Examples of minerals with the halite structure include most alkali halides;

oxides and sulfides of alkali-earth elements, with the exception of Be; some oxides of divalent transition metals; and galena [PbS].

The CsCl structure is simple cubic (Fig. 2-3). The Cl^- ions occupy the corners of the cubic primitive cell and the Cs^+ ion occupies the center. There

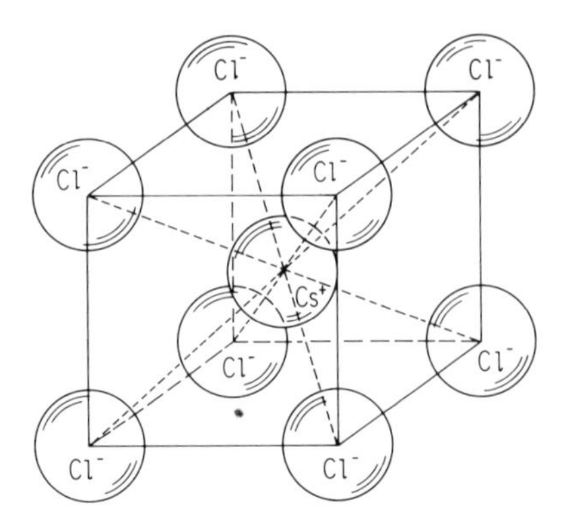

Figure 2-3 CsCl structure. This structure is made up of two interpenetrating PC lattices. The unit cell of CsCl may be shown with either Cs^+ or Cl^- in the PC positions. All ions are in cubic coordination.

are, thus, one Cs^+ ion and one Cl^- ion per unit cell. Again, it is immaterial whether the Cl^- or the Cs^+ ion occupies the corner position, as the structure is made up of two interpenetrating simple cubic arrays.

Each Cs^+ ion is surrounded by eight Cl^- ions and each Cl^- ion by eight Cs^+ ions. As has already been explained, pressure tends to increase cation coordination numbers. Under very high pressure, 18,000 bars for halite itself, minerals with halite structure invert to the CsCl structure.

At this point in the discussion, a slight digression on the subject of terminology is warranted. Halite is a mineral with the specific composition NaCl. It has the structure illustrated in Fig. 2-2. All minerals and crystalline compounds having this structure are said to have *halite structure*, regardless of their chemical composition. All structures are named for their chief example. Whether or not it is wise to have two related but different meanings for a mineral name is a question which belongs more properly to philosophy than to mineralogy. Let it suffice to note that the two meanings for the same word exist side by side in modern mineralogic usage, and that the meaning should be clear from the context.

The two preceding crystal structures had chemical bonds that were primarily ionic in character. The remaining two AX structure examples have bonds that are more covalent in character. They are similar in that both are exemplified by polymorphs of ZnS and both are based on close packing of the anions (Fig. 2-4).

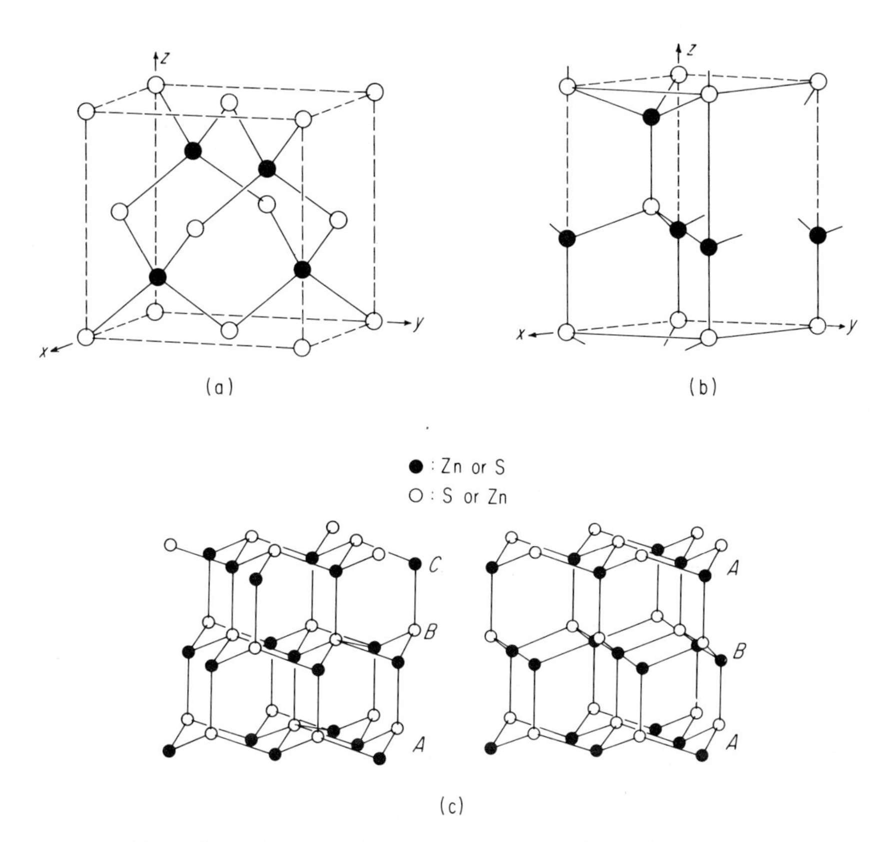

Figure 2-4 Sphalerite and wurtzite structures. (a) Sphalerite structure with S^{2-} ions in an FCC array and Zn^{2+} ions occupying interior positions. Compare this structure with the diamond structure (Fig. 2-1). (b) Wurtzite structure. (c) Sphalerite and wurtzite structures compared. The closest-packed layers are parallel to the bottom in both drawings. Each species of ion is in a CCP array in sphalerite and in an HCP array in wurtzite. All ions are in fourfold coordination.

In the sphalerite structure, the anions are in CCP array and the cations occupy one half the tetrahedral voids. Chalcopyrite [$CuFeS_2$] is a half-breed derivative of sphalerite, with Cu^{2+} and Fe^{2+}, alternatively, replacing Zn^{2+} in the tetrahedral site, which results in doubling the size of the unit cell.

The wurtzite structure has anions in an HCP array with the cations occupying one half the tetrahedral voids.

2-3. AX_2 Structures

The three important mineral structures that have a cation-to-anion ratio which can be expressed as AX_2 are fluorite [CaF_2], rutile [TiO_2], and quartz [SiO_2]. The cation CN's are VIII, VI, and IV, respectively, reflecting a decrease in cation size. The fluorite structure is based on a FCC array of Ca^{2+} ions, which yields a unit cell of four cations. The anions occur along the body diagonals of the unit-cell cube, and, as can be seen in Fig. 2-5, there are eight

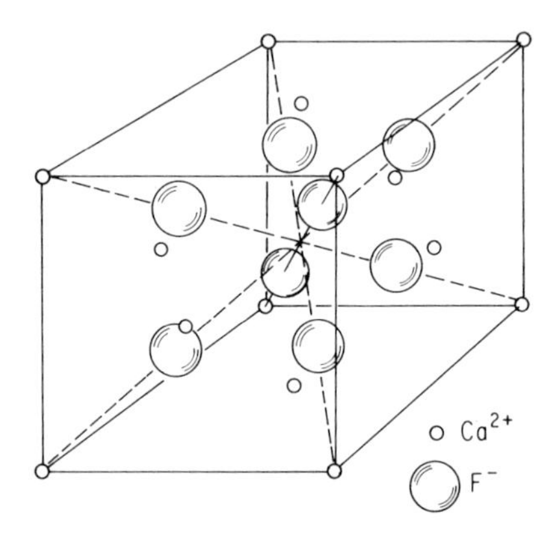

Figure 2-5 Fluorite structure. The Ca^{2+} ions occupy the cube corners and the centers of the faces. The F^- ions occur along the body diagonals of the cube. Each Ca^{2+} ion is in eightfold coordination and each F^- ion in fourfold coördination.

of them, all interior to the unit-cell cube. Each cation coordinates eight anions, and each anion coordinates four cations. It is worth noting, at this point, that the octahedral cleavage of fluorite follows planes composed of Ca^{2+} ions in a closest-packed layer (Section 4-4). The minerals thorianite [ThO_2] and cerianite [CeO_2] have the fluorite structure.

The rutile structure (Fig. 2-6) can be described as having cations in a distorted BCC array, for a total of two cations per unit cell. Four anions lie in pairs on two opposite faces, and two are interior to the unit cell, making a total of four anions per unit cell. This maintains the AX_2 ratio. Cassiterite [SnO_2], polianite [MnO_2], stishovite [SiO_2], and rutile [TiO_2] have this structure.

The quartz structure is based on the all-important silica tetrahedron. Each cation is surrounded by four anions and each anion is bonded to two cations. That is, the coordination polyhedra share corners. The subject of

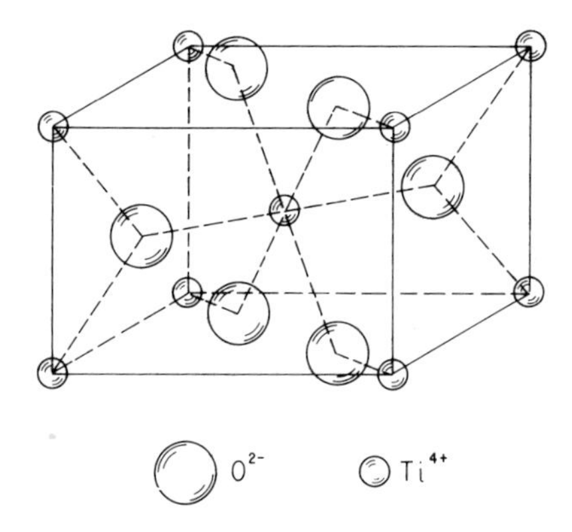

Figure 2-6 Rutile structure. The unit cell is square in outline as viewed from the top, but the dimension shown vertically in the drawing is shorter than the side of the square. The Ti^{4+} ions are at the corners and the center of the unit cell. The O^{2-} ions are on two of the faces plus two in the interior of the unit cell. Each Ti^{4+} ion is surrounded by six O^{2-} ions in a distorted octahedron. These octahedra share edges between adjacent unit cells, causing the shortening of the shared elements of the coordination polyhedra (Section 1-12). Each O^{2-} is in threefold coordination.

silica structures is of such importance mineralogically that it is covered in a separate section (Section 2-12).

2-4. A_2X_3 Structures

There are a large number of A_2X_3 structures, but the most common one among the natural minerals is the corundum structure. The corundum structure (Fig. 2-7) is based on hexagonal close packing of O^{2-} ions, with Al^{3+} ions occupying two thirds of the available octahedral voids. The Al^{3+} ions are not randomly distributed, but are in a very ordered array. Because one third of the octahedral voids are vacant, the O^{2-} ions are distorted slightly from perfect HCP.

Hematite has the corundum structure with Fe^{3+} in place of Al^{3+}. The sesquioxides of Cr^{3+} and V^{3+} also have the corundum structure.

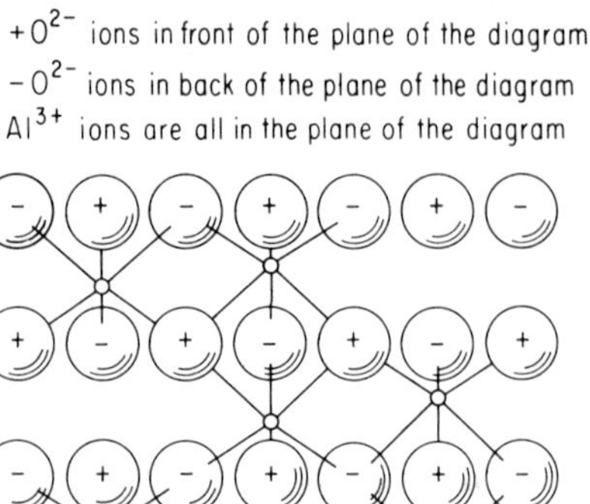

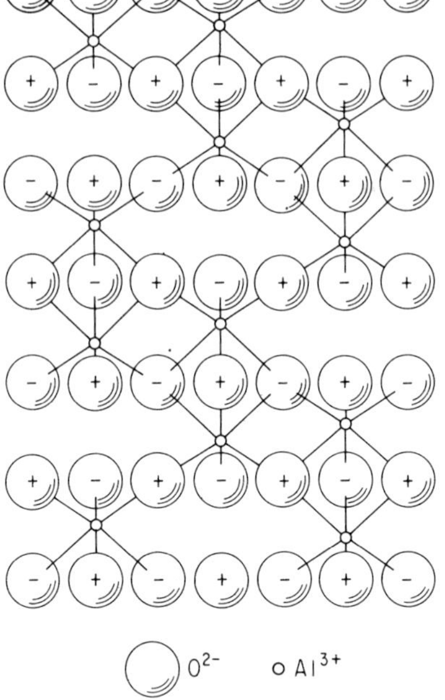

Figure 2-7 Corundum structure. The O^{2-} ions are in an HCP array with each row in the figure representing a closest-packed layer. Four sequential O^{2-} ions form the rhombus outlined in Fig. 1-7. The Al^{3+} ions are in octahedral coordination between the closest-packed layers, two thirds of the octahedral voids being occupied.

Two additional structures similar to the corundum structure are gibbsite [$Al(OH)_3$] and brucite [$Mg\ (OH)_2$] (Fig. 2-8). Both structures are based on HCP hydroxide layers in sheets of two. In the brucite structure, Mg^{2+} ions fill all the octahedral voids between the two OH^- sheets. In the gibbsite structure, Al^{3+} ions fill two thirds of the available octahedral voids. In addition to being important minerals in their own right, the gibbsite and brucite layers are of great importance in describing the micas and other layered silicates (Section 2-11). These layers are also referred to as *tri*octahedral and *di*octahedral, depending upon whether three thirds or two thirds of the octahedral voids are filled.

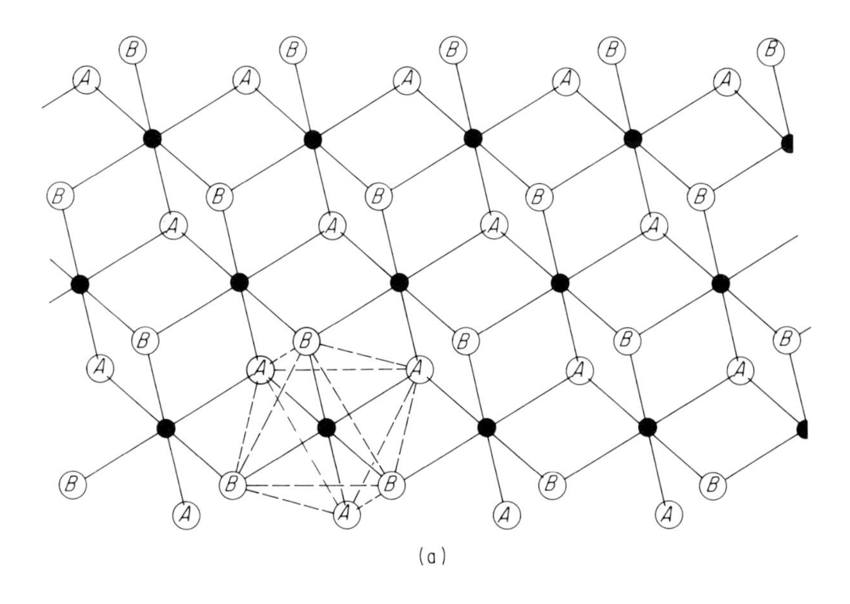

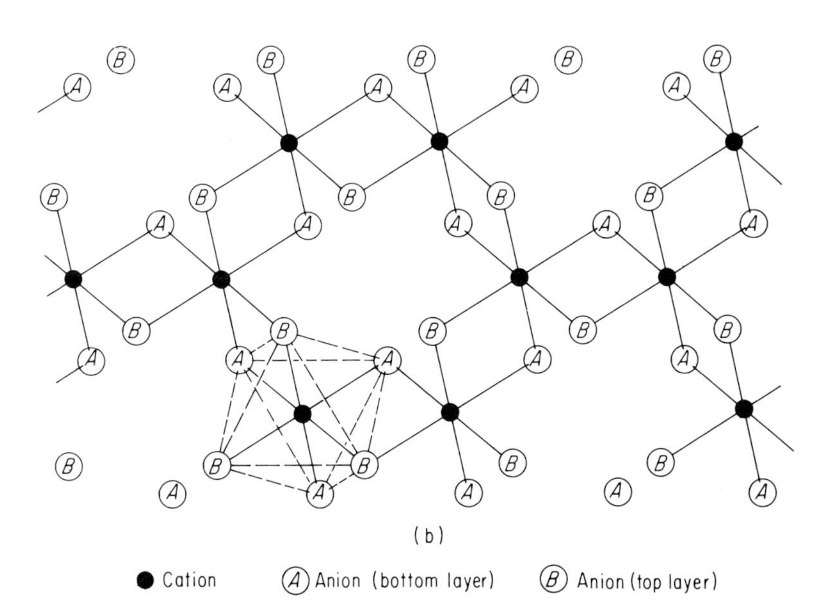

Figure 2-8 Brucite and gibbsite structures. Both structures have two closest-packed layers of OH^- radicals. (Compare Fig. 1-8a.) (a) Brucite structure with all interlayer octahedral voids filled by Mg^{2+} ions. (b) Gibbsite structure with two thirds of the octahedral voids filled with Al^{3+} ions. Van der Waals bonds hold the two-layer sheets together in both structures.

2-5. ABX_3 Structures

There are four important crystalline structures having the general chemical formula ABX_3: calcite, aragonite, ilmenite, and perovskite. In the calcite structure the A ion has octahedral coordination, and the B ion has trigonal planar coordination. The calcite structure (Fig. 2-9a) may be visualized as

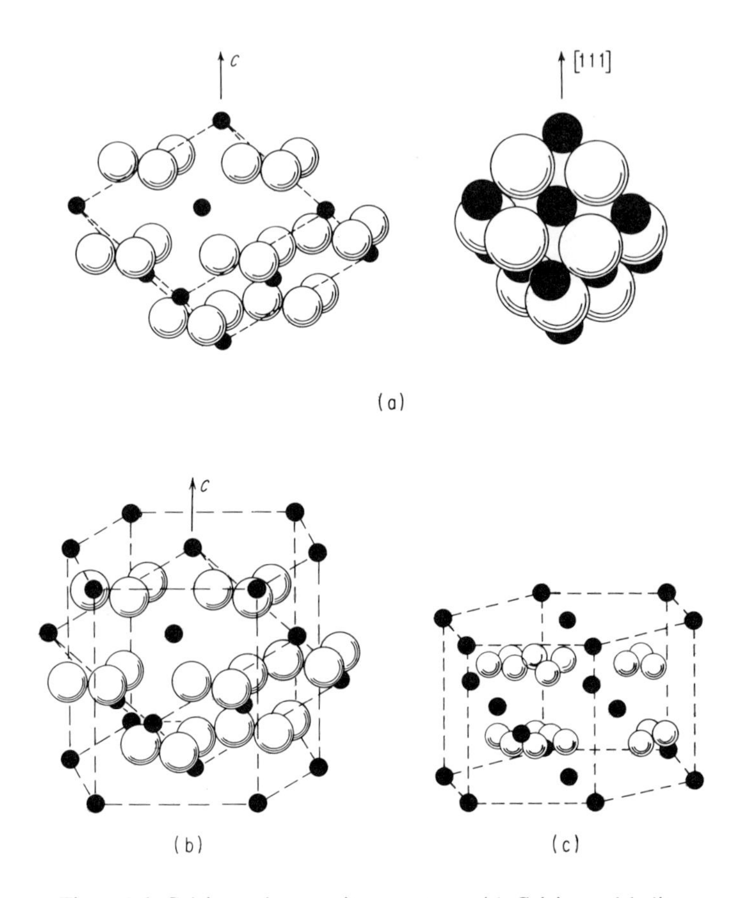

Figure 2-9 Calcite and aragonite structures. (a) Calcite and halite structures compared. These structures are analogous, the Na^+ of the halite being replaced by the Ca^{2+} of the calcite and the Cl^- by CO_3^{2-}. The nonspherical shape of the CO_3^{2-} radical distorts the halite cube to a rhombohedron. (b) Calcite structure. Both the Ca^{2+} ions and the CO_3^{2-} radicals are in distorted CCP arrays. (c) Aragonite structure. Both the Ca^{2+} ions and the CO_3^{2-} radicals are in distorted HCP arrays.

derived from a distorted halite structure, with Ca^{2+} ions in the place of Na^+, and with CO_3^{2-} radicals in place of Cl^- ions. The unit cell is a rhombohedron rather than a cube, however. All carbonate minerals having *A* ions smaller than Ca^{2+} (Table 1-5) crystallize with the calcite structure (Fig. 2-9b). They are calcite [$CaCO_3$], magnesite [$MgCO_3$], siderite [$FeCO_3$], rhodochrosite [$MnCO_3$], and smithsonite [$ZnCO_3$]. Nitratine [$NaNO_3$] has the calcite structure.

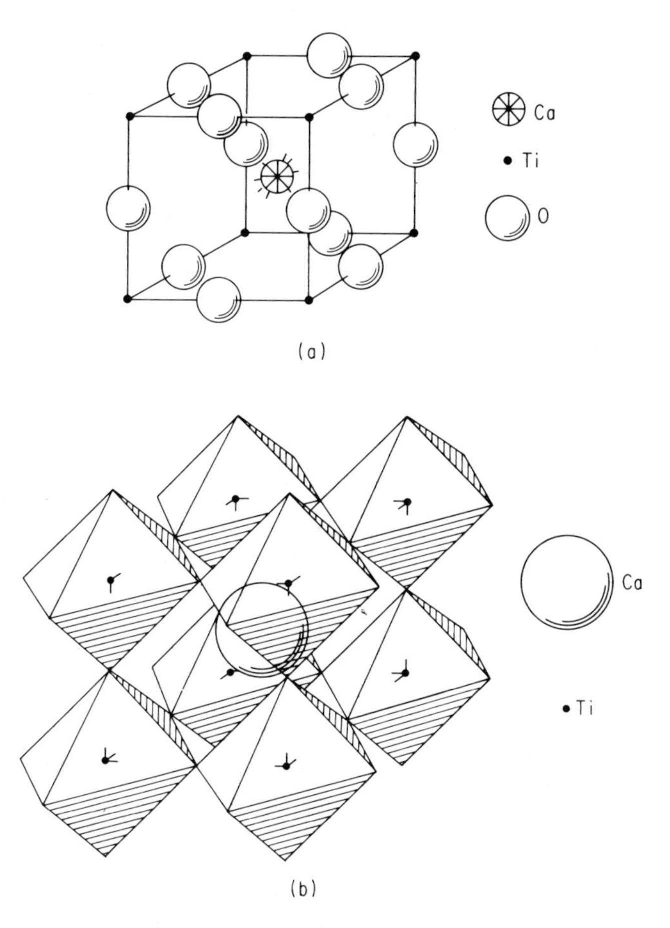

Figure 2-10 Perovskite structure. (a) Unit cell showing Ti^{4+} ions at each corner, O^{2-} ions along each cell edge, and Ca^{2+} ion at the cube center. (b) Each Ti^{4+} ion is at the center of a TiO_6^{8-} octahedron, which shares all corners with adjacent octahedra. The Ca^{2+} ions occupy the spaces among the octahedra and are in twelvefold coordination.

Calcium carbonate also crystallizes as a high-pressure or biogenetic polymorph, aragonite. In the aragonite structure (Fig. 2-9c), the *A* ion is in ninefold coordination and the *B* ion is in trigonal planar coordination, as in the calcite structure. The calcite structure is described as a modified halite structure. In the halite structure the Na^{+} ions are in a FCC array, which is equivalent to a CCP array. The aragonite structure is similarly based on the Ca^{2+} ions being in a distorted HCP array, with CO_3^{2-} radicals occupying half the octahedral voids. Note that in both the calcite and aragonite structures, each Ca^{2+} ion is surrounded by six CO_3^{2-} radical groups. In the calcite structure, these are arranged so that each Ca^{2+} ion is also surrounded by six O^{2-} ions. However, in the aragonite structure the carbonate radicals are oriented such that each Ca^{2+} ion is surrounded by nine O^{2-} ions. All carbonate minerals having *A* ions larger than Ca^{2+} crystallize with the aragonite structure. They are aragonite [$CaCO_3$], witherite [$BaCO_3$], strontianite [$SrCO_3$], and cerrusite [$PbCO_3$]. Niter [KNO_3] has the aragonite structure.

The ilmenite [$FeTiO_3$] structure is a half-breed derivative of the corundum structure. All the Al^{3+} ions between one pair of closest-packed O^{2-} layers are relaced by Fe^{2+} ions, and all of the Al^{3+} ions in the next layer are replaced by Ti^{4+} ions.

The last ABX_3 structure is the perovskite [$CaTiO_3$] structure (Fig. 2-10). Each Ca^{2+} ion is in twelvefold coordination and each Ti^{4+} ion is in octahedral coordination.

This structure may be visualized in two different ways. Consider a cube with a Ti^{4+} ion in the center, six O^{2-} ions at the center of each face, and eight Ca^{2+} ions at the corners. Alternatively, consider a cube with a $(TiO_6)^{8-}$ octahedron at each corner and a Ca^{2+} ion in the center (Fig. 2-10).

2-6. ABX_4 Structures

The ilmenite structure is a half-breed derivative of corundum with Fe^{2+} and Ti^{4+} replacing Al^{3+} in alternation. Most ABX_4 structures are half-breed derivatives of AX_2 structures.

There are many half-breed derivatives based on the several silica structures. Although the silica structures are summarized in Table 2-2 and discussed in Section 2-12, a few observations here can be made about their derivatives. Berlinite [$AlPO_4$] is a half-breed derivative structure of quartz, with half the Si^{4+} ions replaced by Al^{3+} ions and half by P^{5+} ions. Eucryptite [$LiAlSiO_4$] is a stuffed derivative of quartz, with half the Si^{4+} ions replaced by Al^{3+}, and an equal number of Li^{+} ions stuffed into the structure to maintain electrostatic neutrality.

The zircon structure (Fig. 2-11) has *A* ions in cubic coordination and *B* in tetrahedral coordination. The *X* ions about *A* do not all occur at the same

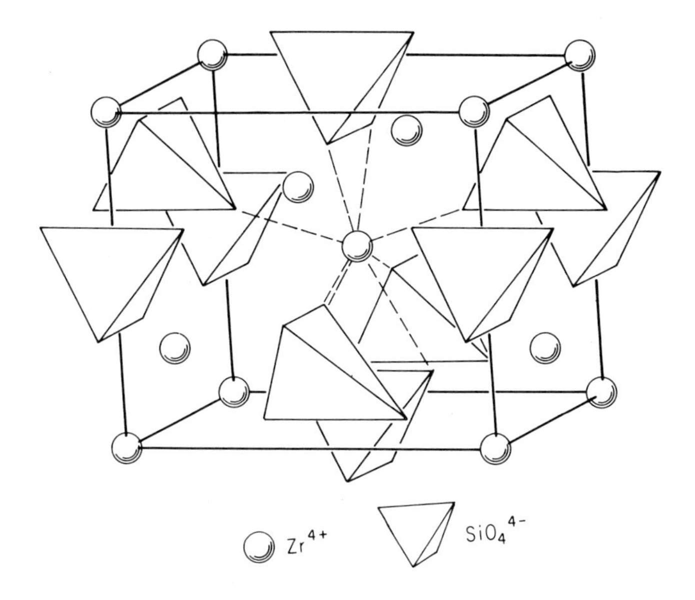

Figure 2-11 Zircon structure. The tetrahedra represent SiO_4^{4-} coordination polyhedra. The Zr^{4+} ions lie at the cell corners, the cell centers, and on the cell faces. Each Zr^{4+} ion is surrounded by eight O^{2-} ions, as shown by the dashed lines.

distance, there being four nearer to A and four a little farther away. The minerals zircon [$ZrSiO_4$], thorite [$ThSiO_4$], and xenotime [YPO_4] crystallize with this structure.

The monazite structure has A ions in cubic coordination and B ions in tetrahedral coordination. As in the zircon structure, the eight O^{2-} ions about A are not equidistant. Minerals crystallizing with this structure are monazite [$CePO_4$] and huttonite [$ThSiO_4$].

2-7. A_2BX_4 Structures

One of the most important two-cation families of minerals has the general chemical formula A_2BX_4. Olivines, which are important both petrologically and as refractory materials; spinels, which are important in refractories and in ferrimagnetic materials; and Ca_2SiO_4, which is important in refractories and in portland cement, are included.

The olivine structure (Fig. 2-12) is based on HCP of the O^{2-} ions. The A ions occupy octahedral voids, and the B ions occupy tetrahedral voids. Minerals having the olivine structure include forsterite [Mg_2SiO_4], fayalite [Fe_2SiO_4], and tephroite [Mn_2SiO_4]. Monticellite [$CaMgSiO_4$] is a half-breed

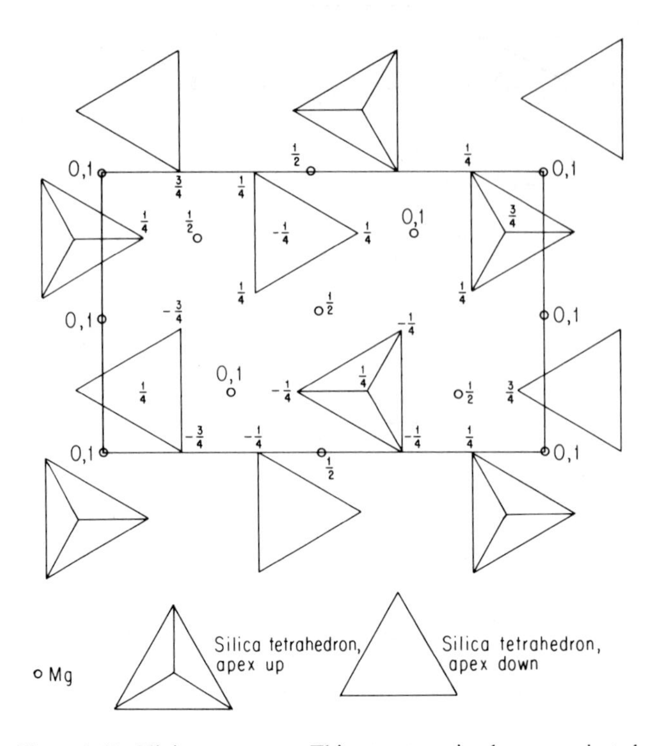

Figure 2-12 Olivine structure. This structure is shown projected onto one face of the unit cell. The SiO_4^{4-} groups are shown as tetrahedra, half with an apex up and half with an apex down. Fractions represent the distance above the opposite side of the unit cell as fractions of the total cell dimension. The O^{2-} ions are in a distorted HCP array, with Mg^{2+} occupying octahedral voids and Si^{4+} occupying tetrahedral voids.

derivitive of the olivine structure. One of the polymorphs of Ca_2SiO_4 also has the olivine structure. Because of their similar ionic size, Fe^{2+} and Mg^{2+} ions freely substitute one for the other in the olivine minerals (Table 1-8).

The spinel structure (Fig. 2-13) is based on CCP of the O^{2-} ions. Since the numbers of octahedral and tetrahedral voids are the same in CCP and in HCP structures, there is no basic change in coordination in going from olivine to spinel structures. In both cases, two thirds of the ions are in octahedral coordination and one third in tetrahedral. A spinel is called normal if all the *A* ions are in octahedral sites and all the *B* ions in tetrahedral sites. It is called inverse if half the *A* ions are in octahedral coordination, half the *A* ions in tetrahedral coordination, and all the *B* ions in octahedral coordination. In general, the Al^{3+} and Cr^{3+} spinels are normal, and the Fe^{3+} spinels are inverse. Because of their ferrimagnetic properties, many totally new spinels

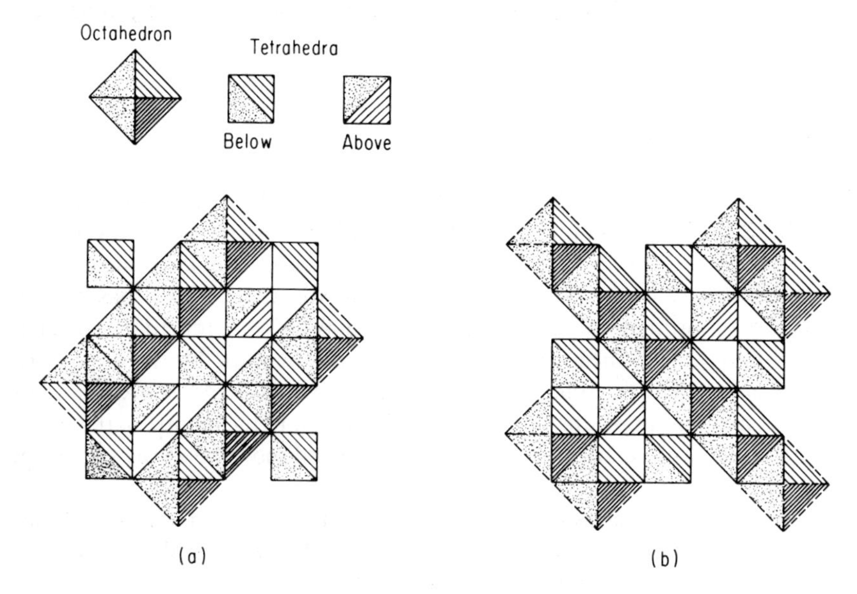

Figure 2-13 Spinel structure. Coordination polyhedra are projected onto the cube face of the unit cell. The chains of octahedra, sharing edges, are at the same level. The cross-linking tetrahedra are alternately above and below the plane of the octahedra (note that superposed tetrahedral corners are not shared). The octahedral chains in the first layer (a) are at right angles to the chains in the second layer (b). Four such layers make up the complete unit cell of spinel. Each O^{2-} (represented by the apices of the polyhedra) is shared by two octahedra and one tetrahedron.

Table 2-1

MINERALS WITH SPINEL STRUCTURE

Normal spinels		
Spinel series (Al)	Spinel	$MgAl_2O_4$
	Hercynite	$FeAl_2O_4$
	Gahnite	$ZnAl_2O_4$
	Galaxite	$MnAl_2O_4$
Chromite series (Cr)	Chromite	$FeCr_2O_4$
	Magnesiochromite	$MgCr_2O_4$
Inverse spinels		
Magnetite series (Fe)	Magnetite	$FeFe_2O_4$
	Magnesioferrite	$MgFe_2O_4$
	Franklinite	$ZnFe_2O_4$
	Jacobsite	$MnFe_2O_4$
	Trevorite	$NiFe_2O_4$

have been synthesized in the past decades. The naturally occurring spinels are listed in Table 2-1.

2-8. Orthosilicates

Having discussed some of the principles of crystal chemistry and seen how they have applied to simple ionic structures, it is now possible to study, in some detail, that group of minerals which makes up the bulk of the crust of the earth, the silicates. With the exception of stishovite, which has Si^{4+} in octahedral coordination, all silicate minerals have Si^{4+} in fourfold coordination with O^{2-}. The silica tetrahedron is the basic building block of the crust of the earth. There is some experimental evidence that Si^{4+} in sixfold coordination may be important at depths below 500 km, in the mantle of the earth, but no mineral phase from these depths has been discovered at the surface of the earth.

The silicates are classified on the basis of the manner in which the silica tetrahedra are linked together, which, in turn, is based upon the Si : O ratio. This classification is summarized in Table 2-2.

The first class of silicates to be considered is that in which the silica tetrahedra are independent of one another (Fig. 2-15). That is, there are no O^{2-} ions shared between two adjacent tetrahedra. The silica tetrahedra are bonded together by cations that lie between them. The minerals in this group are called *orthosilicates*, and include the olivines (discussed in Section 2-7), the garnets, the zircon group (discussed in Section 2-6), the phenakite group, the aluminosilicates, sphene, and topaz.

The garnet group of minerals includes a wide variety of naturally occurring garnets and an even larger group of synthetic garnets, which have no natural counterpart. Some of the synthetic garnets have very interesting electrical and magnetic properties, and materials with garnet structure are used in microwave transmission. Although natural garnets are all silicates, it is possible to synthesize crystals that have the garnet structure, but which are totally free of silicon in the tetrahedral sites.

The general formula for garnet is $R_3^{VIII} R_2^{VI} X_3^{IV} O_{12}$. The subscripts refer to ionic ratios, as usual, and the superscripts refer to the CN of the various ionic sites. The ions commonly found in these sites in natural garnets and the common garnets are listed in Table 2-3.

The O^{2-} ions in the garnet structure are arranged, roughly, in a BCC array, and the unit cell is rather large and complex. In terms of coordination polyhedra, the silica tetrahedra are linked in a three-dimensional array with R^{VI} octahedra. The R^{VIII} ions are distributed among the polyhedra.

The aluminosilicates comprise the minerals sillimanite, andalusite, and kyanite, all of which have the chemical formula Al_2SiO_5. All three minerals

Table 2-2

CLASSIFICATION OF SILICATE STRUCTURES

Class	*Arrangement of tetrahedra*	*Shared corners*	*Repeat unit*	*Si: O*	*Example*
Orthosilicates	Independent tetrahedra	0	SiO_4^{4-}	1 : 4	Garnet
Sorosilicates	Pair of tetrahedra sharing corner	1	$Si_2O_7^{6-}$	1 : $3\frac{1}{2}$	Hemimorphite
Cyclosilicates	Closed rings of tetrahedra each sharing two corners (single rings)	2	SiO_3^{2-}	1 : 3	Tourmaline
	or three corners (double rings)	3	$Si_{12}O_{30}^{12-}$	1 : $2\frac{1}{2}$	Osumilite
Inosilicates	Infinite single chains of tetrahedra each sharing two corners	2	SiO_3^{2-}	1 : 3	Pyroxene
	Infinite double chains of tetrahedra alternately sharing two and three corners	$2\frac{1}{2}$	$Si_4O_{11}^{6-}$	1 : $2\frac{3}{4}$	Amphibole
Phyllosilicates	Infinite sheets of tetrahedra each sharing three corners	3	$Si_2O_5^{2-}$	1 : $2\frac{1}{2}$	Mica
Tektosilicates	Unbounded framework of tetrahedra each sharing four corners	4	SiO_2	1 : 2	Quartz

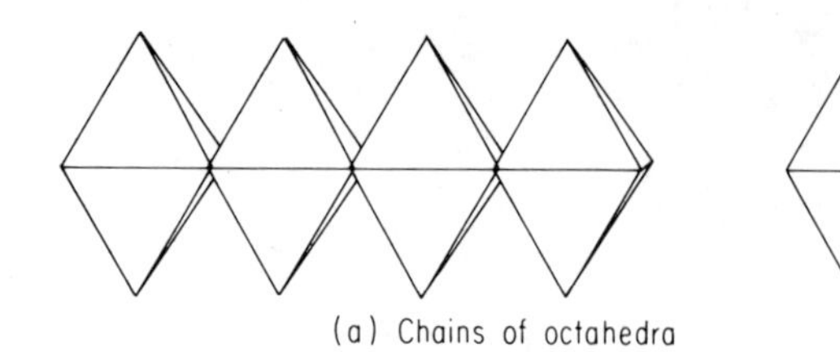

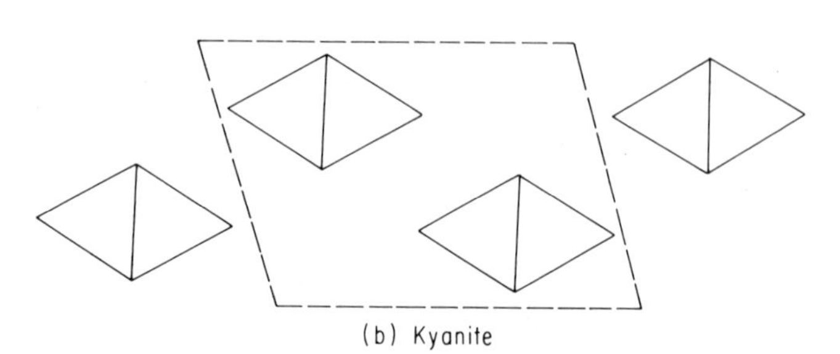

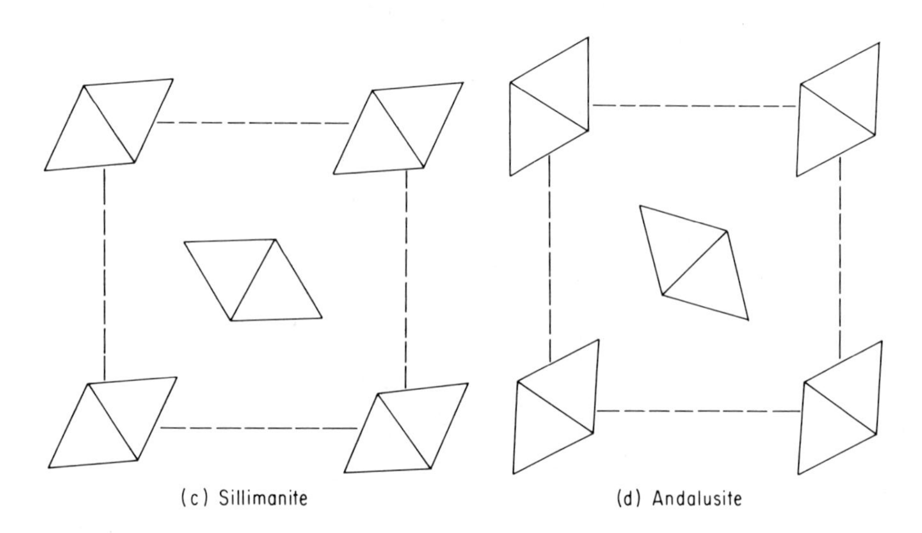

Figure 2-14 Aluminosilicate structures. These structures are composed of chains of octahedra that are cross-linked with Al^{3+} and Si^{4+} ions. The Si^{4+} ions are always in tetrahedral coordination. (a) Single chain of edge-sharing AlO_6 octahedra, side and end view. (b) Kyanite structure. The interchain Al^{3+} ions (not shown) are in octahedral coordination. (c) Sillimanite structure. The interchain Al^{3+} ions are in tetrahedral coordination. (d) Andalusite structure. The interchain Al^{3+} ions are in fivefold coordination.

Table 2-3
GARNET COMPOSITIONS

Structural formula: $R_3^{VIII}R_2^{VI}X_3^{IV}O_{12}$

Site	*Occupants*
R^{VIII}	Ca^{2+}, Mg^{2+}, Fe^{2+}, Mn^{2+}, Y^{3+}, trivalent lanthanides
R^{VI}	Al^{3+}, Cr^{3+}, Fe^{3+}, Ti^{4+}
X^{IV}	Si^{4+}, Al^{3+}, Fe^{3+}, Ti^{4+}

Major naturally occurring garnet end members:
Pyrope, $Mg_3Al_2Si_3O_{12}$
Almandine, $Fe_3Al_2Si_3O_{12}$
Grossular, $Ca_3Al_2Si_3O_{12}$
Andradite, $Ca_3Fe_2Si_3O_{12}$

have a structure that consists of chains of Al octahedra, which are formed by sharing edges with two adjacent octahedra (Fig. 2-14). The chains are cross-linked with Si^{4+} and Al^{3+} ions. The Si^{4+} ions are all in tetrahedral coordination. The difference between the structures lies in the coordination polyhedron of the cross-linking Al^{3+} ions. In sillimanite, these Al^{3+} ions are in fourfold coordination, in andalusite they are in fivefold coordination, and in kyanite they are in sixfold coordination. Andalusite is the only common rock-forming mineral that has Al^{3+} ions in fivefold coordination. The O^{2-} ions in kyanite are in CCP, which gives it the highest density of the three. The phase relationships among the aluminosilicates are given in Fig. 4-1b. In sillimanite, the Si^{4+} and Al^{3+} ions are in a disordered array in the tetrahedral sites.

In topaz [$Al_2SiO_4(OH, F)_2$], the anions approach a close-packing array, which is neither cubic nor hexagonal. The separate silica tetrahedra are linked together by alumina octahedra, with one fifth the O^{2-} ions replaced by OH^- and by F^-.

Sphene [$CaTiSiO_5$] has independent silica tetrahedra linked by Ti^{4+} octahedra. The Ca^{2+} ions are in sevenfold coordination.

2-9. Silicates with Independent Tetrahedral Groups

The linkage between silica tetrahedra that gives independent tetrahedral groups is shown in Fig. 2-15. One corner shared between two tetrahedra yields an $(Si_2O_7)^{6-}$ group. The mineral lawsonite [$CaAl_2(Si_2O_7)(OH)_2 \cdot H_2O$] is made up of these double tetrahedra groups. In the epidote minerals [$(X_2^{VIII}Y_3^{VI})(Si_2O_7)(SiO_4)O(OH)$], both double and single tetrahedral groups occur in the same structure. The epidote minerals zoisite and clinozoisite have Ca^{2+} in eightfold coordination and Al^{3+} in sixfold. The mineral epidote itself has Fe^{3+} replacing some of the Al^{3+} in the octahedral sites. Allanite may

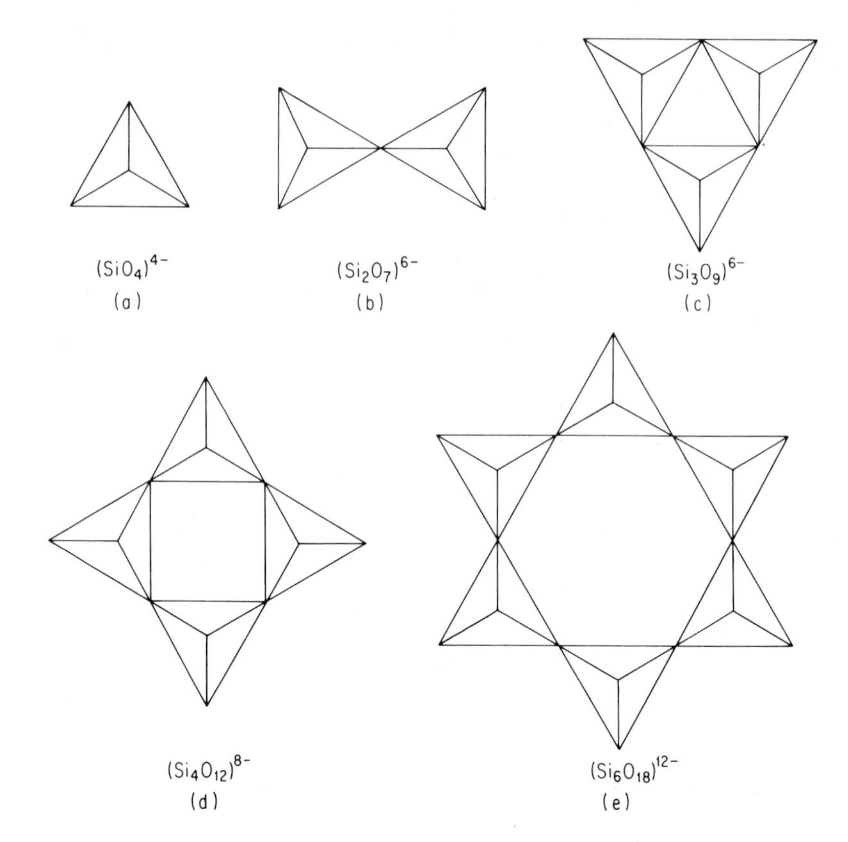

Figure 2-15 Silicate structures with independent groups. (a) Independent SiO_4 tetrahedral unit, as in garnet and zircon. (b) Double Si_2O_7 group with one corner shared, as in the epidote minerals. (c) Three-member Si_3O_9 ring with each tetrahedron sharing two corners. (d) Four-member Si_4O_{12} ring. (e) Six-member Si_6O_{18} ring, as in the tourmaline minerals. Another independent unit is formed by apical sharing between two six-member rings, one upon the other.

be conceived from epidote by a coupled substitution of trivalent rare-earth ions for Ca^{2+} in the eightfold site, and Mg^{2+} for Al^{3+} and Fe^{3+} in the octahedral site.

Silica tetrahedra, sharing two corners with two of their neighbors, can form rings of tetrahedra, as shown in Fig. 2-15. Independent rings of three $(Si_3O_9)^{6-}$, four $(Si_4O_{12})^{8-}$, six $(Si_6O_{18})^{12-}$, eight $(Si_8O_{24})^{16-}$, and double-six $(Si_{12}O_{30})^{12-}$ members exist in naturally occurring minerals. The double-six independent ring structure consists of two $(Si_6O_{18})^{12-}$ groups that share apical O^{2-} ions, so that three tetrahedral corners are shared. All but the independent six-member rings are rare.

Tourmaline $[Na(Mg, Fe)_3Al_6(BO_3)_3(Si_6O_{18})(OH)_4]$ has six-membered rings of silica tetrahedra stacked alternately with the borate groups. These are bonded together by alkalis, alkali earths, Al^{3+}, and Fe^{2+}.

2-10. Chain Silicates

Two major mineral groups, the pyroxenes and the amphiboles, have silica tetrahedra linked to form infinite chains (Fig. 2-16). Although the chemical

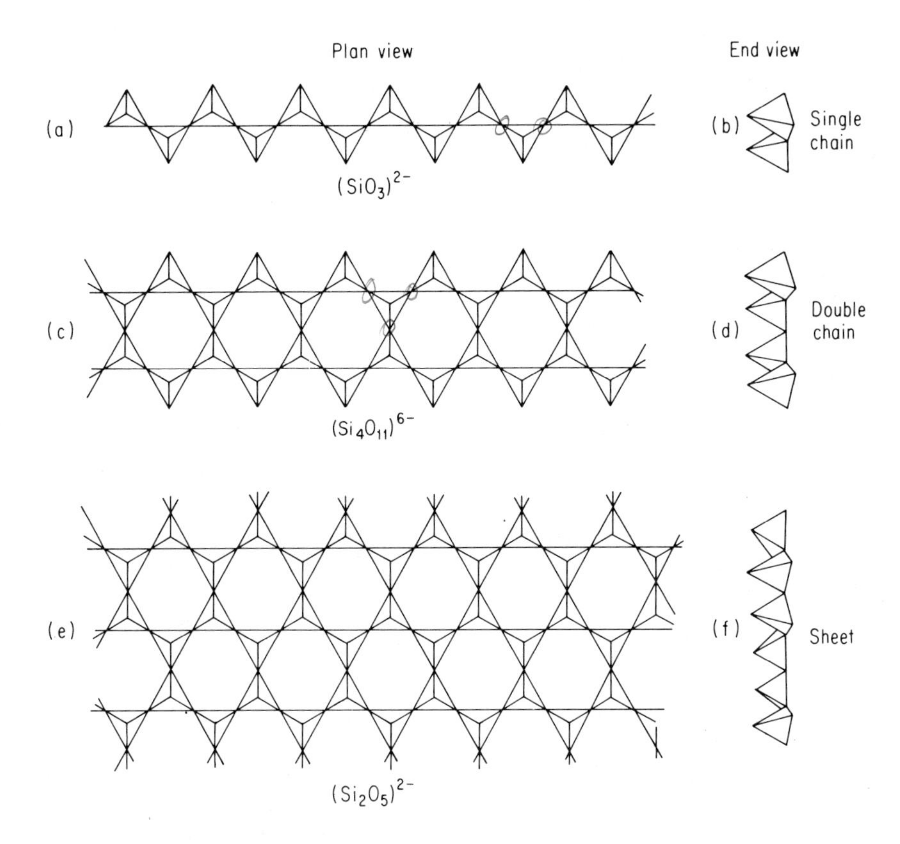

Figure 2-16 Silicate structures with infinite linkages. (a) Single chain formed by tetrahedra sharing two corners, as in the pyroxene minerals. (b) End view of single chain. (c) Double chain formed by tetrahedra sharing two and three corners, as in the amphibole minerals. This chain can be visualized either as two single chains with cross-linkage or as six-member rings linked with individual tetrahedra. (d) End view of double chain. (e) Sheet formed by tetrahedra sharing three corners, as in clays and micas. This sheet may be visualized as six-membered rings sharing outside corners or as side-linked double chains. (f) Edge view of sheet.

Table 2-4

CHAIN SILICATE MINERALS

Pyroxene structural formula: $X^{VIII}Y^{VI}Z_2^{IV}O_6$

Site	*Occupants*
X^{VIII}	Ca^{2+}, Na^{+}
Y^{VI}	Mg^{2+}, Fe^{2+}, Mn^{2+}, Ni^{2+}, Li^{+}, Al^{3+}, Ti^{4+}, Fe^{3+}
Z^{IV}	Si^{4+}, Al^{3+}, Fe^{3+}

Amphibole structural formula: $\square_{1-0}X_{2-3}^{VIII}Y_5^{VI}Z_8^{IV}O_{22}(OH)_2$

Site	*Occupants*
$\square$	K^{+}, Na^{+}
X^{VIII}	Mg^{2+}, Fe^{2+}, Na^{+}, Ca^{2+}, K^{+}, Ba^{2+}, Li^{+}, Mn^{2+}
Y^{VI}	Mg^{2+}, Fe^{2+}, Fe^{3+}, Al^{3+}, Mn^{2+}, Mn^{3+}, Ti^{4+}, Cr^{3+}, Ni^{2+}, Zn^{2+}
Z^{IV}	Si^{4+}, Al^{3+}, Fe^{3+}

Comparison of pyroxene and amphibole compositions:

Pyroxene group	*Amphibole group*
Orthorhombic	
Enstatite, $MgSiO_3$	Anthophyllite, $(Mg, Fe)_7Si_8O_{22}(OH)_2$
Hypersthene, $(Mg, Fe)SiO_3$	
Orthoferrosilite, $FeSiO_3$	
Monoclinic	
Clinoenstatite, $MgSiO_3$	Cummingtonite, $(Mg, Fe)_7Si_8O_{22}(OH)_2$
Ferrosilite, $FeSiO_3$	
Diopside, $CaMgSi_2O_6$	Tremolite, $Ca_2Mg_5Si_8O_{22}(OH)_2$
Pigeonite, $(Ca, Mg, Fe)_2Si_2O_6$	Actinolite, $Ca_2(Mg, Fe^{2+})_5Si_8O_{22}(OH)_2$
Hedenbergite, $CaFe^{2+}Si_2O_6$	Soda-tremolite, $CaNa_2Mg_5Si_8O_{22}(OH)_2$
Johannsenite, $CaMnSi_2O_6$	Arfvedsonite, $Na_3Mg_4AlSi_8O_{22}(OH)_2$
Augite, $(Ca, Mg, Fe^{2+}, Fe^{3+}, Al)_2(Si, Al)_2O_6$	Riebeckite, $Na_2Fe_3^{2+}Fe_2^{3+}Si_8O_{22}(OH)_2$
Aegirine, $NaFe^{3+}Si_2O_6$	Glaucophane, $Na_2(Mg, Fe^{2+})_3(Al, Fe^{3+})_2Si_8O_{22}(OH)_2$
Jadeite, $NaAlSi_2O_6$	Hornblende, $(Ca, Na, K)_{2-3}(Mg, Fe^{2+}, Fe^{3+}, Al)_5(Si, Al)_8O_{22}(OH)_2$
Spodumene, $LiAlSi_2O_6$	

compositions of the minerals in these groups vary widely (see Table 2-4), their physical properties and their natural associations have led mineralogists to group them together. This grouping has been verified by detailed X-ray analysis of their structures. The basic structure of the pyroxene group (Figs. 2-16 and 2-17) is silica tetrahedra that share two of their corners with neighboring tetrahedra so as to produce infinite chains with the composition $(SiO_3)_n^{2-}$. These parallel chains are linked together with cations, which lie between pairs of them. The amphibole group differs in that two chains of silica tetrahedra are cross-linked at every other tetrahedron (Figs. 2-16 and 2-18) with a chain composition $(Si_4O_{11})_n^{6-}$. Parallel double chains are also linked together by cations, which lie between them and which give the structure electrostatic neutrality.

The chain silicates are further subdivided on the basis of crystallographic symmetry, a subject discussed in Chapter 3. The basic difference is whether or not the unit cell can be described in terms of three coordinates all at right angles. If the unit cell is so described, it is called *orthorhombic*, and the pyroxenes, *orthopyroxenes*. If, however, the unit cell is described in terms of three coordinates, one of which is not at right angles to one of the other two, it is *monoclinic*, and the pyroxenes, *clinopyroxenes*. (If the unit cell requires that the coordinates all lie at angles other than right angles, it is called *triclinic*. See Section 3-7.) The major pyroxene and amphibole minerals are compared in Table 2-4.

The basic chemical composition of the pyroxene minerals may be referred

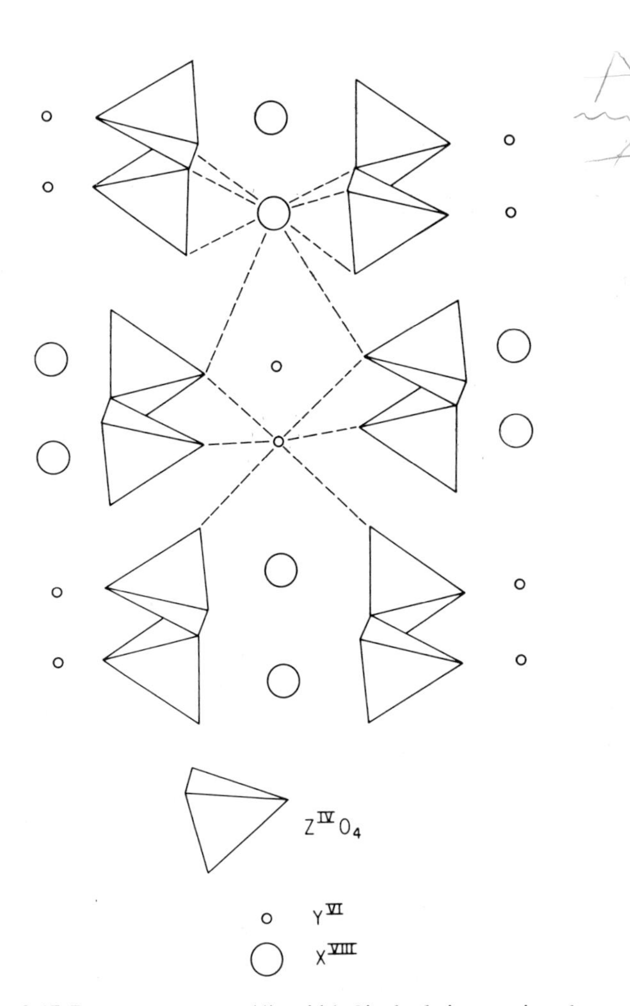

Figure 2-17 Pyroxene structure (diopside). Single chains are viewed from the end. The bonding of interchain cations to chain O^{2-} ions is shown with dashed lines.

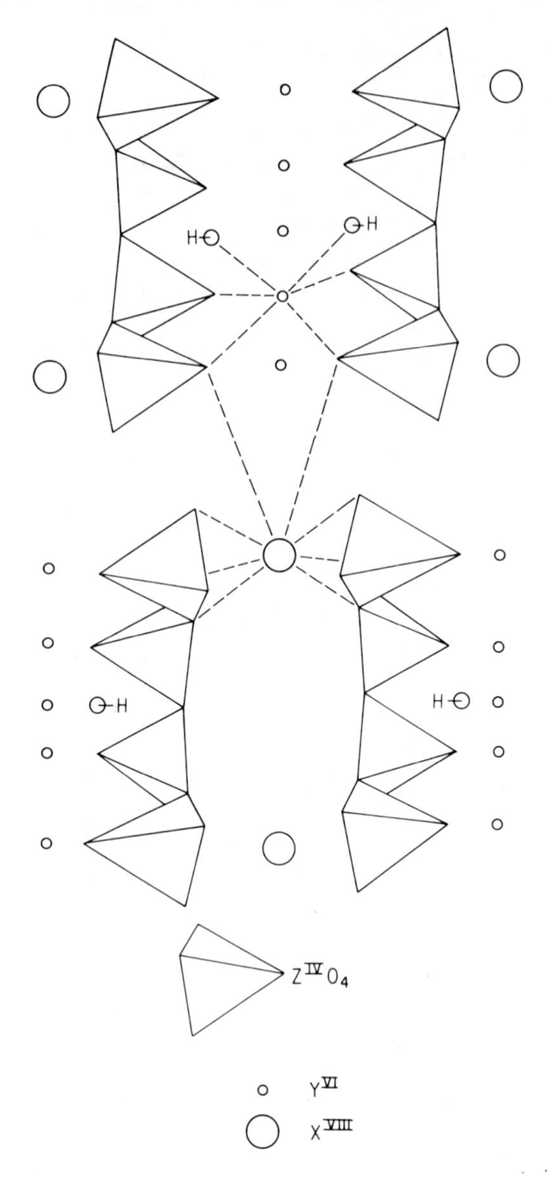

Figure 2-18 Amphibole structure (tremolite). Double chains are viewed from the end. The OH^- radicals are found in the center of the hexagonal rings, at the level of the apical O^{2-} ions. Bond directions for one X^8 and one Y^6 are shown.

to the structural formula $X^{VIII}Y^{VI}Z_2^{IV}O_6$ (Fig. 2-17). The superscripts again refer to cation CN. The eightfold coordination about X is not fixed rigidly, and may be reduced to six when small ions occupy the site. Also, the X site may or may not be occupied, depending upon the size and charge requirements of the ions involved. The Y^{VI} site contains Mg^{2+}, divalent transition metals, and Li^+, with minor amounts of Fe^{3+}, Al^{3+}, Cr^{3+}, and Ti^{4+} ions. The tetrahedral site Z^{IV} contains mostly Si^{4+}, with minor amounts of Al^{3+} and Fe^{3+}.

Although a general structural formula for the amphibole group is possible, ionic substitution is more varied and the relationships among the different minerals are more complex than those of the pyroxenes. Like the pyroxenes, amphibole unit cells may be orthorhombic or monoclinic. One general structural formula has the form

$$\square_{1-0}X_{2-3}^{VIII}Y_5^{VI}Z_8^{IV}O_{22}(OH)_2$$

The $\square$ in the formula refers to a structural vacancy between the chains that is similar to the vacancy occupied by K^+ in the mica minerals, which are described in Section 2-11. The CN of this vacancy is probably larger than VIII and smaller than XII. It is the site of some of the Na^+ and K^+ in the amphiboles. The number of occupants of the X site varies inversely with the occupancy of the vacancy. The various ions that can occupy each of these sites are listed in Table 2-4. In addition, the (OH) group may be replaced by F^-, Cl^-, or O^{2-}.

One major distinguishing feature between the pyroxenes and the amphiboles is their cleavage, that of pyroxene being prismatic at 93° and that of amphibole being prismatic with an angle of 54° between the two cleavage directions. The prismatic cleavage is to be expected because of the relative strength of the chemical bonds in the crystal. The Si–O bonds in the chains are substantially stronger than the other cation-to-oxygen bonds between the chains, and the cleavages do not cut across the chains. The difference in the angle between the cleavage planes derives from the different widths of the silica chains, as is illustrated in Figs. 2-17 and 2-18 (see also Section 4-4).

2-11. Layered Silicates

Layers of silica tetrahedra are formed when each tetrahedron shares three corners with three other tetrahedra in a two-dimensional continuous array (Fig. 2-16). Each tetrahedron has a net −1 charge located at the unshared apical O^{2-} ion. The formula for the layer is $(Si_2O_5)_n^{2-}$. The layered silicate minerals differ one from another in the way in which their Si_2O_5 layers are stacked up and bonded together by octahedral layers.

The simplest structure is that of the clay mineral kaolinite, which is a layered mineral made up of one silicate layer and one gibbsite layer (Figs.

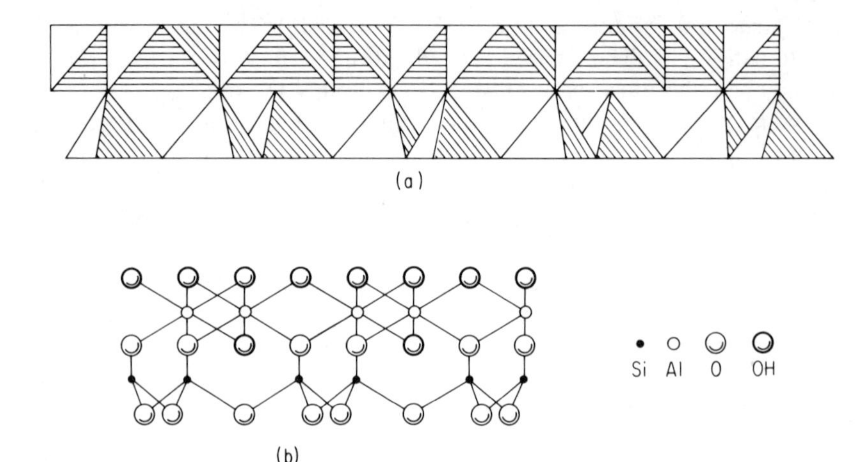

Figure 2-19 Kaolinite structure. (a) Coordination polyhedra. (b) Ions and bond directions. Both representations are to the same scale. The kaolinite layer may be thought of as a gibbsite octahedral sheet (Fig. 2-8b) imposed upon a tetrahedral silicate sheet. Layers are bonded together by van der Waals forces.

2-8 and 2-19). As can be seen from Fig. 2-16, the silica layers are made up of hexagonal units of six tetrahedra each, with all unbonded apices pointing in the same direction. If one OH^- group is added at the center of each hexagonal ring at the level of the tetrahedral apices, an approximate hexagonal closest-packed sheet is created. This sheet is the base of the gibbsite layer (Fig. 2-8). The top of the gibbsite layer is a normal HCP sheet of OH^- ions. It is called dioctahedral because two thirds of the octahedral voids are filled with Al^{3+} ions. Thus, the kaolinite layered structure is formed. There are no chemical bonds other than hydrogen bonds between the top of the gibbsite layer and the bottom of the next silica layer. The formula of the kaolinite unit cell is $Al_4Si_4O_{10}(OH)_8$.

Because the kaolinite layer is 7.13 Å in thickness, kaolinite and other minerals with the same basic structure are called 7-Å layered silicates. They are also 1 : 1 layered silicates, because there is one tetrahedral layer to one octahedral layer. Minerals of the kaolinite group differ in the way in which the successive kaolinite layers are stacked. Kaolinite has a repeat distance of one, dickite a repeat distance of two, and nacrite a repeat distance of six layers (see Fig. 2-22). The mineral halloysite differs from kaolinite in that there is a water layer one molecule thick between the kaolinite layers. In general, there is very little substitution for either Al^{3+} or Si^{4+} in the kaolinite-group minerals, and they are chemically very pure.

When divalent cations fill all the available octahedral voids in the hydroxide layer, it is called trioctahedral. The trioctahedral 7-Å layered silicates are the septechlorites. The structure of 7-Å serpentine [$Mg_6Si_4O_{10}(OH)_8$] is the same as that of kaolinite, with the four Al^{3+} ions replaced in the octahedral layer by six Mg^{2+} ions. Greenalite [$Fe_6Si_4O_{10}(OH)_8$] has the same structure, but with Fe^{2+} in place of Mg^{2+} ions. Other members of the septechlorite group exist, but are relatively rare.

The OH^- ion spacing in the octahedral layer is slightly different from the O^{2-} ion spacing in the tetrahedral layer. For this reason, structural stresses between the layers are built up over relatively short distances. In the kaolinite group, this results in very small crystals, which are typical of clays in general. In the septechlorite group, the result is a curvature of the layer so that crystals of chrysotile occur in tubes approximately 100 Å in radius. Megascopically, it appears as fibrous crystals, as in asbestos. The mineral antigorite has chemical composition and structure similar to chrysotile, but the intraplaner stress is relieved by corrugation rather than by curling.

The talc–pyrophyllite layered structure may be regarded as a sandwich formed by two tetrahedral layers with their apices pointed at an interposed octahedral layer (Fig. 2-20). The octahedral layer is chemically bonded to each tetrahedral layer in the manner described by the bonding in kaolinite. Only van der Waals bonding between silica bases holds these three-layer

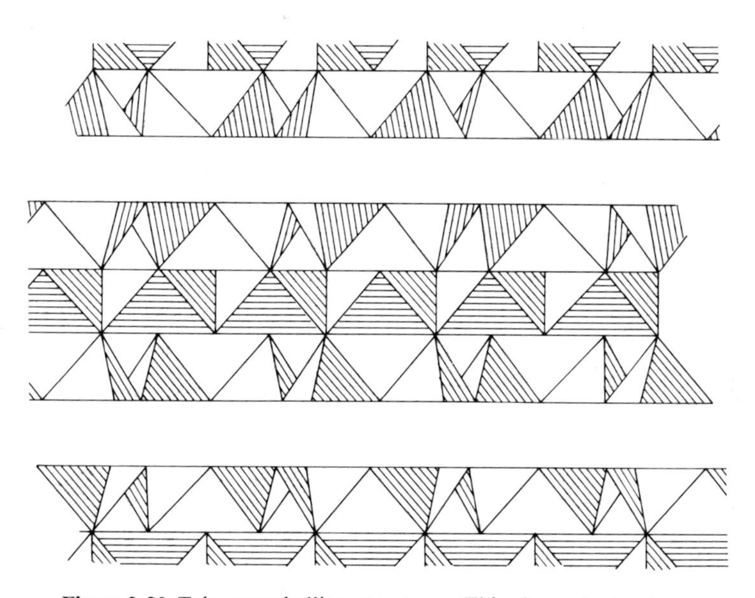

Figure 2-20 Talc–pyrophyllite structure. This layered structure may be visualized as a gibbsite or brucite octahedral sheet (Fig. 2-8) between two opposing tetrahedral silicate sheets (Fig. 2-16). Van der Waals forces bond the layers together.

units together. Because of the thickness of the talc–pyrophyllite unit, this group is called the 9-Å group. Because of the ratio of tetrahedral to octahedral layers, it is designated the 2 : 1 group.

If the octahedral layer contains Mg^{2+} ions, the mineral is talc [$Mg_3Si_4O_{10}(OH)_2$]; if Al^{3+} ions, pyrophyllite [$Al_2Si_4O_{10}(OH)_2$]; and if Fe^{2+}, minnesotaite [$Fe_3Si_4O_{10}(OH)_2$]. Talc and minnesotaite are trioctahedral, and pyrophyllite is dioctahedral.

The mica-group minerals (Fig. 2-21) may be derived from the talc group by removing every fourth Si^{4+} ion from the tetrahedral layer and substituting

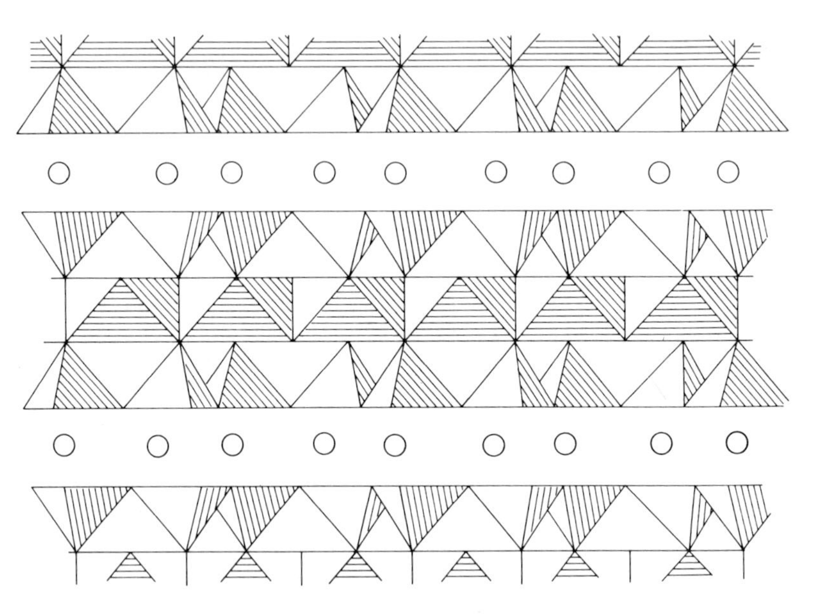

Figure 2-21 Mica structure. This layered structure may be visualized as a talc–pyrophyllite structure in which one fourth of the tetrahedral Si^{4+} is replaced by Al^{3+}, with electrostatic neutrality being maintained by the introduction of K^+ between the layers.

an Al^{3+} ion for it. Electrostatic neutrality is maintained by putting Na^+, K^+, Ca^{2+}, or H_3O^+ between the layers. The interlayer cation has a six-member tetrahedral ring on each side of it, giving it a coordination number of 12. This is equivalent to the □ position in the amphiboles.

Muscovite [$KAl_2(AlSi_3O_{10})(OH)_2$] is thus derived from pyrophyllite and has K^+ in the interlayer position. Paragonite is similar, with Na^+ in the interlayer position. Phlogopite [$KMg_3(AlSi_3O_{10})(OH)_2$] is similarly derived from talc, and annite from minnesotaite. Few trioctahedral micas are pure end members, however, and biotite is intermediate between phlogopite and annite. Vermiculite has a biotite structure in which some, or all, of its K^+ is replaced by hydronium (H_3O^+).

Lepidolite [$KLi_2Al(Si_4O_{10})(OH)_2$] is a lithium-bearing mica found in some pegmatite deposits. In this structure, there is no Al^{3+} substituting for Si^{4+} in the tetrahedral layer. Two Li^+ ions replace one Al^{3+} ion in the octahedral layer, resulting in a type of trioctahedral mica. The charge deficiency thus comes from the octahedral layer and is balanced by interlayer K^+ ions.

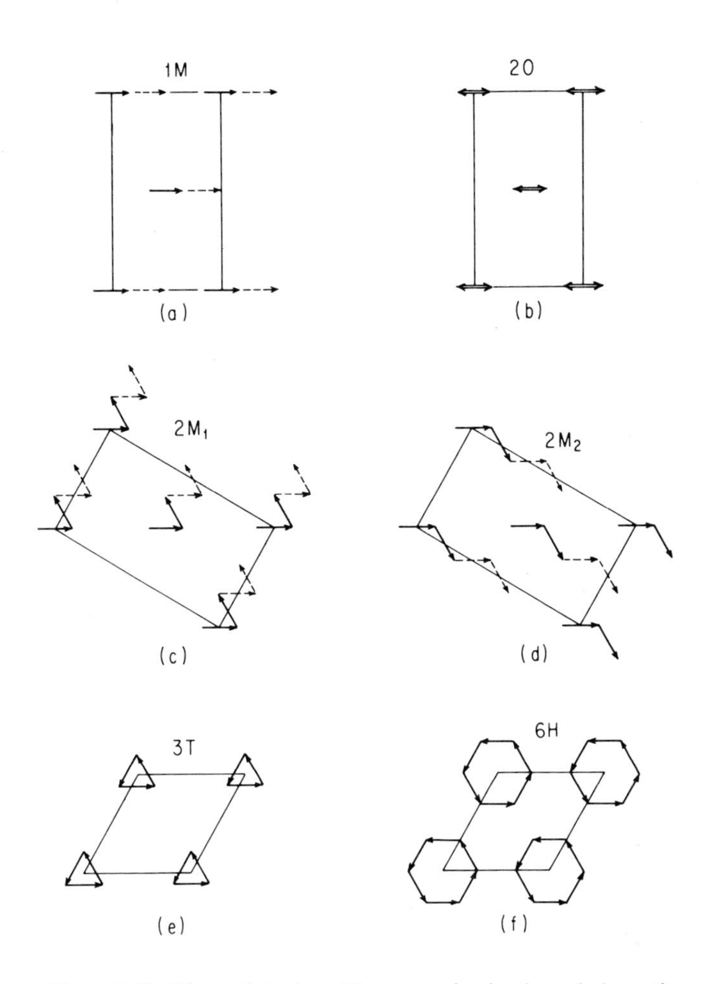

Figure 2-22 Mica polytypism. There are six simple and theoretically possible ways of stacking mica layers in an ordered manner. Arrows connect identical points in two adjacent layers. (a) One-layer unit cell with monoclinic symmetry (symmetries are defined in Chapter 3). (b) Two-layer unit cell with orthorhombic symmetry (c) Two-layer unit cell with monoclinic symmetry. (d) Alternative two-layer unit cell with monoclinic symmetry. (e) Three-layer unit cell with trigonal symmetry. (f) Six-layer unit cell with hexagonal symmetry. Stacking may also be disordered with monoclinic symmetry. (After J. V. Smith and H. S. Yoder, 1956, *Min. Mag. 31*, p. 209.)

The composition of glauconite may be derived from that of muscovite by replacing some of the octahedral Al^{3+} with Fe^{2+} and Fe^{3+}, by the tetrahedral Al^{3+} being variable but in less than one fourth of the tetrahedra, and by having the interlayer K^+ partially replaced by Na^+, Ca^{2+}, and H_3O^+.

The composition of montmorillonite may be derived from muscovite by replacing some of the Al^{3+} with Mg^{2+}. The amount of Al^{3+} in the tetrahedral layer is small, generally occurring in considerably fewer than one fourth of the tetrahedra. The interlayer ions are Na^+ and Ca^{2+} in small and variable amounts.

Sericite is predominantly very fine grained muscovite. Illite is a mixture of alternately stacked muscovite and montmorillonite sheets. There is ready substitution of F^- for OH^- in the structure of all the mica-group minerals.

The structure of the chlorite group of minerals may be derived from the talc structure by adding a complete brucite layer between the talc sheets, thereby making it into a 14-Å structure. The chlorite minerals are mostly trioctahedral, but the proportions of Fe^{2+} and Mg^{2+} are variable. There is some substitution of Al^{3+} and Fe^{3+} in the octahedral layers and of Al^{3+} in the tetrahedral layer, so that there is ionic bonding between the talc and brucite layers. The 14-Å true chlorites differ from the 7-Å septechlorites in that septechlorites have the tetrahedral sheets all pointing their apices in the same direction, whereas the true chlorites have their apical directions alternating. The septechlorites are the stable polymorphs at low temperatures.

Polytypism is common in the mica group. It may be exemplified by the five different types of stacking shown for muscovite in Fig. 2-22. In addition, the stacking may be completely disordered, can contain different types of layers, or both. This mixed layering and stacking disorder, coupled with the fine grain size that occurs in sedimentary materials, presents a formidable challenge in the analysis and identification of these minerals. There is much disagreement in the classification of layered silicates as a result of systems based on different methods of analysis, which emphasize different chemical and structural similarities.

2-12. Framework Silicates

If all four corners of the silica tetrahedra are shared and each O^{2-} is bonded to two Si^{4+} ions, the result is a three-dimensional network of linked tetrahedra with the formula SiO_2. The different silica minerals—quartz, tridymite, cristobalite, and coesite—have tetrahedra linked together in different fashions. The exception to this tetrahedral coordination of Si^{4+} ions occurs in stishovite, a mineral formed by meteorite impact in sandstone, which has the rutile structure (Fig. 2-6).

Quartz, tridymite, and cristobalite have high- and low-temperature polymorphs that result from a slight change in bond angle between tetrahedra, and a resulting change in symmetry. The high–low polymorphic inversion is rapid and nonquenchable, since it involves no rupture of Si–O bonds. The polymorphic transitions between the silica mineral species, however, are sluggish and quenchable because they involve rupture and rearrangement of the Si–O bonds.

The stable form of silica at the surface of the earth is quartz, but the high-temperature polymorphs tridymite and cristobalite, and the high-pressure polymorphs coesite and stishovite, are found in rocks and are preserved in the laboratory. The three-dimensional linkage of the silica tetrahedra in the silica minerals is complex enough so that they are best studied with models.

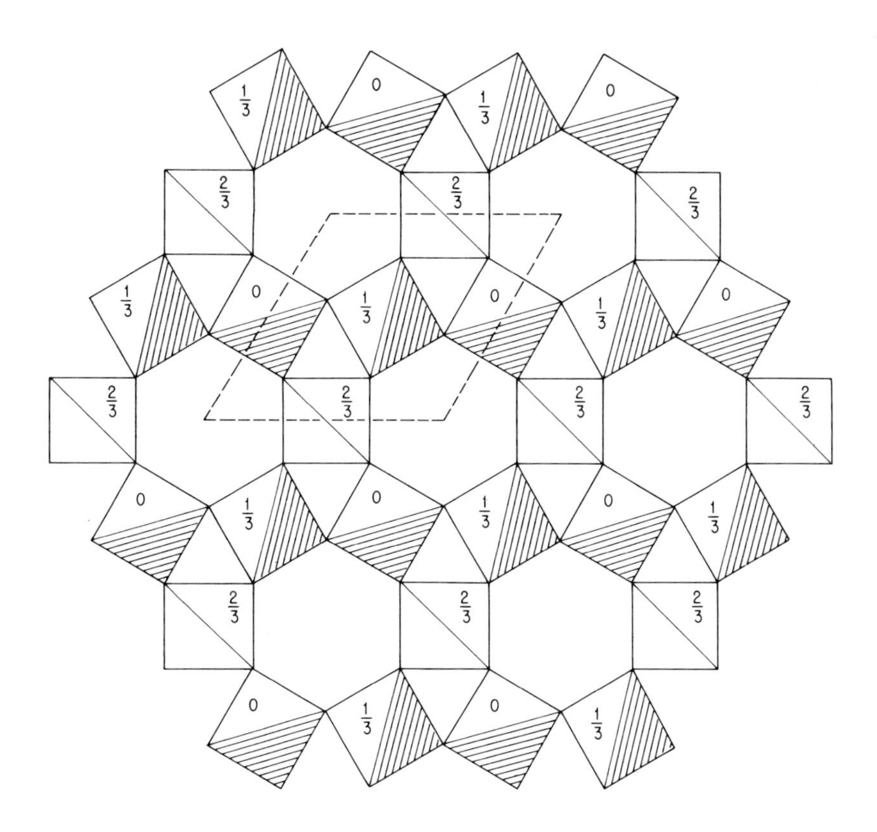

Figure 2-23 Quartz structure. Tetrahedra are shown in silhouette with one edge down. The structure may be visualized in terms of interconnected double helices of tetrahedra that share all four corners. The elevation of the individual tetrahedra above the base is indicated by fractions, such that each helix proceeds 0, $\frac{1}{3}$, $\frac{2}{3}$, above 0, above $\frac{1}{3}$. . . . The dashed line outlines the unit cell.

The structure of quartz (Fig. 2-23) is based on a three-dimensional linkage of silica tetrahedra so that they are arranged in cross-linked double helices. The helices of silica tetrahedra are shown in projection in Fig. 2-23, with the relative heights of the tetrahedra indicated by fractions. An individual helix begins at a 0 tetrahedron and continues around the hexagonal ring to the $\frac{1}{3}$ and the $\frac{2}{3}$ tetrahedra. The fourth tetrahedron in a helix is directly over the next 0 tetrahedron and the fifth over the next $\frac{1}{3}$ tetrahedron. These helices may appear as rotating either to the right (as in Fig. 2-23) or to the left, the two being mirror images of one another (see Section 3-10).

The structure of tridymite (Fig. 2-24) is a more open three-dimensional network of silica tetrahedra than that of quartz. The tridymite structure may be visualized in terms of sheets of silica tetrahedra arranged in six-membered rings, but with adjacent tetrahedra in the sheet pointing in opposite directions. The second sheet in the stack is a mirror image of the first and joined to it

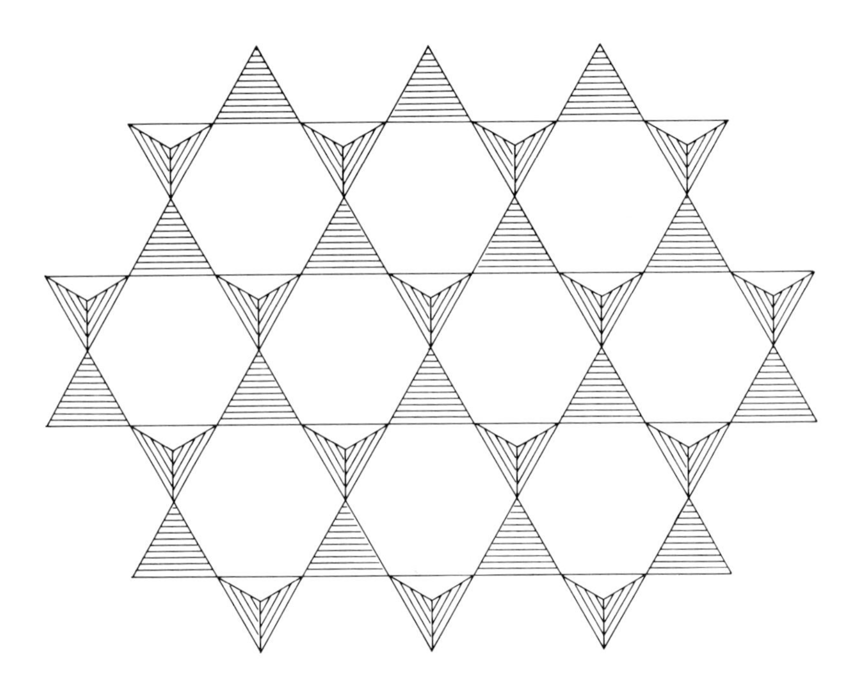

Figure 2-24 Tridymite structure. This structure may be visualized as layers of hexagonal rings of tetrahedra with alternate apices pointing in opposite directions. Layers are added such that the upward-pointing apices of the lower layer are shared with the downward-pointing apices of the upper layer. All four corners of each tetrahedron are shared. Layers are stacked such that the hexagonal rings are aligned.

where the tetrahedral apices meet. As with quartz, large channels run through the structure, but, unlike in quartz, these are not helical.

The structure of cristobalite (Fig. 2-25) is related to that of tridymite in that it may be visualized in terms of sheets of six-membered rings with

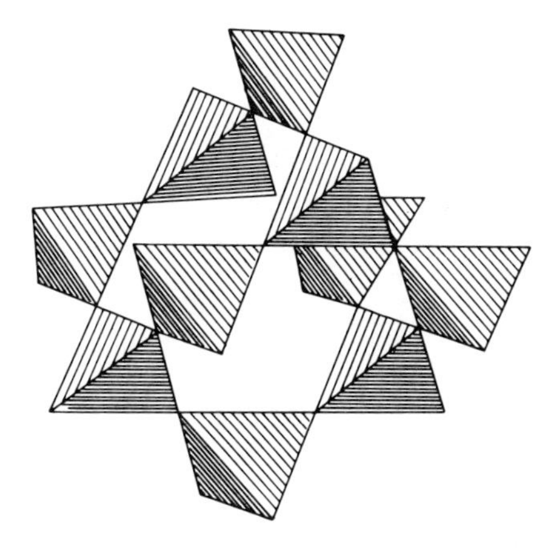

Figure 2-25 Cristobalite structure. This structure may be visualized as layers of hexagonal rings of tetrahedra with alternate apices pointing in opposite directions. Layers are added such that the upward-pointing apices of the lower layer are shared with the downward-pointing apices of the upper layer. All four corners of each tetrahedron are shared. Layers are stacked such that the hexagonal rings are staggered between layers.

adjacent tetrahedra pointing in opposite directions. The stacking of the sheets is different, in that the rings are offset between sheets. However, open channels are still present in the structure.

The helices of quartz are large enough to house ions no larger than Li^+, whereas the channels of tridymite and cristobalite are large enough to tolerate ions as large as Na^+ and even K^+.

The high-pressure, high-temperature form of silica, coesite, was first synthesized and described in 1953. Subsequently, it was found as a rare mineral, along with stishovite, in sandstones that had been severely shocked by meteorite impact. The structure of coesite (Fig. 2-26) is of interest in that it shows considerable similarity with the feldspar structures. In coesite, the silica tetrahedra are linked in two kinds of four-member rings. Each such ring has eight corners not shared within the ring. Two of these corners are linked with adjacent rings to form chains (Fig. 2-28). The remaining unshared corners are cross-linked to adjacent layers of parallel chains.

The feldspar minerals have structures based upon the coesite structure,

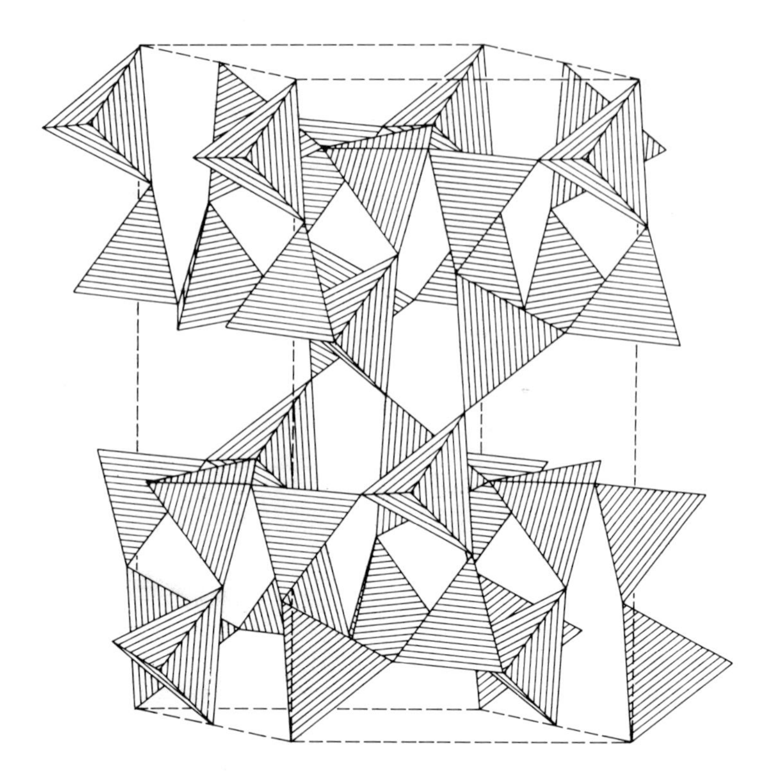

Figure 2-26 Coesite structure. Tetrahedra, each sharing all four corners, are linked together into four-member rings. See Figs. 2-28 and 2-29 for details. (After T. Zoltai and M. J. Buerger, 1959, *Zeits. Krist. 111*, p. 129.)

but with one fourth to one half the tetrahedral Si^{4+} ions replaced by Al^{3+} ions, and electrostatic neutrality maintained by stuffing K^+, Na^+, Ba^{2+}, and Ca^{2+} ions between the linked chains of tetrahedra. The various feldspars and their crystal solution series are shown in Fig. 2-27.

As can be seen from Fig. 2-26, spaces exist between the linked chains in the coesite structure. These spaces house the alkali and alkali-earth ions, which maintain electrostatic neutrality in the structure when Al^{3+} replaces Si^{4+} in the feldspar structures. Thus, these structures are stuffed derivatives of the coesite structure. They are divided into two groups based upon the size of the stuffing ion, Ba^{2+} and K^+ being substantially larger than Na^+ and Ca^{2+}. Within these groups, they may also be divided on the basis of the ordering of the Al^{3+} and Si^{4+} ions in different tetrahedral sites.

The basic structure of celsian [$BaAl_2Si_2O_8$], orthoclase, and microcline may be illustrated with the structure of high sanidine (Figs. 2-28 and 2-29).

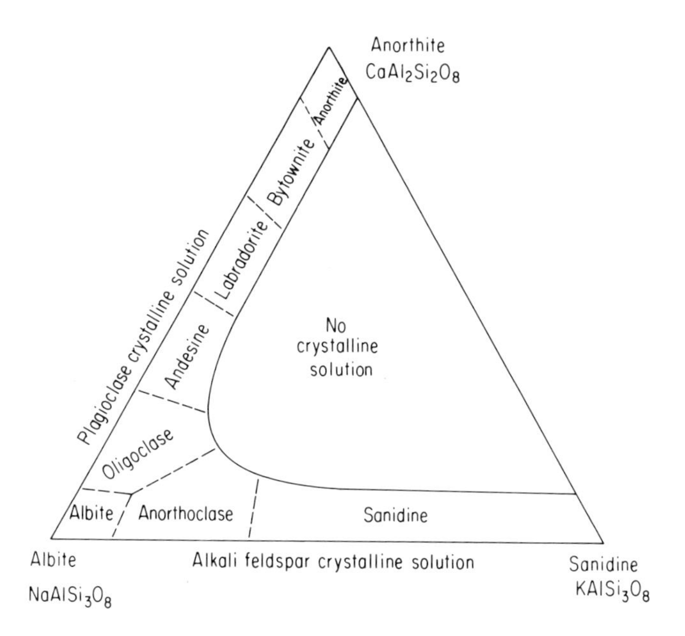

Figure 2-27 Feldspar minerals. Complete crystalline solution exists between sanidine and albite and between albite and anorthite at high temperatures. At low temperatures orthoclase and albite are immiscible, and the plagioclase crystalline solution series becomes immiscible in a complex and, as yet, poorly understood fashion.

As in the coesite structure (Fig. 2-26), the basic structural unit is a four-membered ring with one half the tetrahedra pointing up and one half pointing down (Fig. 2-28a). The elevations of the apices of the tetrahedra are indicated by fractions. Tetrahedron *A* is superimposed on tetrahedron *B* where they are joined at the apices numbered $\frac{1}{2}$. These four-membered rings are thus stacked into chains (Fig. 2-28b), with the remaining unshared apices available to cross-link the chains into a three-dimensional network of tetrahedra (Fig. 2-29). In this structure the Al^{3+} and Si^{4+} ions are distributed completely at random in the tetrahedral sites. The K^{+} ions are in ninefold coordination with one O^{2-} ion measurably closer than the others. Some Na^{+} commonly substitutes for K^{+} in this mineral.

Orthoclase has a similar structure, but the overall symmetry is different because of a partial ordering of the Al^{3+} and Si^{4+} ions between distinct types of tetrahedral sites. Microcline results from the complete ordering of Al^{3+} and Si^{4+} ions, so that there are distinctly different kinds of tetrahedral coordination polyhedra in the structure. In the high-temperature disordered

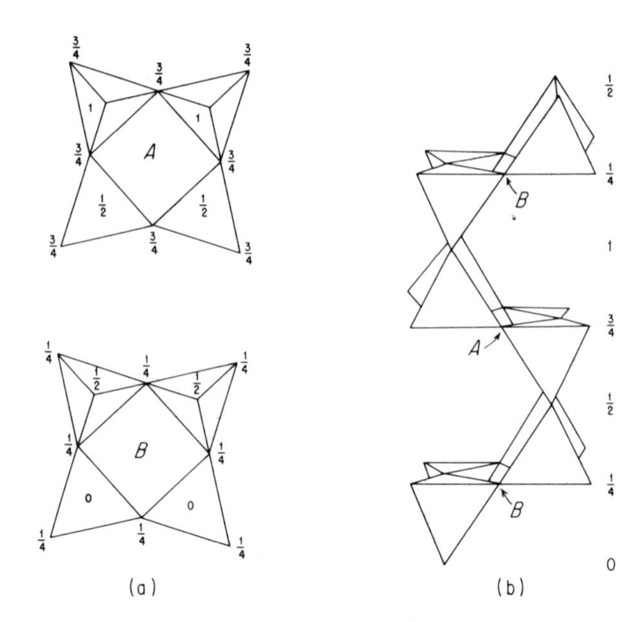

Figure 2-28 Tetrahedral linkage in the sanidine structure. This structure is fundamentally similar to the coesite structure. (a) Two four-member rings of tetrahedra. The elevations of the apices are indicated by fractions. The apices labeled $\frac{1}{2}$ in the *A* ring are shared with the apices labeled $\frac{1}{2}$ in the *B* ring. (b) Side view of chain of four-member rings. The fractions and the ring labels are the same as in (a). The unshared apices are shared with adjacent chains to form a three-dimensional array.

forms, there is complete crystalline solution between K-feldspar and Na-feldspar, and between K-feldspar and Ba-feldspar. In the low-temperature ordered form, the degree of crystalline solution is restricted.

The basic difference between the plagioclase structures and the sanidine structure is a marked shortening of the unit cell, owing to a partial collapse of the chain around the smaller Na^+ and Ca^{2+} ions. Again, there is a high-temperature disordered form and a low-temperature ordered form for the plagioclase minerals. There is complete crystalline solution between high albite and high anorthite. The degree and kinds of crystalline solutions between the low-temperature end members are quite complex and only poorly understood at present. The inversion from high- to low-temperature structures is sluggish, to the extent that low-temperature ones have not been synthesized from chemical components in the laboratory.

The nepheline group of minerals (Fig. 2-30) has a structure based on the tridymite structure (Fig. 2-24). Half the Si^{4+} ions have been replaced by Al^{3+}

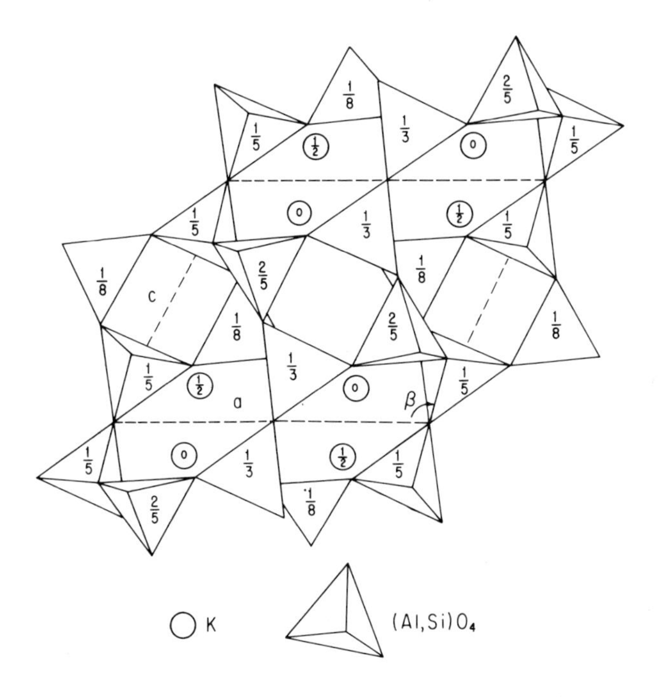

Figure 2-29 Sanidine structure. Chains of four-member rings linked in a three-dimensional array. One fourth of the tetrahedra are occupied by Al^{3+} and three fourths by Si^{4+}. Electrostatic neutrality is maintained by stuffing K^+ ions (circles) between the chains. Fractions indicate elevations of cations above the base. Compare the sanidine structure with the coesite structure (Fig. 2-26). Other feldspar structures are modifications of the sanidine structure.

ions, and electrostatic neutrality is maintained by Na^+ and K^+ ions occupying the voids in the tridymite network of tetrahedra. Ideal nepheline has three Na^+ ions for each K^+ ion. This ratio is maintained because introduction of the alkali ions into the tridymite structure distorts it; three fourths of the cavities are collapsed into an oblong form and one fourth retain a more regular shape. The smaller Na^+ ions go into the collapsed cavities with a CN of VIII, and the larger K^+ ions into the regular ones with a CN of IX. In natural nepheline, there is a coupled deficiency from stoichiometry of K^+ and Al^{3+}.

Kalsilite [$KAlSiO_4$] also has a structure based on that of tridymite, but it is not isostructural with nepheline. Kalsilite is relatively rare, but important as an end-member mineral in experimental studies.

One of the most important characteristics of the feldspar structures is the substitution of Al^{3+} for Si^{4+} in tetrahedral coordination, with the resulting

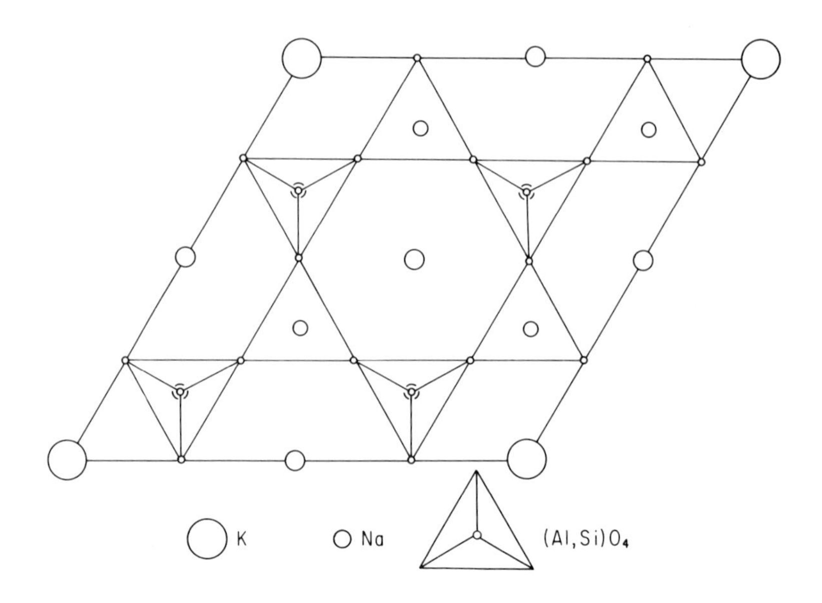

Figure 2-30 Nepheline structure. This is a stuffed derivative of the tridymite structure (Fig. 2-24). One half the tetrahedra have Al^{3+} substituting for Si^{4+}, and electrostatic neutrality is maintained by stuffing alkali ions into the spaces among the tetrahedra.

charge imbalance in the lattice being balanced by other, nonlattice cations. Other ions that can proxy for Si^{4+} in tetrahedral coordination in the tektosilicates are Be^{2+}, P^{5+}, Ge^{3+}, Ga^{4+}, and Fe^{3+}.

Although formerly classified as an independent ring silicate, beryl [$Be_3Al_2Si_6O_{18}$] more logically belongs with the framework silicates in that the six-member rings of silica tetrahedra are linked by berylia tetrahedra in such a manner as to form a three-dimensional tetrahedral network. Cordierite [$Al_3(Mg, Fe^{2+})(AlSi_5O_{18})$] contains six-member rings composed of five silica tetrahedra and one alumina tetrahedron, which are linked into a three-dimensional network by more alumina tetrahedra.

The zeolites are a large group of aluminosilicates of considerable structural complexity. Although less important geologically than the feldspars, they have a wide application in industry as molecular sieves and as ion-exchange materials. As with the garnets, this industrial interest has led to the synthesis of a number of zeolites that have no natural counterpart.

The zeolite structures may be characterized as three-dimensional lace patterns composed of linked four-, five-, six-, eight-, and twelve-membered rings of tetrahedra. The large cavities in this structure may contain water and organic or inorganic molecules, as well as various ions and ion groups. In many zeolites, volatile cavity-filling materials can be driven off by heating and reintroduced or replaced by soaking in the appropriate solution. The rings of tetrahedra form windows that vary in size according to the number of members in the ring. These windows permit the passage of molecules up to a certain size and block the passage of anything larger. As in the feldspars, electrostatic neutrality is maintained, in many zeolites, by the presence of alkali and alkali-earth ions. Many of these are quite easily replaced. The commonly used water softener is a zeolite that contains Na^+ which it exchanges for the Ca^{2+} ions in hard water.

The structure and properties of the zeolites, in general, may be illustrated by a synthetic zeolite called Linde molecular sieve type A $[Na_{12}(Al_{12}Si_{12}O_{24})NaAlO_2 \cdot 29H_2O]$ (Fig. 2-31), which is made up of four-, six-, and eight-membered rings. A feature of the structure is a pair of four-membered rings joined together in such a fashion as to form a cube (Fig. 2-31a). The cubes lie along the edge centers of the unit cell (Fig. 2-31b). The

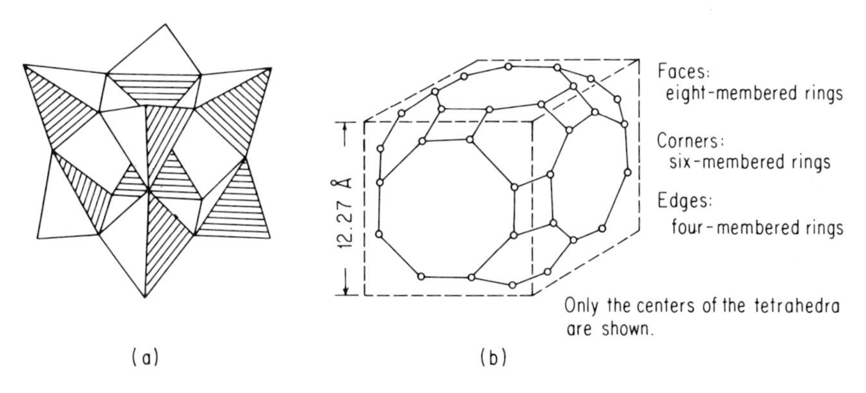

Figure 2-31 Structure of the synthetic zeolite Linde molecular sieve type A. (a) Two rings of four tetrahedra, each of which are linked together into a cube. (b) Zeolite unit cell. The cubes shown in (a) occur at the edge of the unit cell and are linked together in six-member rings about each corner and eight-member rings about each face. Only the centers of the tetrahedra are shown.

cubes are joined together as six-member rings at the unit-cell corners and as eight-member rings at the unit-cell faces. The diameter of the central cavity is 11.4 Å and the diameter of the window formed by the eight-member ring is 4.2 Å. In addition, there are smaller cavities at the corners of the unit cell, which are 6.6 Å across.

The amount of water in the structure varies from 20 to 30 molecules per unit formula. The Na^+ ions can be replaced in the structure by Li^+, K^+, Tl^+,

NH^{4+}, Ca^{2+}, Sr^{2+}, Ba^{2+}, Mg^{2+}, Zn^{2+}, Pb^{2+}, Ni^{2+}, and Co^{2+} ions. It is interesting to note that all 13 Na^+ ions may be replaced by Ag^+, whereas only 12 Na^+ ions can be replaced by Tl^+. The last Na^+ ion appears to occupy a site different from the other 12 and isolated by an access space with a diameter between 2.52 and 2.94 Å.

The minerals analcime [$NaAlSi_2O_6 \cdot H_2O$] and leucite [$KAlSi_2O_6$] have structures that are similar to Linde molecular sieve type A, but are more complicated. Other zeolites are quite complex in structure, but, in principle, they are built the same way.

FURTHER READINGS

BRAGG, W. L., G. F. CLARINGBULL, and W. H. TAYLOR, *Crystal Structures of Minerals.* Ithaca, N.Y.: Cornell University Press, 1965.

DEER, W. A., R. A. HOWIE, and J. ZUSSMAN, *Rock-Forming Minerals*, vols. 1–5. New York: John Wiley & Sons, Inc., 1962–1963.

———, *An Introduction to the Rock-Forming Minerals.* New York: John Wiley & Sons, Inc., 1966.

WELLS, A. F., *Structural Inorganic Chemistry*, 3rd ed. London: Oxford University Press, Inc., 1962.

3

Crystal Symmetry

Crystallography, a branch of the study of minerals, began in the seventeenth century. As late as 1600, popular opinion held that mineral crystals grew in the same manner as any other living thing, and that transparent gem crystals were ice which somehow had been congealed by extreme cold and under the influence of some ill-defined substance. Indeed, the word crystal (κρ́ηστaλλος) derives from the Greek word for ice. It was usually applied to rock crystal, or quartz.

As a modern science, crystallography began with the demonstration, made in 1669 by Niels Stensen (Nicolaus Steno), that the angles between the same two faces of quartz crystals are the same for all quartz crystals, regardless of the size of the crystal or the relative sizes of the two faces. The constancy of interfacial angles has been called *Steno's law*. This observation opened the first period of crystallographic investigation, which lasted through most of the seventeenth and eighteenth centuries. The activities of crystallographers of this period consisted primarily of systematic observation, of classification, and of speculation about the meaning of these observations.

Without any direct evidence (see Section 1-6), a concept was evolved that crystals were built up of atoms. Johannes Kepler, in 1611, used the idea of packing of spheres to account for the shapes of snowflakes, and Robert Hooke extended the idea to other crystals in 1665. In 1690, Christian Huygens postulated ellipsoidal atoms to account for the rhombic cleavage and double refraction of calcite. Robert Boyle and Isaac Newton proposed, on the other hand, that the atoms which make up crystals must be small polyhedra, so as to avoid the necessity of void spaces between atoms.

In the late eighteenth century, Romé de l'Isle sought to classify crystalline materials on the basis of crystal form. One of his students invented, in 1783, a contact goniometer for the measurement of interfacial angles. This led to the enumeration of over 100 different crystal forms.

The leading school of crystallography at the end of the eighteenth and the beginning of the nineteenth centuries was headed by René Haüy, who devoted most of his effort to the analysis by cleavage of the shapes of the ultimate building blocks of matter. He concluded that there were but three shapes of atoms: parallelepipeds, triangular prisms, and tetrahedra. One of his students, Christian Weiss, found that a more convenient way of visualizing crystal forms was with reference to a set of three coordinate axes. This line of study began the second period of crystallographic investigation, one that lasted from 1815, when Weiss succeeded in identifying four different crystal systems, until 1912, when von Laue supervised the first X-ray diffraction by a crystal structure. This period is characterized by the development of the theoretical foundations of crystallography on a mathematical basis.

Weiss also proposed that crystal faces could easily be represented in terms of their intercepts on the crystal axes at integral multiples of their lengths along the axes, and that certain crystal faces could be related by symmetry operations.

Within a decade, Friedrich Mohs, whose name identifies the hardness scale used in mineralogy, discovered two more crystal systems. The 32 point groups, which are all the possible combinations of symmetry operations, were derived in 1830. The 14 space lattices, which are arrays of points in a geometrical pattern, were derived by Auguste Bravais in 1848. The 230 space groups, which are arrays of all the symmetry elements in all possible locations in the 14 space lattices, were derived independently by E. S. Federov and Artur Schonflies in 1891. Geometrical crystallography was complete.

The third period of crystallographic investigation began in Munich, in 1912, with the discovery that crystalline substances diffract X rays. The development of this tool, which could *see* the internal structure of crystals, made it possible for crystallographers to verify many of the speculations of the previous periods. The year after von Laue supervised the first diffraction experiment, W. L. Bragg, the son, began a systematic determination of crystal structures at Cambridge University. The structures of sphalerite, halite, and CsCl had been postulated already, and Bragg quickly verified them. At the University of Leeds, W.H. Bragg, the father, constructed an X-ray spectrometer with which he was able to measure the characteristic radiation of many metals. This led to the possibility of X-ray tubes that could produce monochromatic radiation. At the same time, the son continued his studies of the crystal structure of many minerals.

Since that time, X-ray diffraction has been used to determine the internal crystal structure of more than 15,000 crystalline compounds, both inorganic and organic.

3-1. Symmetry Operations

The unit cell of a crystal structure is the smallest volume that, when conceptually repeated indefinitely in three dimensions, contains all the different ions of the structure and all the interrelationships among those ions. The unit cell, thus, is the building block of the crystal structure. Transformations that relate individual ions within the unit cell to one another and to all other ions in the structures are called *symmetry operations.*

A symmetry operation is an imaginary transformation of ions, such that when the transformation is completed the resulting configuration is indistinguishable from the original, that is, identical with it. Ions within the unit cell that are interchangeable by symmetry operations are symmetrically identical, and the ions of different unit cells within the structure are also symmetrically identical. Therefore, the entire crystal lattice can be built from the unit cell by these conceptual transformations.

The first operation of symmetry to be considered is one that was assumed tacitly throughout Chapter 2, the operation of *translation*. The three translations necessary to build up a crystal structure from the unit cell are shown in Fig. 3-1. Each translation (**t**) is parallel in direction to a set of unit cell edges, and is equal in magnitude to the length of that set of edges. These translations need not be of equal length. In the unit cell of CsCl, which is shown in Fig. 3-1, they are equal, but in the unit cell of rutile (Figs. 2-6 and 3-2) one is clearly different from the other two. All three may be different.

Neither do the three translations need to be orthogonal. It is possible to fill space with unit cells that are parallelepipeds, in which case the translations are not at right angles to one another. An example is the unit cell of calcite, which is shown in Fig. 2-9.

The second symmetry operation to consider is that of rotation. *Rotational symmetry* occurs when the unit cell can be rotated about an imaginary axis for some fraction of a complete revolution to reveal a structure that is indistinguishable from the initial one. In addition, after completion of a rotational symmetry operation, an entire lattice, which may have been generated by the operation of translation, is also returned to a configuration which is identical to that observed initially.

The simplest rotation operation is the *1-fold rotation*. The *fold* of a rotation is the number of times that it must be repeated to get a total rotation of 360°. A 1-fold operation is a complete revolution. Clearly, the structure thus is back into identity with itself.

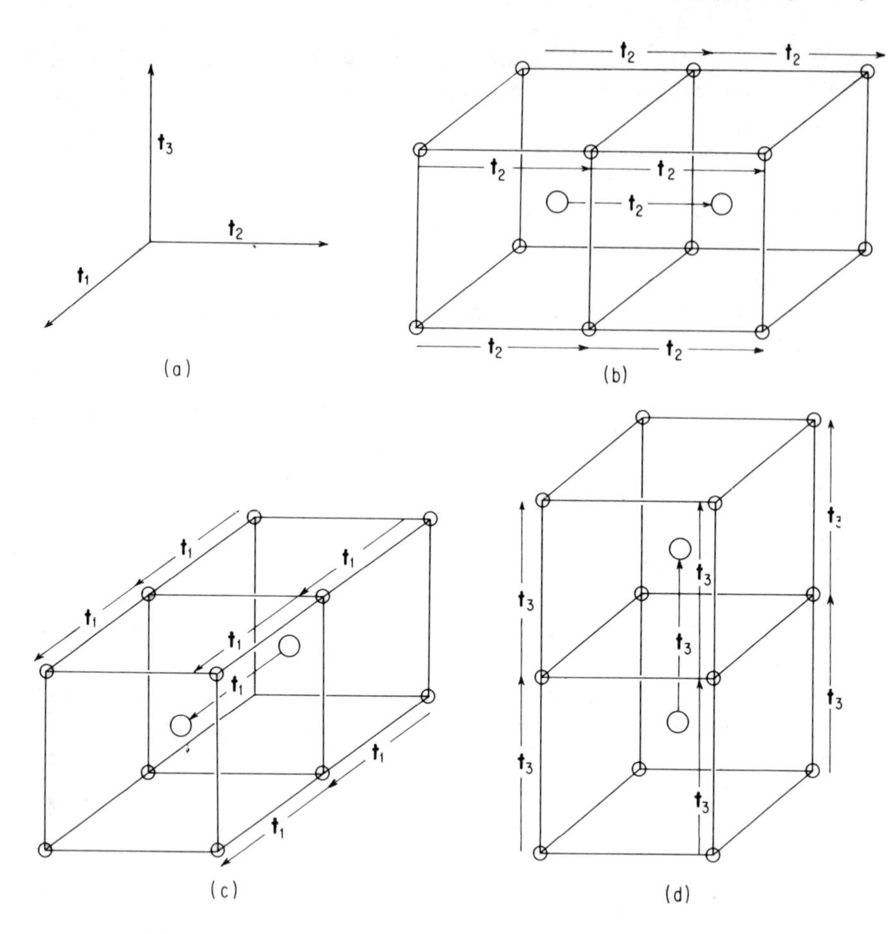

Figure 3-1 Translations of a unit cell. (a) Three translation vectors that are parallel to the unit-cell edges and of length equal to the unit-cell edges. (b) Unit cell of CsCl translated parallel to $\mathbf{t}_2$. (c) Unit cell translated parallel to $\mathbf{t}_1$. (d) Unit cell translated parallel to $\mathbf{t}_3$. The translations need not be of equal length or orthogonal.

The next rotation is the *2-fold*, that is, a rotation of 180°. Figure 3-2 shows two different 2-fold rotation axes that are possible with the rutile unit cell. The ions are numbered for clarity, although they are structurally indistinguishable. Figure 3-2b shows the resultant position of the ions after they have been rotated 180° about an axis that runs diagonally across the body of the unit cell. The ions that are on the axis are clearly not shifted by the rotation. The other ions exchange positions as indicated by the numbers in the diagram. Figure 3-2c shows a second possible 2-fold rotation axis at right angles to the one shown in Fig. 3-2b. The change of position from that indicated in Fig.

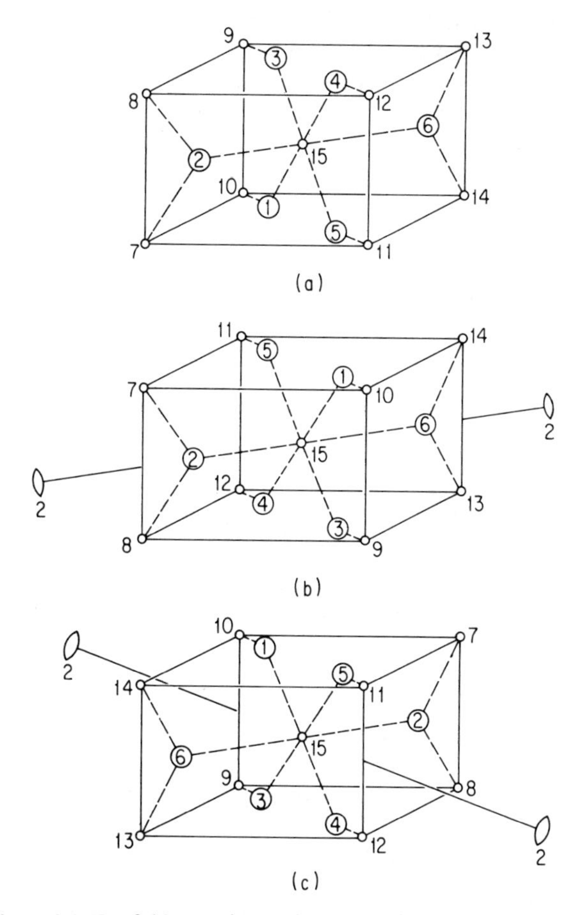

Figure 3-2 Twofold rotation axis. (a) Unit cell of rutile with individual ions numbered. (b) Same unit cell operated on by 2-fold rotation axis. (c) Rutile unit cell operated on by different 2-fold rotation axis. Numbered ions are the same in all three drawings. A second operation of each axis returns numbered ions to their original position.

3-2a is shown by the numbers in part c. A second rotation about either of these 2-fold axes will return the numbered ions to the position that they occupied in Fig. 3-2a.

A *3-fold rotation* is one of 120°. Figure 3-3 shows three layers of a CCP structure. The *A* layer is dashed, the *B* layer shown with solid circles, and the *C* layer shown by position only. A rotation of 120° about the 3-fold axis, which is perpendicular to the page of the figure through the ion labeled *C*7, moves the ions to an indistinguishable configuration, as shown by the numbers in Fig. 3-3b. Two additional operations about this axis bring the ions back to the original position shown in Fig. 3-3a.

In any unit cell in which the three translations are both equal and at right angles, there are four 3-fold axes that pass through the opposite corners of the cube (Fig. 3-4). The numbers in Fig. 3-4b show the position of the ions

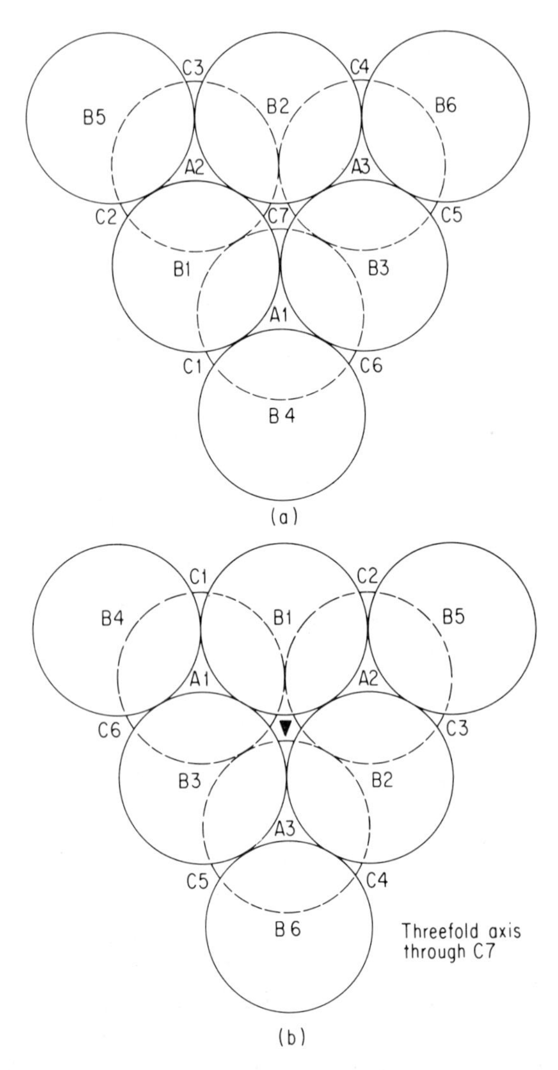

Figure 3-3 Threefold rotation axis. (a) Three layers in CCP array (compare with Fig. 1-8). *C* layer shown by position only. (b) Rotation about 3-fold axis passing throught ion *C*7. A second rotation would produce an identical configuration, but with numbered ions moved. A third rotation returns numbered ions to original positions.

after a rotation of 120° about a 3-fold axis that passes through the corners numbered 4 and 6. Figure 3-4c shows the ionic positions after a second rotation about this axis. A third rotation returns the ions to their original positions, as shown in Fig. 3-4a. Figures 3-4d, e, and f show the locations of the other three axes and the positions of the ions after one operation.

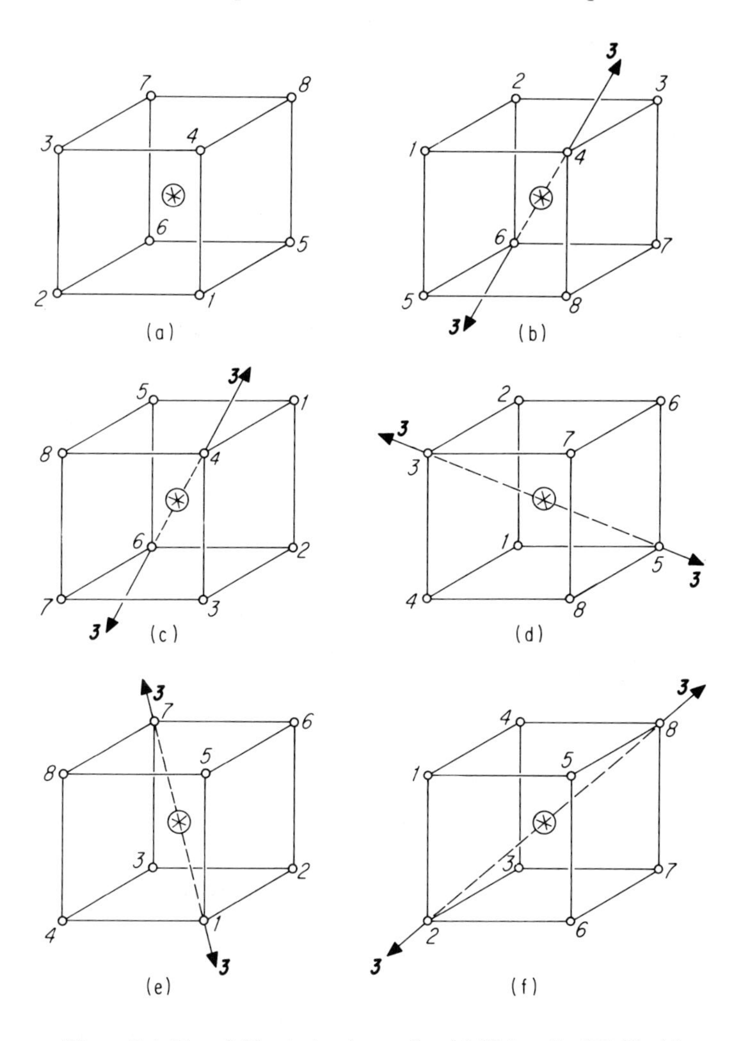

Figure 3-4 Threefold rotation in a cube. (a) Unit cell of CsCl with corner ions numbered. (b) Same unit cell operated on by a 3-fold rotation axis passing through ions *4* and *6*. (c) Ionic positions after second operation of the 3-fold axis passing through ions *4* and *6*. A third rotation about this axis returns the numbered ions to their original positions. (d) Unit cell of CsCl operated on by a 3-fold rotation axis passing through ions *3* and *5*. (e) and (f) Same unit cell operated on by axes passing through *1* and *7* and through *2* and *8*.

A *4-fold axis* of symmetry is one that returns to identity after a 90° rotation. Figure 3-5 shows the three such axes in the CsCl unit cell. The numbers in Fig. 3-5b show the first operation from part a; part c shows the second operation; part d shows the third operation; and a fourth rotation of 90° returns the ions to the original positions as shown in part a. Figures 3-5e and f show the other two axes, each operated once from the positions shown by the numbers in part a.

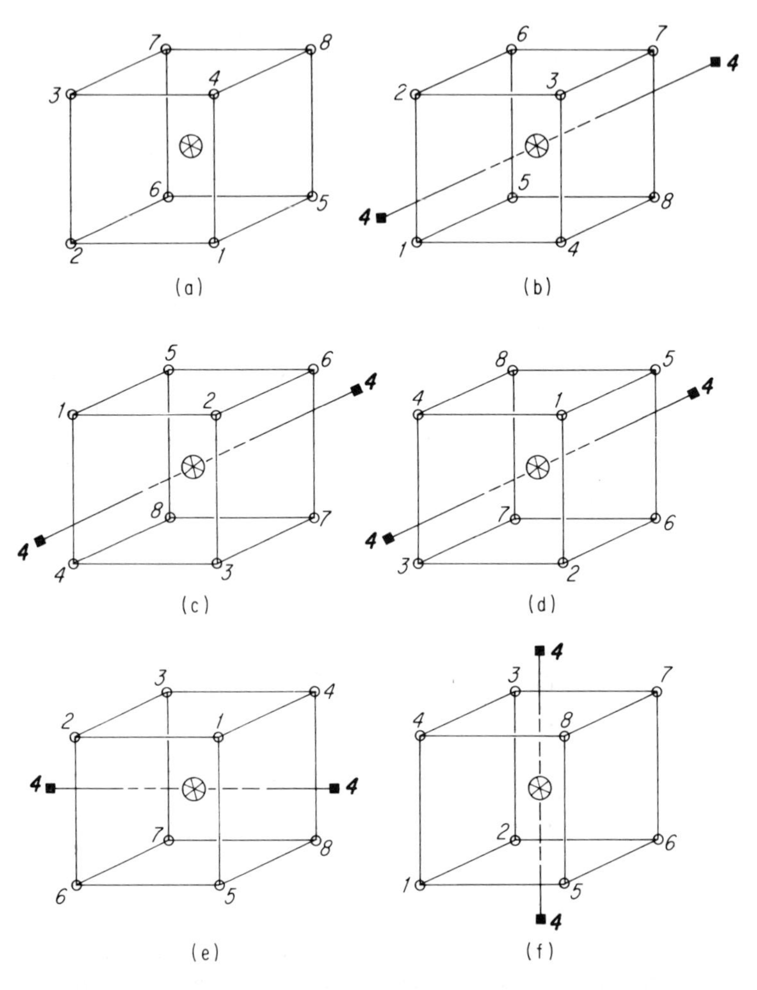

Figure 3-5 Fourfold rotation axes. (a) Unit cell of CsCl with corner ions numbered. (b) Same unit cell operated on once by a 4-fold rotation axis passing through the centers of the front and back faces. (c) Second operation of axis shown in (b). (d) Third operation of axis shown in (b) and (c). (e) and (f) Two other 4-fold rotation axes possible in CsCl unit cell.

A *5-fold rotation* about an axis is possible, but pentagonal unit cells cannot be fitted together so as to fill space.

A *6-fold rotation* is one of 60°. The numbers in Fig. 3-6 show the operation of a 6-fold axis in the graphite structure.

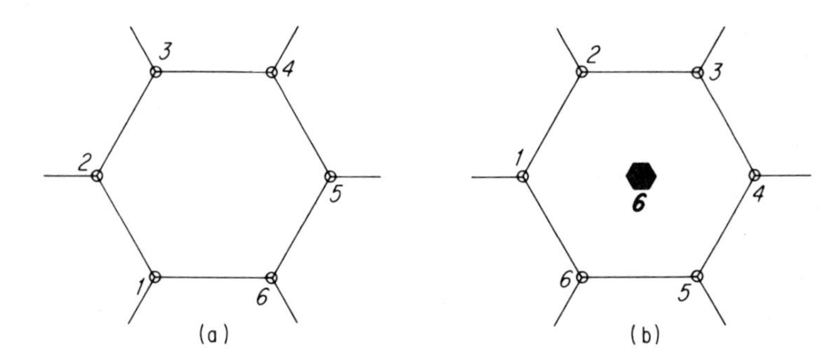

Figure 3-6 Sixfold rotation axis. (a) Part of the graphite structure with C ions numbered. (b) One operation of a 6-fold rotation axis passing through the center of the hexagon.

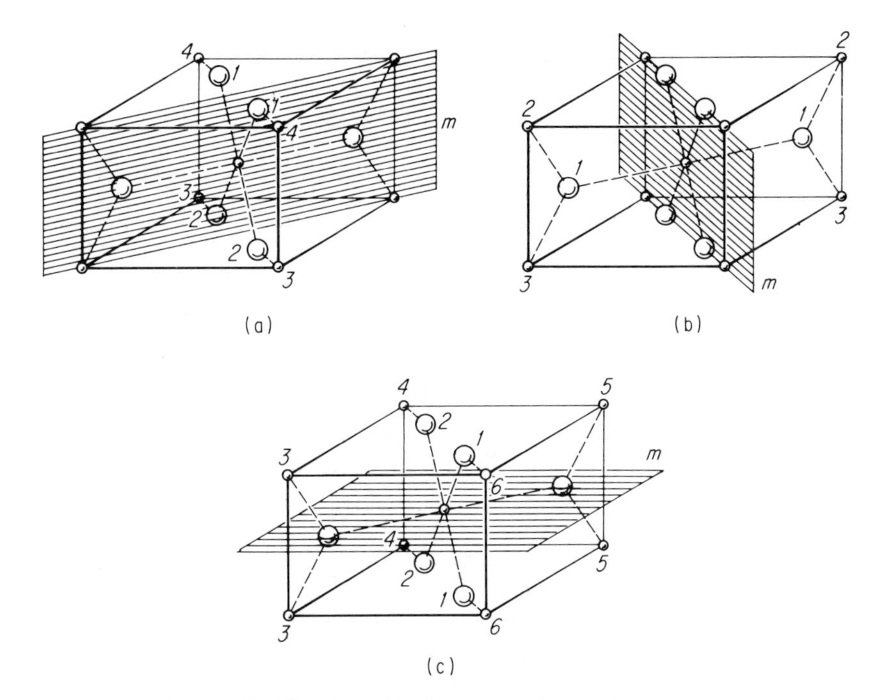

Figure 3-7 Mirror planes of symmetry. Three diagonal mirror planes of symmetry are shown for the rutile structure. Mirror symmetrical ions are indicated by numbered pairs.

The third symmetry operation to be considered is that of a mirror plane. A *mirror plane* bisects the structure in such a way that each half is a mirror image of the other half. Three possible mirror planes for the rutile structure are shown in Fig. 3-7. The paired ions on opposite sides of the plane are indicated by giving them the same numbers.

The fourth symmetry operation to consider is that of *inversion*, or a *center of symmetry*. In this operation, each ion in a unit cell has another ion paired with it on a straight line equidistant through the center (Fig. 3-8). The paired ions on opposite sides of the center are indicated by giving them the same numbers.

The combination of two operations into one may be used to describe some symmetry relations in crystal structures. Figure 3-9a illustrates, schematically, what happens when a 4-fold rotation axis is combined with a mirror plane to form the *rotoreflection* operation. Any allowable rotation may be combined with a mirror plane at right angles to the rotation axis to form a rotoreflection. The fold of the axis is indicated by a number, and the reflection is indicated by placing a tilde over the number.

Combination of a rotation with an inversion gives the *rotoinversion* operation (Fig. 3-9b). Again, any fold of rotation may be combined with inversion through the center. A bar over the number indicates that a rotation axis is one of rotoinversion.

Rotoinversion and rotoreflection are called *improper* rotations to set them

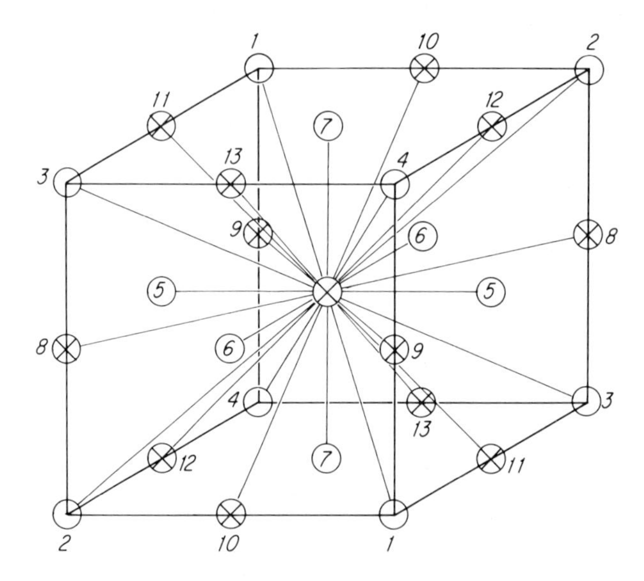

Figure 3-8 Inversion in a unit cell. The centrosymmetric ions in the unit cell of halite are indicated by paired numbered ions.

$2 + m = \tilde{2}$

$3 + m = \tilde{3}$

$4 + m = \tilde{4}$

$6 + m = \tilde{6}$

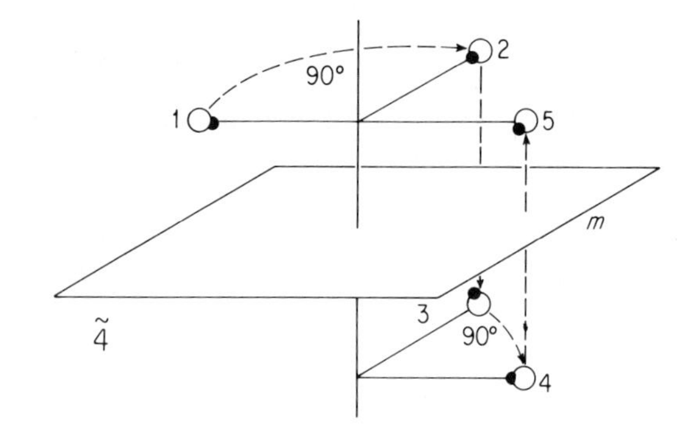

(b) Rotoinversion

$2 + i = \bar{2}$

$3 + i = \bar{3}$

$4 + i = \bar{4}$

$6 + i = \bar{6}$

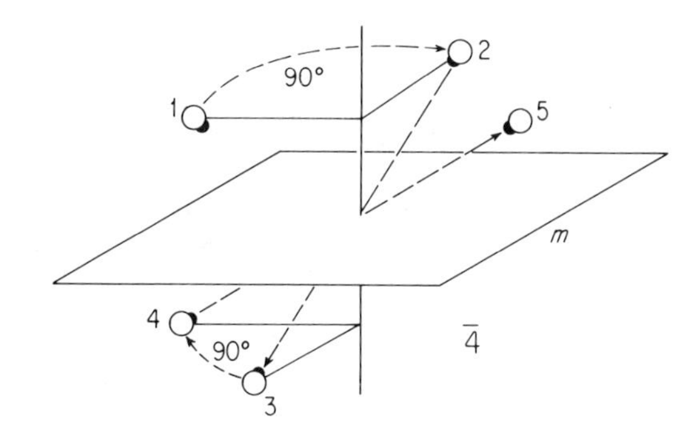

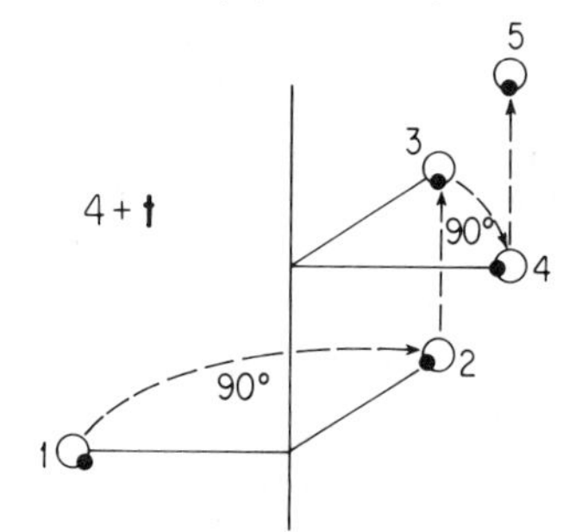

(d) Glide plane

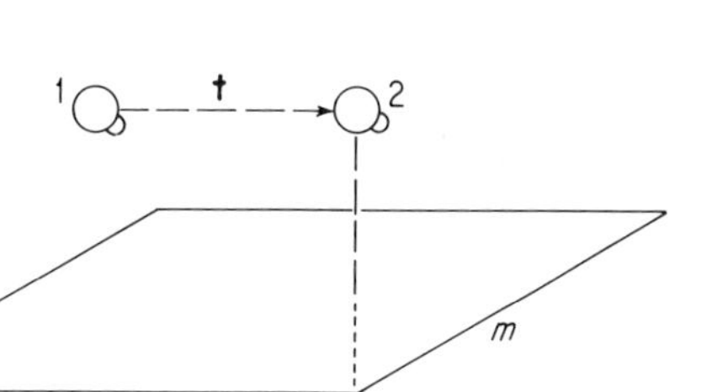

Figure 3-9 Combined symmetry operations. (a) Rotoreflection. A 4-fold rotation followed by reflection across a mirror plane that is normal to the rotation axis is shown in steps 1–3. (b) Rotoinversion. A 4-fold rotation is followed by inversion through a center of symmetry in steps 1–3. (c) Screw axis. A 4-fold rotation operation is followed by translation parallel to the axis of rotation in steps 1–3. A second operation is shown in steps 3–5. Reflection, inversion, and translation may be combined with 1-, 2-, 3-, and 6-fold rotations as well. (d) Glide plane. A translation is followed by reflection across a mirror plane parallel to the translation.

apart from *proper*, or simple, rotation operations. Improper rotations show a reversal of right- to left-handedness, whereas proper rotations do not. Each improper rotation is equivalent to some other operation or combination of operations, so that the same symmetry may be described by more than one operation. Operational equivalences and preferred notations are given in Table 3-1.

Table 3-1

EQUIVALENCE OF SYMMETRY AXES

Improper axis	*Decomposed into rotation and reflection or inversion*	*Equivalent improper axis*	*Preferred designation*
$\bar{1}$	$1 \cdot i$	$\tilde{2}$	$\bar{1}$
$\bar{2}$	$1 \cdot m$	$\tilde{1}$	m
$\bar{3}$	$3 \cdot i$	$\tilde{6}$	$\bar{3}$
$\bar{4}$	—	$\tilde{4}$	$\bar{4}$
$\bar{6}$	$3 \cdot m$	$\tilde{3}$	$\frac{3}{m}$
$\tilde{1}$	$1 \cdot m$	$\bar{2}$	m
$\tilde{2}$	$1 \cdot i$	$\bar{1}$	$\bar{1}$
$\tilde{3}$	$3 \cdot m$	$\bar{6}$	$\frac{3}{m}$
$\tilde{4}$	—	$\bar{4}$	$\bar{4}$
$\tilde{6}$	$3 \cdot i$	$\bar{3}$	$\bar{3}$

When dealing with crystal structures at the ionic level, two more combined operations are used. These operations do not show the symmetrical identity of crystal faces or the geometric shape of the unit cell, but, rather, show a symmetrical identity of two ions. Hence, they are not apparent in macroscopic crystals. The first such operation is the *screw axis*, which involves rotation about an n-fold axis, followed by translation parallel to the axis of rotation, the resulting motion being like that of a screw (Fig. 3-9c). The second such operation is the *glide plane*, which involves translation followed by reflection through a mirror plane that is parallel to the translation (Fig. 3-9d).

The ions that illustrate the several operations in Fig. 3-9 are shown as asymmetrical radicals, the way in which OH^- might be represented. In all symmetry operations, the shape of the radical or molecule that occupies a particular position must be accounted for in the operation, as well as the position itself. This becomes very significant in crystalline structures that contain radicals or molecules.

3-2. Directions and Planes in Crystal Structures

There are several methods of indicating directions with respect to various coordinate systems. The method most used in crystallography utilizes the three possible translations along the three sets of unit-cell edges. Figure 3-10f shows the standard orientation of space coordinates that are used in mineralogy. The ***a*** axis ($\mathbf{t}_1$) is positive out of the paper, the ***b*** axis ($\mathbf{t}_2$) is positive to the right, and the ***c*** axis ($\mathbf{t}_3$) is positive up. Remember that these coordinates do not necessarily have to be at right angles to one another; nor do the unit translations along the axes have to be of equal magnitude.

Directions are stated in terms of multiples of the translation distances in each of the coordinate directions. Figure 3-10a shows the [111] direction as the resultant of unit translations first along the ***a*** axis, second parallel to the ***b*** axis, and finally parallel to the ***c*** axis. The distance of the final point from the origin is of no importance, but the direction is defined as the [111] direction (read: one, one, one). Figure 3-10b shows the [110] (read: one, one, zero; or: one, one, oh) direction, where translation along the ***c*** axis is zero. Figure 3-10c shows the [010] direction with both the ***a*** and ***c*** translations zero.

Since only direction, and not magnitude, is of interest, the [222] direction, which results from the addition of two units of translation along each axis, is shown as crystallographically equal to the [111] direction in Fig. 3-10d.

Any direction of crystallographic interest can be expressed in this system, such as the $[\bar{2}\bar{1}\bar{3}]$ direction (read: bar two, bar one, bar three), which is shown in Fig. 3-10e. Directions in all eight octants of space can be designated by this method.

The directions of planes in a crystal are indicated by the direction of their poles. The *pole* of a plane is the direction of the line that is normal (at right angles) to it. Clearly, there are only two straight-line directions normal to a plane, the ones shown in Fig. 3-11 and those in the exact opposite direction. The plane normal to the [100] direction is the (100) plane. This is strictly true only for systems described in terms of orthogonal coordinates. Again, position is of no importance, but direction is.

These numbers, which indicate the directions of crystallographic planes, are known as the *Miller indices* of a plane. Note that, although the numbers are the same as the directional notation, the Miller indices apply only to the direction of the plane and not to the direction of a line. Miller indices of a plane are always shown in parentheses, such as (110). Directions of a line are always shown in brackets, such as [110].

Another method of designating the direction of a plane is in terms of

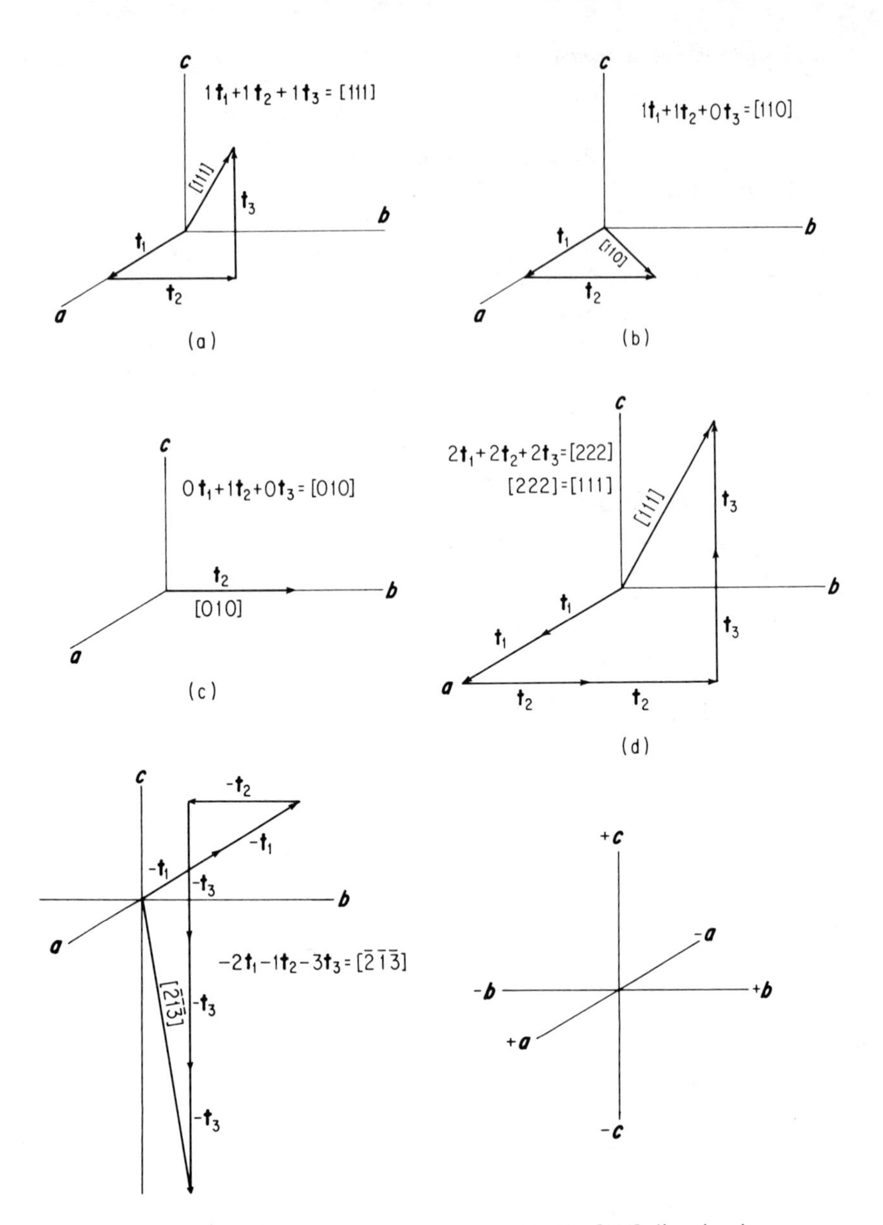

Figure 3-10 Directions in crystal space. (a) The [111] direction is arrived at by successive unit translations parallel to $\mathbf{t}_1$, $\mathbf{t}_2$, and $\mathbf{t}_3$. (b) The [110] direction is arrived at by unit translations parallel to $\mathbf{t}_1$ and $\mathbf{t}_2$, with no translation parallel to $\mathbf{t}_3$. (c) The [010] direction is parallel to the $\mathbf{t}_2$ translation. (d) Translations of two units in each direction produce the [111] direction. (e) The $[\bar{2}\bar{1}\bar{3}]$ direction is arrived at by two unit translations in the negative $\mathbf{t}_1$ direction, one unit translation in the negative $\mathbf{t}_2$ direction, and three unit translations in the negative $\mathbf{t}_3$ direction. (f) Coordinate axes labeled according to crystallographic convention.

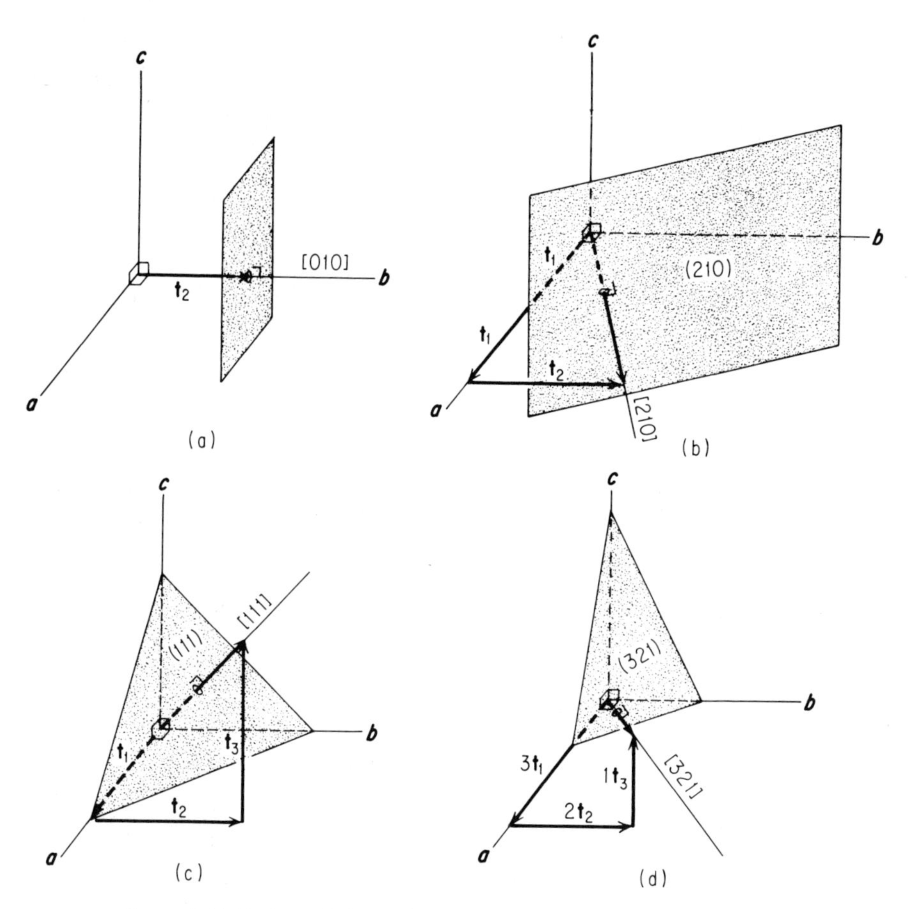

Figure 3-11 Indices of planes. (a) The (010) plane at right angles to the [010] direction. (b) The (210) plane at right angles to the [210] direction. (c) The (111) plane at right angles to the [111] direction. (d) The (321) plane at right angles to the [321] direction. These planes are at right angles to the corresponding direction only in systems where the translations are mutually orthogonal.

Weiss indices, which are the points at which the plane intersects the three crystallographic axes. Several planes that intersect all three axes are shown in Fig. 3-12c. If the distance *OG* is proportional to $\mathbf{t}_1$, the unit-cell length along the ***a*** axis, then *OB* is proportional to one third that edge. Similarly, if *OH* is proportional to $\mathbf{t}_2$, the edge of the unit cell in the ***b*** direction, then *OE* is proportional to one half the unit-cell edge. Furthermore, if *OF* is proportional to $\mathbf{t}_3$, the length of the unit cell in the ***c*** direction, *OC* is proportional to one third of that edge.

Three other planes—*GEF*, *EAF*, and *ABC*—are also plotted in Fig. 3-12. The axial intercepts and the Weiss indices of these planes are tabulated in Table 3-2. Fractions may be eliminated from the plane indices by taking the reciprocal of each and clearing fractions. Thus, $\frac{1}{2}$ becomes 2 and $\frac{1}{3}$ becomes 3,

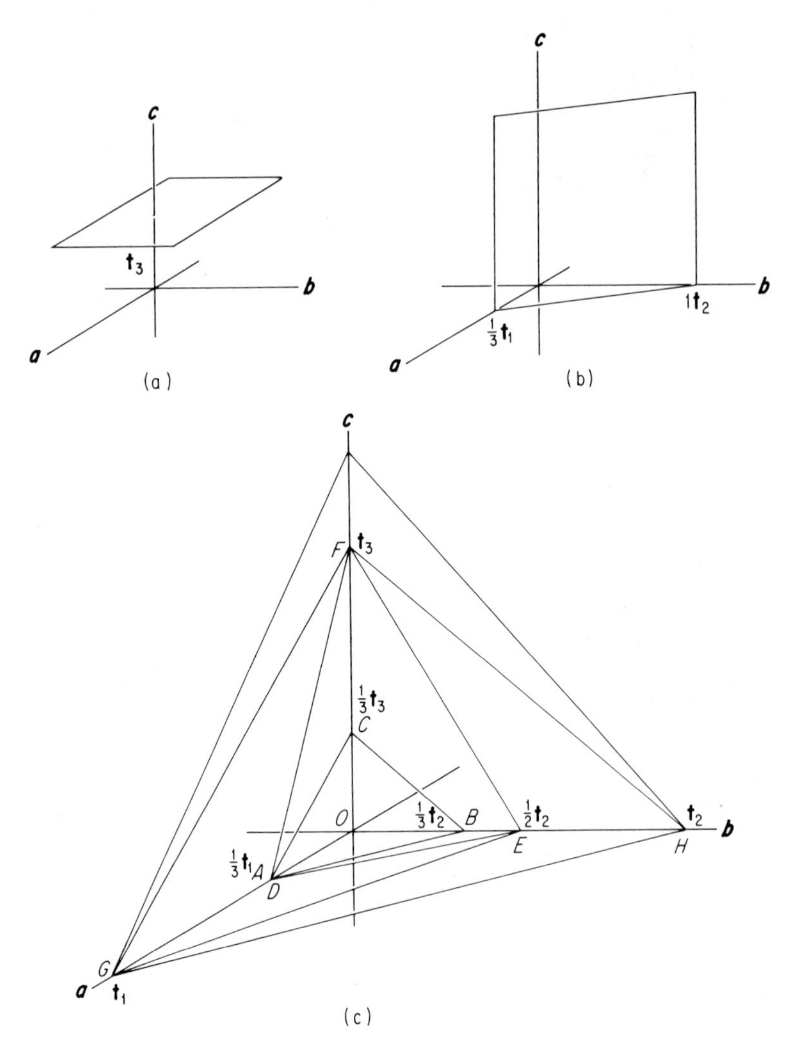

Figure 3-12 Miller indices. (a) Plane intercepts ***c*** axis at unity and ***a*** and ***b*** axes at infinity. Intercept reciprocals are $1/\infty$, $1/\infty$, and $\frac{1}{1}$, which give Miller indices (001). (b) Plane intercepts ***a*** axis at $\frac{1}{3}\mathbf{t}_1$, ***b*** axis at $1\mathbf{t}_2$, and ***c*** axis at infinity. Intercept reciprocals are $\frac{3}{1}$, $\frac{1}{1}$, and $1/\infty$, which give Miller indices (310). (c) Indices of planes: *ABC*, (333) = (111); *FGH*, (111); *AEF*, (321); *EFG*, (121).

while 1 remains 1. Since only direction is important, *ABC* is equivalent to *FGH* because the two planes are parallel. Therefore, it is usual to take the largest factor out of the reciprocals, which in the case of *ABC* is three. Thus, 3 3 3 becomes 1 1 1.

The reciprocals of the intercepts of the plane with the crystallographic

Table 3-2
WEISS AND MILLER INDICES

Plane	*Axial intercepts*	*Weiss indices*	*Intercept reciprocals*	*Miller indices*
FGH	$1\boldsymbol{a}$ $1\boldsymbol{b}$ $1\boldsymbol{c}$	1 1 1	1 1 1	(111)
GEF	$\frac{1}{2}\boldsymbol{a}$ $1\boldsymbol{b}$ $1\boldsymbol{c}$	$\frac{1}{2}$ 1 1	2 1 1	(211)
EAF	$\frac{1}{2}\boldsymbol{a}$ $\frac{1}{3}\boldsymbol{b}$ $1\boldsymbol{c}$	$\frac{1}{2}$ $\frac{1}{3}$ 1	2 3 1	(231)
ABC	$\frac{1}{3}\boldsymbol{a}$ $\frac{1}{3}\boldsymbol{b}$ $\frac{1}{3}\boldsymbol{c}$	$\frac{1}{3}$ $\frac{1}{3}$ $\frac{1}{3}$	3 3 3	(111)

axes are the Miller indices of the plane. Proof of mathematical identity of the indices of a pole and the reciprocals of the intercepts for orthogonal coordinates is to be found in any good analytical geometry textbook. For unit cells that are not described by orthogonal translations this relationship does not hold, and Miller indices must be found from reciprocals of the intercepts.

Figure 3-12b shows a plane that is parallel to the $\boldsymbol{c}$ axis and which has intercepts at $1\boldsymbol{a}$, $\frac{1}{3}\boldsymbol{b}$, and $\infty\boldsymbol{c}$. The reciprocals are $\frac{1}{1}$, $\frac{3}{1}$ and $\frac{1}{\infty}$, which gives the Miller indices (130). Figure 3-12c shows a plane parallel to both the $\boldsymbol{a}$ and $\boldsymbol{b}$ axes. Its intercepts are at $\infty\boldsymbol{a}$, $\infty\boldsymbol{b}$, and $1\boldsymbol{c}$. The Miller indices of the plane are (001).

Slightly different indices are used to describe directions of planes and lines in crystal structures that display a 6-fold axis of rotation. Figure 3-13a shows a closest-packed layer with the 6-fold rotation axis normal to the page of the diagram. There are clearly two alternative cells that can be used to describe this structure. The first is a hexagonal prism and connects the centers of six

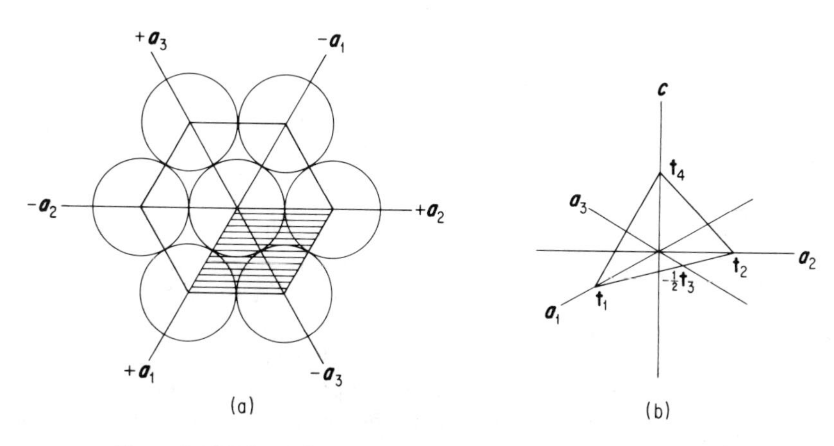

Figure 3-13 Directions in hexagonal system. (a) Closest-packed layer with $\boldsymbol{a}$ axes labeled. The sixfold $\boldsymbol{c}$ axis is normal to the intersection of the $\boldsymbol{a}$ axes. (b) Plane $(11\bar{2}1)$ in hexagonal coordinates.

spheres. The second, which is shaded, is a rhombic prism. The second alternative is clearly simpler, contains fewer spheres, and requires but three axes to describe it. The first, however, is more obviously related to the gross pattern of the structure, the 6-fold axis of rotation being included. For this reason the larger cell is usually employed, even though it is not the unit cell and it requires four coordinate axes to do so. The conventional labeling of these axes is shown in Fig. 3-13.

Figure 3-13b shows a plane that intersects all four coordinate axes at $1\boldsymbol{a}_1$, $1\boldsymbol{a}_2$, $-\frac{1}{2}\boldsymbol{a}_3$, and $1\boldsymbol{c}$. The reciprocals are thus $\frac{1}{1}$, $\frac{1}{1}$, $-\frac{2}{1}$, and $\frac{1}{1}$, and the Miller indices, in this case, $(11\bar{2}1)$. The third number in the hexagonal index is always equal to the negative of the sum of the first two.

The intersection of two planar surfaces is a straight line. A crystal face is a plane of ions. Therefore, the intersection of two crystal faces is a row of ions. The direction of this row can be given indices, just as any direction in a crystal can be given indices. As discussed in the beginning of this section, the indices of a direction, $[hkl]$, are set off in square brackets. A zone is a group of faces whose intersections give rows of ions, all of which have the same direction in a crystal. The row of ions common to the two planes is the zone axis.

3-3. Space Lattices

In dealing with crystal structures, it is usually preferable to deal with the pattern of the structure rather than with the ions themselves. A *space lattice* is a geometrical array of points arranged such that each point has identical surroundings. Rather than considering spherical ions in CCP, which is equivalent to a FCC structure, it is useful to discuss the pattern only, that is, a face-centered array of points. This abstraction permits the consideration of symmetry in a purely abstract sense.

Auguste Bravais, in 1848, demonstrated that there are 14, and only 14, space lattices possible (Fig. 3-14). The names for each of these lattices, along with their dimensions and symmetries, are given in Table 3-3. Other lattices may be conceivable, but further study will show that they can be reduced to one of the 14 given.

A primitive lattice (P) is one in which the unit cell is drawn so that lattice points occur only at the corners (Figs. 3-14a, d, f, g, h, l, and n). In the primitive unit cells, parts a, d, and h, the edges are all at right angles to one another, and the lattices differ in the relative lengths of the edges. In the primitive unit cell of part f the edges are all equal. The internal angles between the edges are equal to each other, but not equal to 90°. The primative cell in part g has two edges equal and one either longer or shorter than the equal pair. The angle between the equal edges is 120°; the other angles are 90° (three unit cells are actually shown in the drawing). The unit cell in part l has edges of unequal

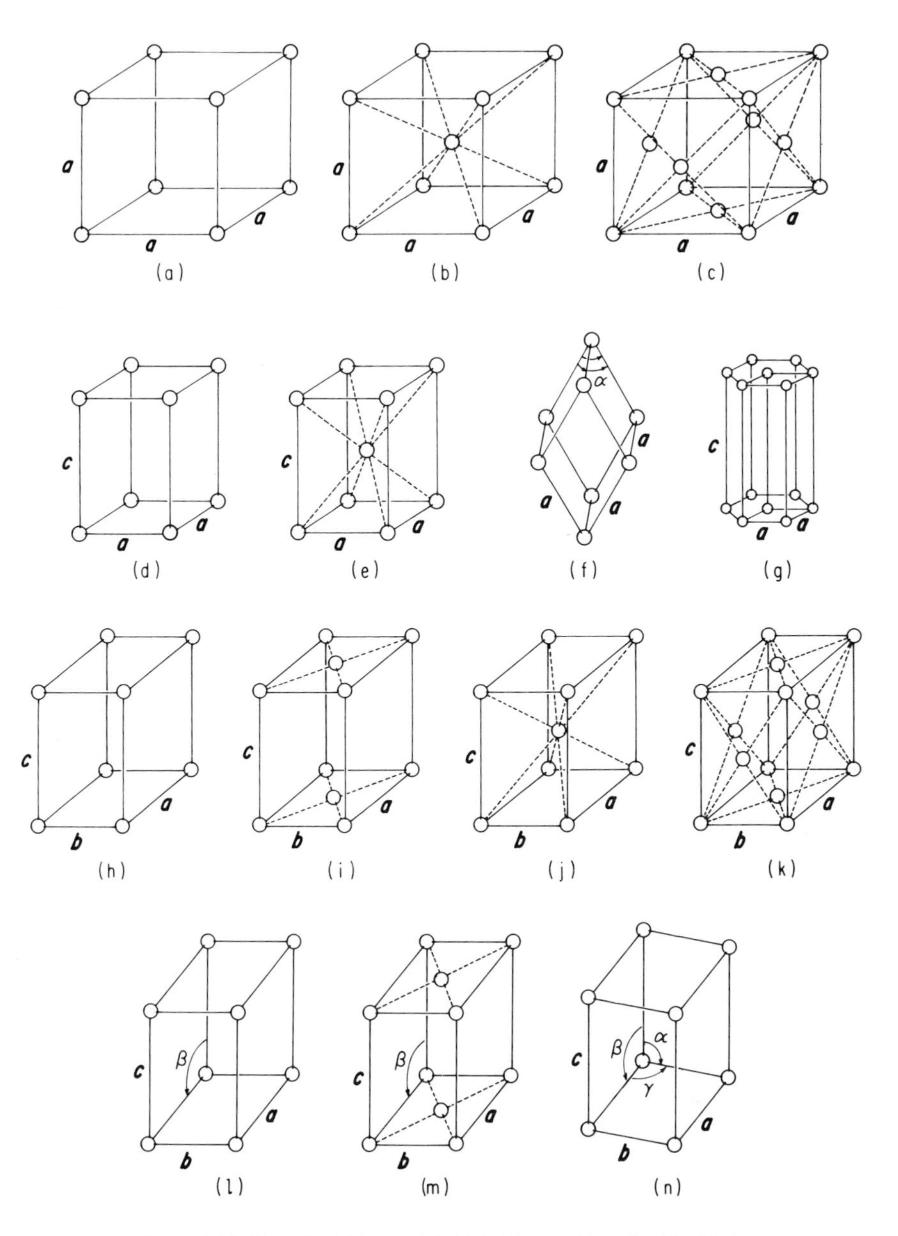

Figure 3-14 Bravais lattices. (a) Primitive cubic (I). (b) Body-centered cubic (BCC). (c) Face-centered cubic (FCC). (d) Primitive tetragonal. (e) Body-centered tetragonal. (f) Rhombohedral (R). (g) Primitive hexagonal. (h) Primitive orthorhombic. (i) End-centered orthorhombic. (j) Body-centered orthorhombic. (k) Face-centered orthorhombic. (l) Primitive monoclinic. (m) End-centered monoclinic. (n) Primitive triclinic (see Table 3-3).

Table 3-3
BRAVAIS LATTICES

Lattice type	*Data needed to describe unit cell completely*	*Symmetry*
Triclinic		
1. Primitive, P	$a\ b\ c\ \alpha\ \beta\ \gamma$	$\bar{1}$
Monoclinic		
2. Primitive, P 3. End centered, C or A	$a\ b\ c\ \beta$	$\frac{2}{m}$
Orthorhombic		
4. Primitive, P 5. End centered, C, B, or A 6. Body centered, I 7. Face centered, F	$a\ b\ c$	$\frac{2}{m}\frac{2}{m}\frac{2}{m}$
Hexagonal and rhombohedral		
8. Hexagonal, P	$a\ c$	$\frac{6}{m}\frac{2}{m}\frac{2}{m}$
9. Rhombohedral, R	$a\ c\ (a_r\alpha)$	$\bar{3}\ \frac{2}{m}$
Tetragonal		
10. Primitive, P 11. Body centered I	$a\ c$	$\frac{4}{m}\frac{2}{m}\frac{2}{m}$
Cubic		
12. Primitive, P 13. Body centered, I 14. Face centered, F	a	$\frac{4}{m}\ \bar{3}\ \frac{2}{m}$

length and one angle that is not 90°. The primitive cell in part n, which is the most general of all and which has the fewest possible symmetry operations, has unequal edges, none of which are at right angles.

A body-centered lattice (I, for interior) has, in addition to a point at each corner, one point in the center of the unit cell (Figs. 3-14b, e, and j). In each of these, the lattice edges are at right angles to one another, but the unit-cell edges are of different relative lengths. There are two lattice points in a body-centered unit cell.

An all face-centered lattice (F) has, in addition to a point at each corner, one point in the center of each face of the unit cell (Figs. 3-14c and k). The first has edges of equal length, which are at 90°. The second has edges at 90°, but of unequal length. There are four lattice points in an all face-centered unit cell.

An end-centered lattice (A, B, or C, depending on which end) has additional lattice points in the centers of two opposite faces (Figs. 3-14i and m). The first has edges of unequal length at right angles to one another. The second has edges of unequal length with one not at right angles to one of the other two. There are two lattice points in an end-centered unit cell.

A convenient method of designating specific lattice points or specific ions in a unit cell is in terms of translations, or fractions of translations, parallel to the lattice or unit-cell edges. The lower, back, left-hand corner is usually designated the origin. In a primative lattice, then, there are lattice points at 0,0,0; 0,0,1; 0,1,0; 1,0,0; 1,1,0; 1,0,1; 0,1,1; and 1,1,1. These numbers represent successive translations along the ***a***, ***b***, and ***c*** axes, respectively. In a body-centered lattice, there is an additional point at $\frac{1}{2}, \frac{1}{2}, \frac{1}{2}$, which is at the center of the cell.

3-4. Stereographic Projections

Accurate and meaningful portrayals of three-dimensional objects on two-dimensional paper have already been encountered in the representation of crystal structures (Chapter 2). Yet another type of representation must now be introduced to show symmetry relationships (Section 3-5) and crystal faces (Section 3-8). The first step in constructing a planar stereographic projection of a crystal face is the construction of a spherical projection. Consider the isometric form of a cube modified by octahedral and dodecahedral faces (Fig. 3-15a). In making a spherical projection of a crystal planc, it is assumed that the crystal is surrounded by a sphere. The center of the crystal, as defined by the intersection of the crystallographic axes, is coincident with the center of the sphere (Fig. 3-15b). Crystal-face normals, which are radii of the sphere, are extended to the surface of the sphere. The intersection of the crystal-face normal with the spherical surface is a point, known as the *pole* of the face. Note that the angular relationships of the faces are completely defined by their poles.

Because it is the angular relationship between faces, lattice planes, or elements of symmetry that is of interest, and not their actual position as determined by translations, after plotting the pole of the face or any other lattice plane, it is permissible and very advantageous to consider all lines and planes to be translated parallel to their poles so as to pass through the center of the sphere. The cube and dodecahedral planes are shown in Fig. 3-16, where each of the forms is treated separately for clarity. The spherical projection of a plane passing through the center of a sphere clearly is a great circle. A great circle has a diameter equal to the diameter of the sphere; a small circle has a lesser diameter. Thus, in terms of terrestrial coordinates, meridians of longitude and the equator are great circles and parallels of latitude are small circles.

The angular relationships between crystallographic planes of a crystal are preserved in the angular relationships between their poles in spherical projection. Note, however, that the angle between two poles is the complement of the angle between the two planes (Fig. 3-17). Although these angular rela-

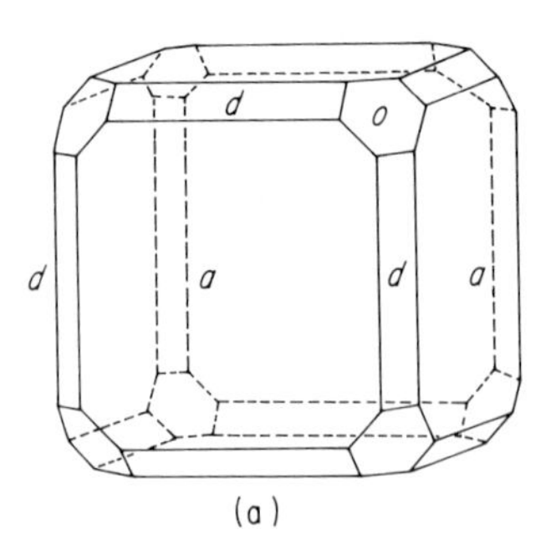

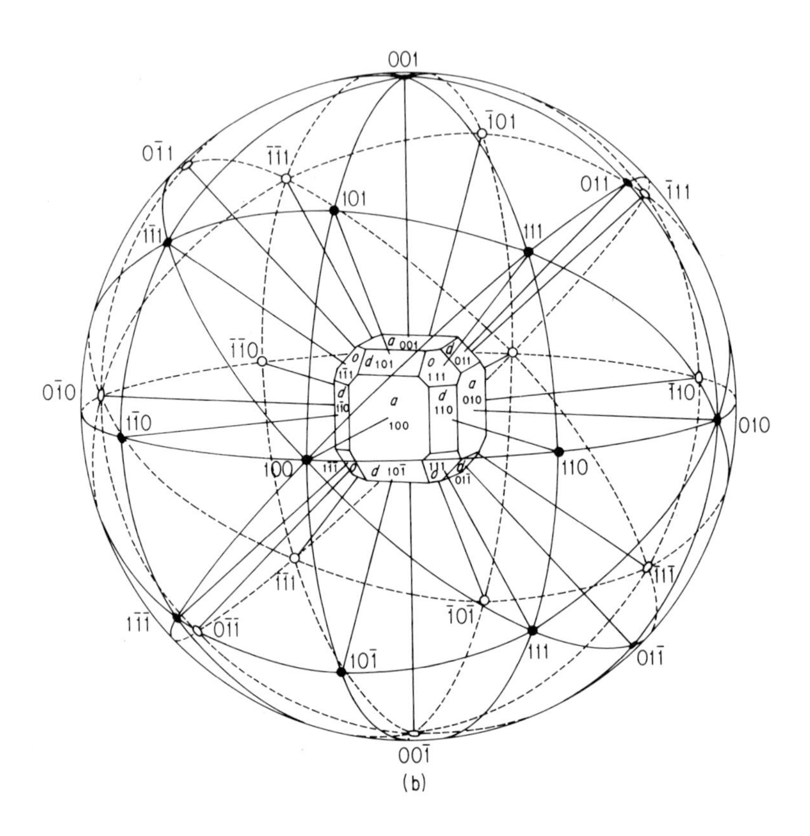

Figure 3-15 Spherical projection of the poles of a cube, an octahedron, and a dodecahedron. (a) Cube *a*, octahedron *o*, and dodecahedron *d* faces. (b) Spherical projection. The poles of the crystal faces project as points on the surface of the sphere.

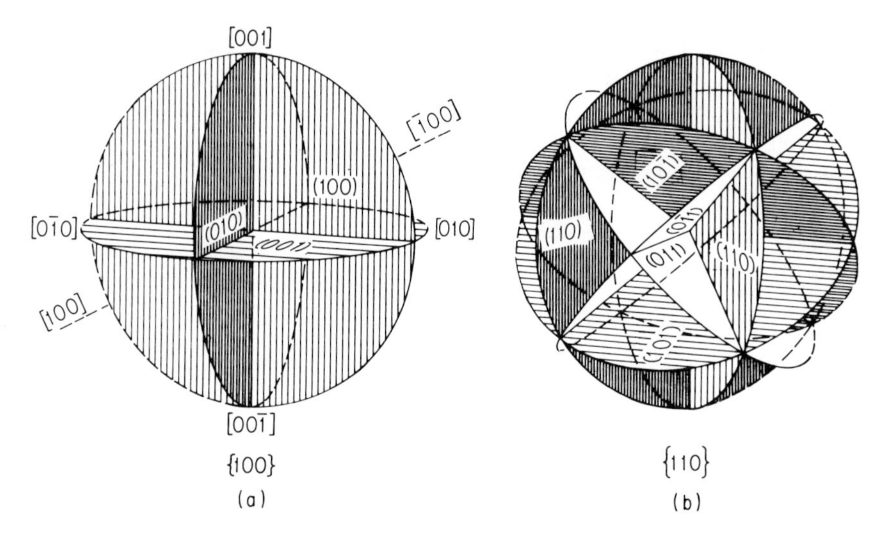

Figure 3-16 Spherical projections of crystallographic planes. The planes are so transposed as to place all of them passing through the center of the sphere. The projection of such a plane is a great circle on the surface of the sphere. (a) Cube. (b) Dodecahedron.

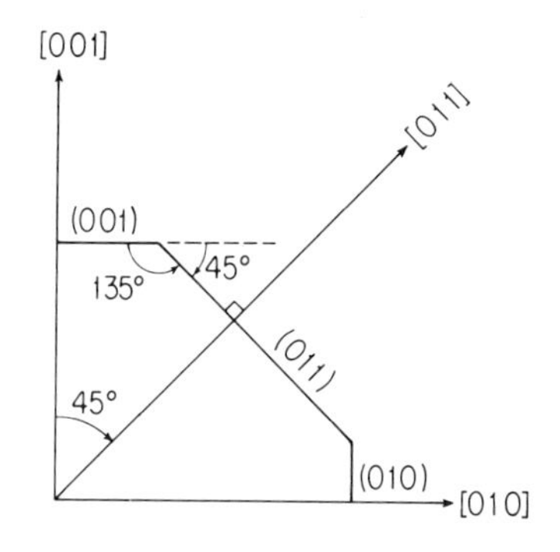

Figure 3-17 Relationship between polar angles and planar angles. The angle between two face poles is the same as the external angle, and it is the complement of the internal angle between the faces.

tionships are preserved in spherical projection, and the relationship between the plane and its pole is clearly demonstrated, the representation of the spherical projection on two-dimensional paper does not permit the direct measurement of these relationships.

The stereographic projection is derived from the spherical projection by connecting each pole in the spherical projection with a line to the south pole of the sphere. Poles in the southern hemisphere are projected similarly from the north pole. The point of intersection of this line with the equatorial plane of the sphere is the stereographic projection (Fig. 3-18) of the pole. Crystallographic planes are similarly projected. Lines from the great circle are connected with the south pole, and the intersection of these lines with the equatorial plane defines the projection of the great circle.

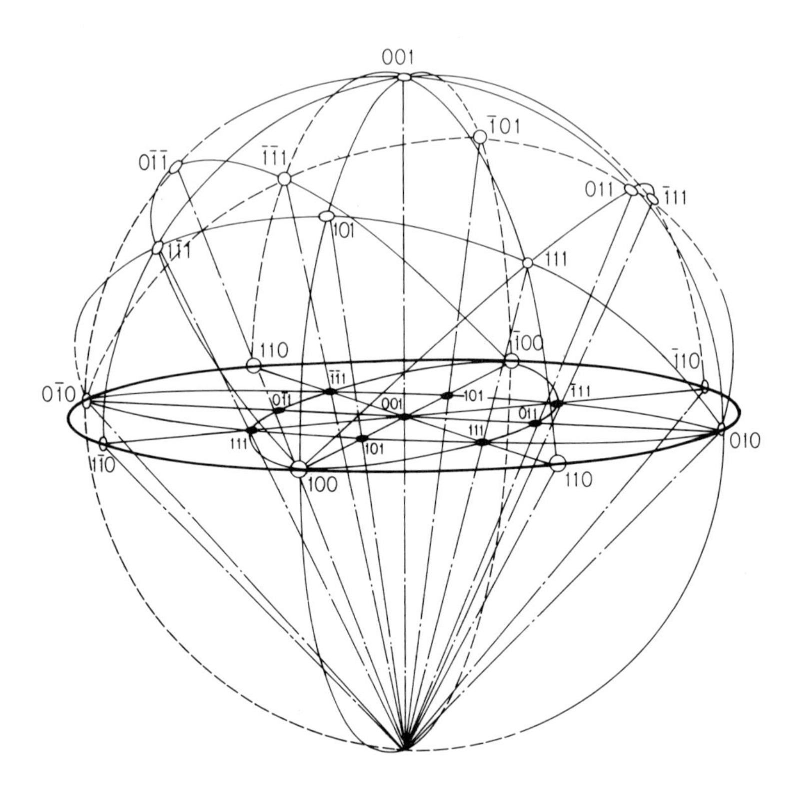

Figure 3-18 Relationship between spherical and stereographic projections. The poles in spherical projection are projected onto the equatorial plane along a line connecting the face pole with the south pole of the sphere. Poles in the southern hemisphere are projected onto the equatorial plane by the north pole.

Two important properties of stereographic projections warrant special mention. First, any circle on the spherical projection, be it a great circle or a small circle, plots as the arc of a circle on the stereographic projection, and, thus, may be plotted with an ordinary compass. Great circles, which pass through the north and south poles of the spherical projection, plot as circles of infinite radius on the stereographic projection, that is, as straight lines passing through the center of the projection. The equatorial circle of the spherical projection is identical with the *primitive* (perimeter) of the stereographic projection.

Second, angular relationships between two or more planes and between two or more poles, as well as between planes and poles, are preserved in stereographic projection. Thus, interfacial angles, which can only be illustrated in the drawing of crystal forms or lattice structures, may be measured and plotted accurately in stereographic projection. Note, however, that the linear value of a degree of arc is not constant on a stereographic projection, but increases from the center outward toward the primitive.

Although the nature of stereographic projection permits construction to be done with the aid of simple drafting tools, these operations are greatly facilitated by the employment of various nets. The most widely used of these devices is known as the *Wulff net* (Fig. 3-19). A Wulff net consists of stereographically projected great circles as lines of equal longitude and small circles as lines of equal latitude. These nets are printed commercially with either a 5- or 7-cm radius and marked at 2° intervals. (A full-scale Wulff net is shown on page 312.)

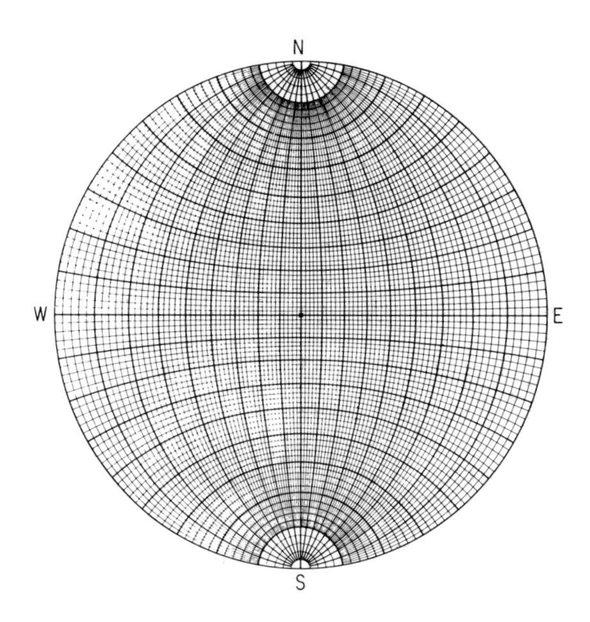

Figure 3-19 Wulff net. The Wulff net is the stereographic projection of meridians and parallels at 2° intervals. Note that the north and south poles, which are used for reference on the Wulff net, are not the same as the north and south poles of the sphere. The north pole of the sphere is at the center of the Wulff net.

A common method of using a Wulff net is to affix to it a sheet of tracing paper with a pin through the center. Mounted in this manner, the paper can be rotated as needed for the desired operation.

Given the pole of a lattice plane or a crystal face, the trace of the plane or face in stereographic projection may be plotted by rotating the paper until the pole is on the equator, counting 90° along the equator across the center point of the net, and tracing in the great circle that is at 90° from the pole. Conversely, given the projection of a plane, its pole may be found by rotating the paper until a great circle of the Wulff net exactly matches it, and counting 90° along the equator.

Angular distances along a great circle are found by matching the great circle to one on the Wulff net and counting the number of degrees between the two points. This procedure is of especial interest in determining the solid angle between two planes in projection. First, the two poles of the two planes are found. Second, the paper is rotated until the two poles lie on the same great circle, where the angle between them can be counted along the great circle.

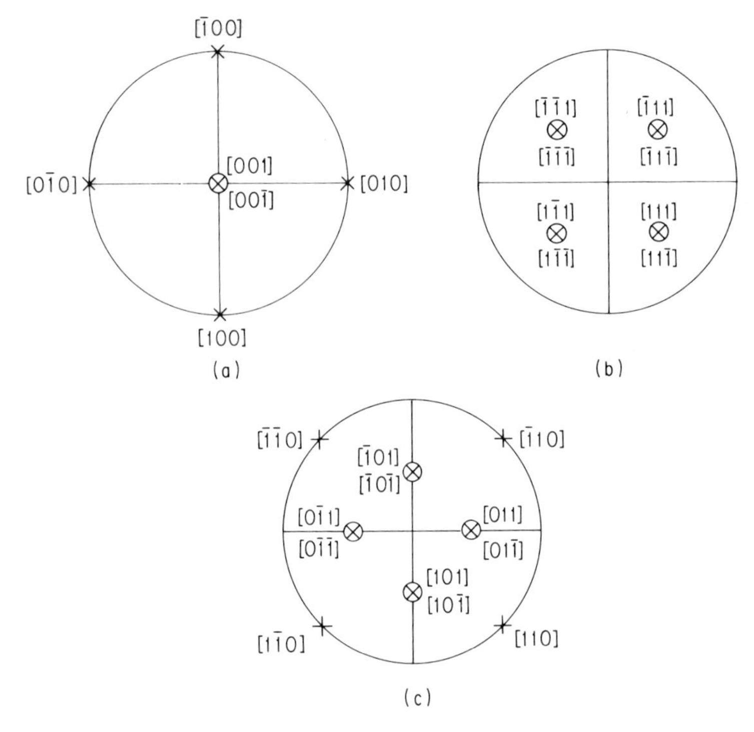

Figure 3-20 Stereographic projections of a cube, an octahedron, and a dodecahedron. Poles projected from the northern hemisphere are shown as ×, and poles from the southern hemisphere as ○. (a) Cube. (b) Octahedron. (c) Dodecahedron.

The stereographic projection is rather simple in concept. However, as has been stated previously, the reader must develop a facility in visualizing the projection from the solid form, lattice plane, or symmetry element, and in visualizing the solid form, lattice plane, or symmetry element from the projection. To this end, it is suggested that the reader measure the interfacial angles on an idealized wooden crystal model and plot the face poles in stereographic projection by means of a Wulff net. As an example, the stereographic projections of a cube, an octahedron, and a dodecahedron are given in Fig. 3-20.

3-5. Point Groups

The atoms in crystals are periodically arrayed as the translation operation indicates. Other operations show that symmetry exists in crystal structures. Therefore, space lattices, which are set up such that points replace atoms, also have symmetry. The different symmetry operations, other than those of translation, have been shown in Section 3-1 as lines and planes that intersect at the center of the unit cell. All the different possible combinations of symmetry axes that can intersect at a given point in space are called *point groups.* There are 32. All crystal structures that have the same point group are members of the same *crystal class.* Remember, in discussing point groups, that all operations containing translation have to be excluded from consideration.

Although it is possible to sketch the symmetry operations, as shown in Section 3-1, for each point group, the results would be confusingly complex. These relationships can be seen, however, more simply and clearly by means of stylized stereographic projections known as *stereograms.*

The 11 point groups that can be generated by proper rotation axes alone or in combination with other proper rotation axes are shown in Fig. 3-21. The point groups 1, 2, 3, 4, and 6 involve a single axis of rotation. The point group 222 has three 2-fold axes at mutual right angles to one another. (These numbers are not related to, nor should they be confused with, similarly appearing Miller indices or lattice directions.)

The point group 32 has one 3-fold axis and three 2-fold axes. Note that symmetrically the 2-fold axes are equivalent by operation of the 3-fold axis. The point group 422 has one 4-fold rotation axis and four 2-fold axes. There are two sets of symmetrically related 2-fold axes. Point group 622 has one 6-fold rotation axis and six 2-fold axes, which are in two sets. Point group 23 has three 2-fold rotation axes at right angles to one another. In addition, there are four 3-fold rotation axes. The orientation of these 3-fold axes is given in Fig. 3-4. The point group 432 has three 4-fold rotation axes at right angles, four 3-fold axes, and six 2-fold axes. The 2-fold axes are those which connect the six pairs of opposite edges of a cube.

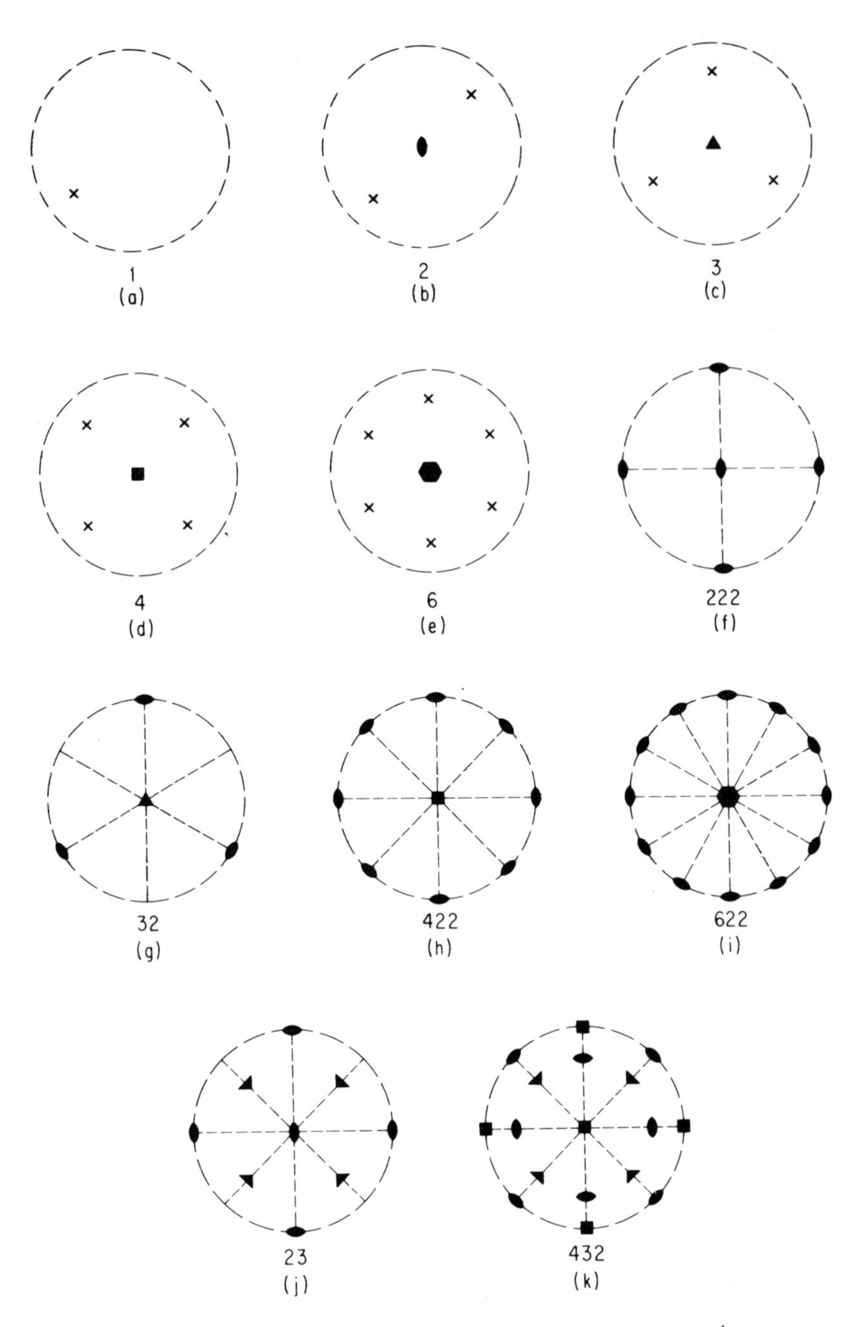

Figure 3-21 Stereograms of point groups with proper rotation axes. 2-fold axes shown with almond symbols; 3-fold axes with triangles; 4-fold with squares; and 6-fold with hexagons. Poles of symmetrically equivalent planes are shown with a cross. Dashed lines are for reference only. Point groups: (a) 1, (b) 2, (c) 3, (d) 4, (e) 6, (f) 222, (g) 32, (h) 422, (i) 622, (j) 23, and (k) 432.

As demonstrated in Section 3-2, the position of a plane in space can be indicated by its pole, the pole direction $[hkl]$ having the same numerals as the Miller indices of the plane (hkl) in orthogonal coordinates. Poles plot as points in stereographic projection. In order to clarify some of the point groups, symmetrically equivalent poles, and, therefore, crystallographic planes, are plotted in the stereograms with a $\times$ signifying a pole in the upper hemisphere and a ○ signifying a pole in the lower hemisphere. The symbol $\otimes$ indicates that there are symmetrically equivalent poles in the upper and lower hemisphere which project to the same point in the stereographic projection.

Point groups involving rotoinversion are shown in Fig. 3-22. The point group $\bar{1}$ (read: bar one) is the same as simple inversion through the center, as shown in Fig. 3-8. As indicated in Table 3-1, this is the usual designation for the center of symmetry. The point group $\bar{2}$, which involves rotation through 180° followed by inversion through the center, is symmetrically equivalent to the mirror operation (Fig. 3-7). The point group $\bar{3}$ generates identical points in such a manner that the three symmetrically related poles above the center are displaced by 60° from the three symmetrically related poles below the

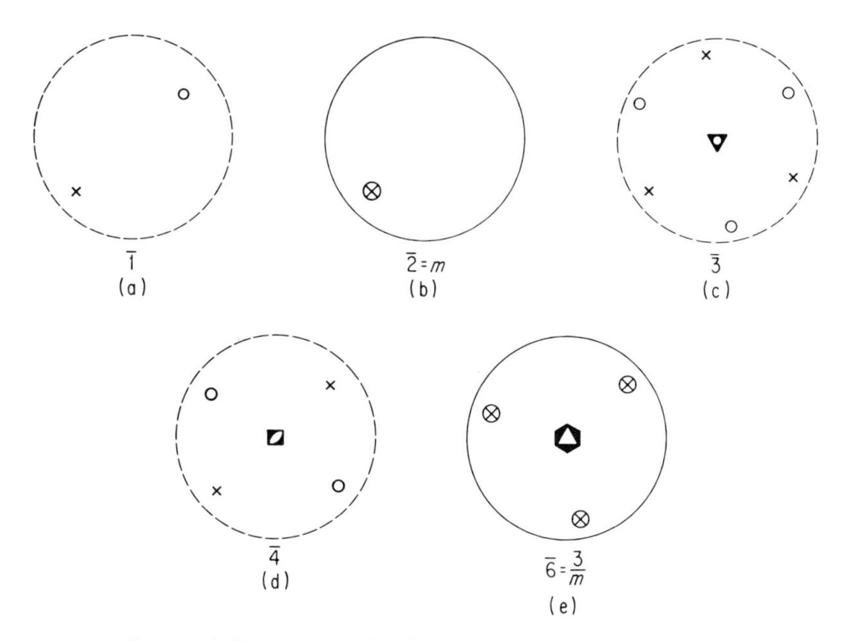

Figure 3-22 Stereograms of point groups with rotoinversion axes. 3-fold axes shown with triangle enclosing circle; 4-fold axes with square enclosing almond symbol; 6-fold with hexagon enclosing triangle. Solid lines indicate mirror planes, and dashed lines are for reference. Cross indicates face pole in northern hemisphere; circle indicates southern face pole. Point groups: (a) $\bar{1}$ = center of symmetry, (b) $\bar{2} = m$, (c) $\bar{3}$, (d) $\bar{4}$, (e) $\bar{6} = 3/m$.

center. The point group $\bar{4}$ has one 4-fold axis of rotoinversion. This axis relates pairs of symmetrically equivalent poles above and below the center. The point group $\bar{6}$ is symmetrically equivalent to a 3-fold rotation axis plus a mirror plane of symmetry, which is at right angles to the axis. It is usually designated $3/m$.

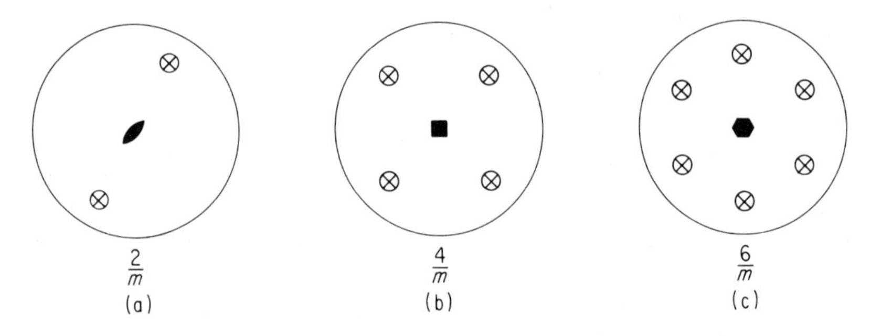

Figure 3-23 Stereograms of point groups with mirror planes normal to proper rotation axes. Symbols as in Figs. 3-21 and 3-22. Point groups: (a) $2/m$, (b) $4/m$, (c) $6/m$. Point group $3/m$ is shown in Fig. 3-22.

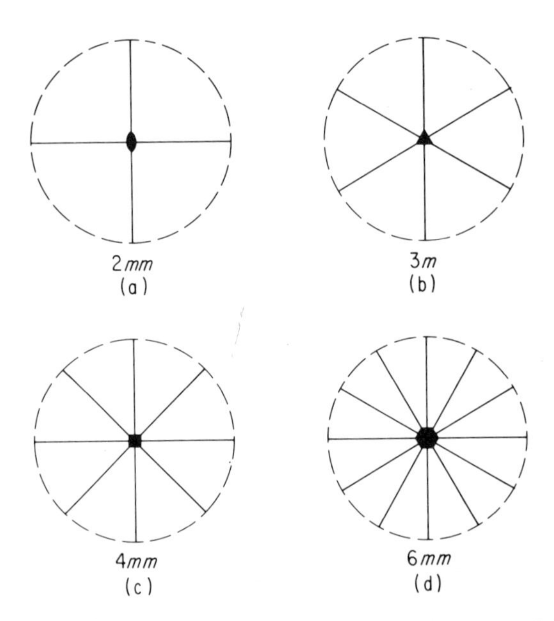

Figure 3-24 Stereograms of point groups with mirrors parallel to proper rotation axes. Symbols as in Fig. 3-21. Point groups: (a) $2mm$, (b) $3m$, (c) $4mm$, (d) $6mm$.

Inspection of Table 3-1 reveals that the improper rotation axes which consist of rotoreflection (Fig. 3-9a) are all equivalent to other axes, axial combinations, or other symmetry operations. Therefore, the operation of rotoreflection, by itself, generates no new point groups.

The next combination of operations to consider is that of a proper rotation axis with a mirror plane normal to it (Fig. 3-23). The combined operation $1/m = m = 2$, as shown in Fig. 3-22 and Table 3-1. This operation creates no new point group. The point groups $2/m$, $4/m$, and $6/m$ each produce a new point group, as shown in Fig. 3-23. The combined operation $3/m$ is identical to $\bar{6}$, as shown in Fig. 3-22. All the three new point groups shown in Fig. 3-23 have, in addition to the elements of symmetry indicated, a center of symmetry.

Four new point groups are generated by adding mirror planes parallel to the proper rotation axes: $2mm$, $3m$, $4mm$, and $6mm$ (Fig. 3-24). The operation $1m$ clearly is equivalent to m, which has already been shown.

The next combination of operations to consider results from the possible existence of mirror planes both normal and parallel to proper rotation axes. Two mirror planes of symmetry, which intersect at right angles, of necessity produce a 2-fold axis of rotation parallel to the intersection of the planes. Therefore, the combination of two such planes with proper axes of rotation is

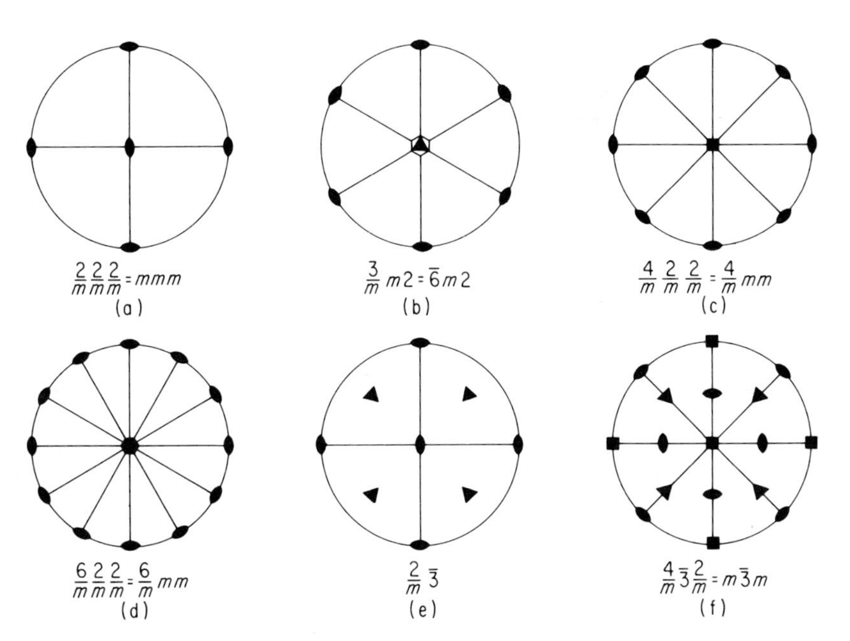

Figure 3-25 Stereograms of point groups that combine proper axes of rotation with mirror planes both parallel and normal to the axis. Symbols as in Fig. 3-21. Point groups: (a) $2/m2/m2/m = mmm$, (b) $3/mm2 = \bar{6}m2$, (c) $4/m2/m2/m = 4/mmm$, (d) $6/m2/m2/m = 6/mmm$, (e) $2/m\bar{3}$, (f) $4/m\bar{3}2/m = m\bar{3}m$.

equivalent to the addition of these planes of symmetry to the point groups shown in Fig. 3-21. The six new point groups thus produced are shown in Fig. 3-25. They are mmm, $\bar{6}m2$, $4/mmm$, $6/mmm$, $2/m\bar{3}$, and $m\bar{3}m$. These symbols are simpler than the full symbols which could be derived from those given in Fig. 3-21. They represent the minimum number of elements, from which all others in the point group can be derived by the operations of symmetry on the symmetry elements themselves. Of the six, only $\bar{6}m2$ lacks a center of symmetry.

The remaining three point groups are obtained by the combination of the $\bar{3}$ and the $\bar{4}$ improper rotation axes with proper rotation axes. They are point groups $\bar{4}2m$, $\bar{3}2/m$, and $\bar{4}3m$, which are shown in Fig. 3-26.

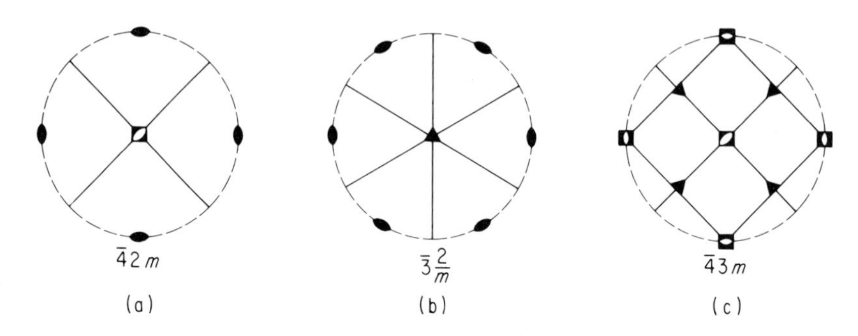

Figure 3-26 Stereograms of point groups that combine proper and improper rotation axes. Symbols as in Fig. 3-21. Point groups: (a) $\bar{4}2m$, (b) $\bar{3}2/m$, (c) $\bar{4}3m$.

3-6. Space Groups

The 14 Bravais space lattices and the 32 point groups are discussed in the preceding two sections. To recapitulate, a *space lattice* is a geometrical array of points arranged in a manner such that each point has identical surroundings. A *point group* contains all the elements of symmetry that intersect at a given point in space. Since each point in a space lattice has identical surroundings, and each point has certain elements of symmetry passing through it, it follows that a whole array of symmetry elements exists. This combination of point groups with space lattices gives arrays known as *space groups*, of which there are 230 possible. Space groups are arrived at by various combinations of translations, glide planes, and screw axes with the 32 point groups. As was pointed out at the beginning of this chapter, the space groups were arrived at mathematically prior to the analysis of crystal structures by means of X-ray diffraction.

Two simple two-dimensional plane groups are illustrated in Fig. 3-27. A plane group is a two-dimensional equivalent of a space group. Figure 3-27a is a rectangular plane group. The group proper is shown with thicker lines and the lattice points occur at the intersections of these lines.

The plane group is made up of a two-dimensional array of points. The translations $\mathbf{t}_1$ and $\mathbf{t}_2$ are of different lengths and at right angles to one another. The lines connecting the group points are clearly lines of mirror symmetry. In addition, lines of mirror symmetry occur midway between the two sets of plane group lines. Each lattice point is the site of a 2-fold axis of rotation. The axis of symmetry, which passes through the center of the diagram, can be seen to operate in a manner such that the unit cell at the origin, when rotated through 180°, comes into coincidence with the unit cell at lower right. There is also a 2-fold rotation axis midway between the lattice points, and one in the center of the unit cell. These axes do not coincide with lattice points.

The plane group shown in Fig. 3-27b is a hexagonal plane group. The two translations are equal and at 120°, so that the unit cell is a rhombus. Axes of 6-fold rotational symmetry occur at the lattice points. Axes of 2-fold rotational symmetry occur midway between lattice points, both along the cell edges and along the cell diagonal. In addition, there are two 3-fold axes of symmetry that lie at one third and two thirds of the distance along the long diagonal of the unit cell. They are, thus, central to an equilateral triangle made by three 6-fold axes of rotation. The lines that connect the lattice points are lines of mirror symmetry, as are the short diagonals of the unit cell. Clearly,

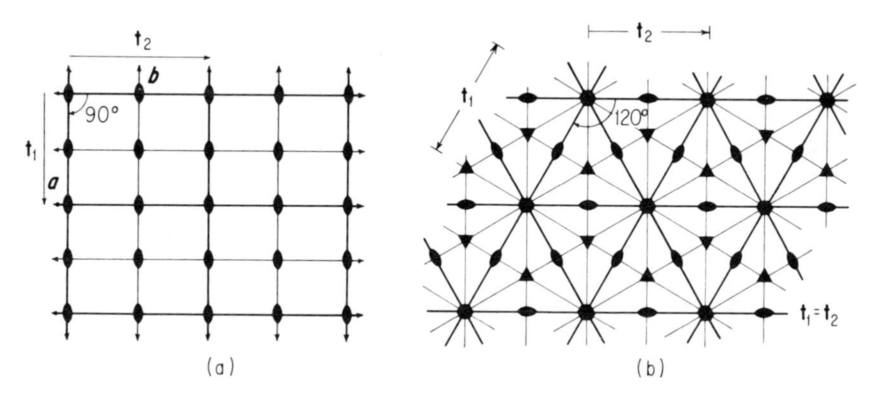

Figure 3-27 Two-dimensional space groups. (a) Rectangular plane group. Heavy lines indicate unit cells, which are also mirror planes; light lines indicate mirror planes; almond symbol indicates 2-fold rotation axis. The unit translations are orthogonal and unequal. (b) Hexagonal plane group. Lines indicate mirror planes of symmetry; heavy lines also indicate real or possible unit cells. Almond symbol indicates a 2-fold axis, triangle a 3-fold axis, and hexagon a 6-fold axis. The unit translations are equal and at 120°.

it is arbitrary whether the present translations are used to define the unit cell, or another set which includes the short diagonal of the rhombus. These cell edges and possible cell edges are shown with thicker lines. The thinner lines show the other mirror lines of symmetry, which lie along actual and possible long diagonals of the unit cell rhombus. Note that the intersections of the lines of mirror symmetry occur at 2-fold and 3-fold axes of rotational symmetry, which do not occur at lattice points.

A space group is a similar array of symmetry elements, but in three dimensions. The description of the 230 space groups is beyond the scope of this book, and the reader is referred to more advanced crystallography texts for a thorough discussion of this topic. Suffice it to state that crystallographic analysis of a mineral is not complete until the space group is known.

3-7. Crystal Systems

The 32 crystal classes, and, therefore, the 14 Bravais lattices, can be divided into 7 *crystal systems*. The relative translation distances and internal angles of the six primitive space lattices, along with the highest fold axes of rotation, provide the basis for this division (Fig. 3-28).

The Bravais lattices are classified according to crystal system in Table 3-3. Only the dimension $\boldsymbol{a}$ is needed to specify the unit cell in the *cubic system*, because, by definition, the cell edges are all of the same length and at right angles to one another. Two dimensions, $\boldsymbol{a}$ and $\boldsymbol{c}$, are needed in the *tetragonal system*, because one unit-cell edge is of length different from the other two. The three cell edges are orthogonal. In the *hexagonal* and *trigonal systems*, the unit cells have two sides of equal length at 120° to each other, and a third of different length is at right angles to the plane of the first two. In the *orthorhombic system*, the unit-cell edges are of different lengths but at mutual right angles. The *monoclinic system* is made up of unit cells that have edges of unequal length, one of which is not at right angles to one of the other two. In the *triclinic system*, the lattice edges are unequal and not at right angles to one another.

The symmetry of the lattice type, which is the maximum symmetry for each system, is also given in Table 3-3. However, because a lattice point may be occupied by more than one ion and, thus, is not necessarily itself symmetrical, the symmetries given in Table 3-3 are not the minimum for each system. The minimum symmetry elements for each crystal system are given in Table 3-4. In the tetragonal, rhombohedral, hexagonal, monoclinic, and triclinic systems, the minimum symmetry required is that of the highest fold axis of rotation or rotoinversion. In the orthorhombic system, the minimum symmetry required is three orthogonal 2-fold axes. Note well that in the cubic

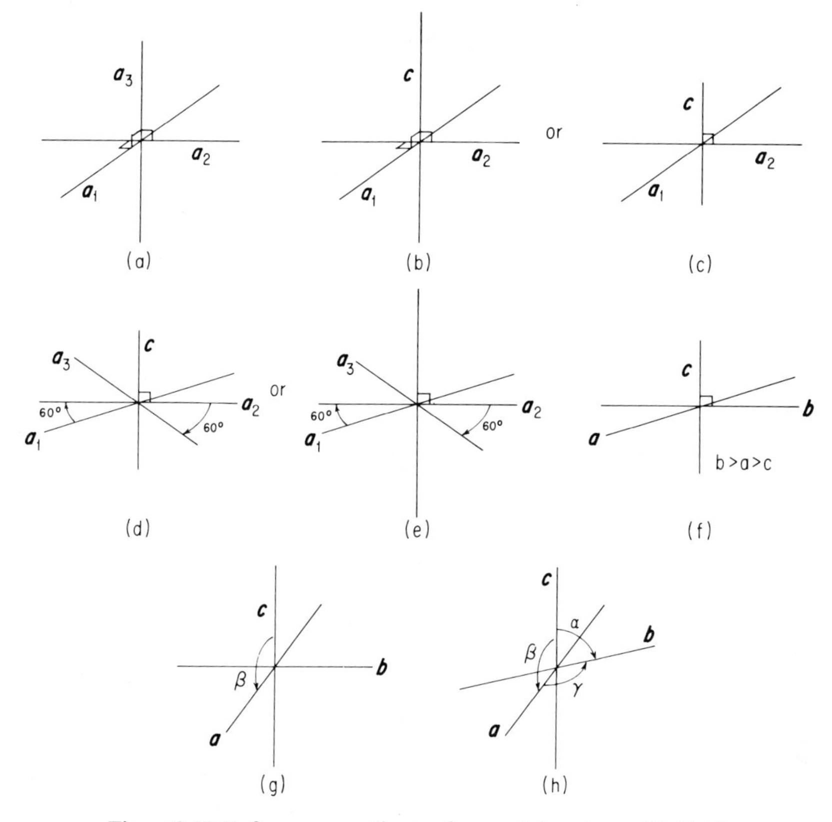

Figure 3-28 Reference coordinates for crystal systems. (a) Cubic coordinates. All orthogonal axes of same unit length. (b) and (c) Tetragonal coordinates. Two axes of the same unit length, and with one longer (b) or shorter (c). All axes orthogonal. (d) and (e) Hexagonal coordinates. Three coplanar axes of the same unit length at 120° angles, and one axis normal to the plane of the other axes and either shorter (d) or longer (e) than those axes. (f) Orthorhombic coordinates. Three mutually orthogonal axes of different unit length. (g) Monoclinic coordinates. Three axes of different unit length, two of which are at mutual right angles and one of which is not at right angles to one of the other two. (h) Triclinic coordinates. Three axes of differing unit length and not mutually orthogonal.

crystal system the minumum symmetry required is four 3-fold axes of rotation, which automatically generate three 2-fold axes, and not the expected 4-fold axes of rotation.

The crystal classes, which are based on the 32 point groups, are named and assigned to their proper crystal systems in Table 3-5. The normal class in each system possesses the highest symmetry of the space lattice (Table 3-3). One

Table 3-4

MINIMUM SYMMETRY ELEMENTS FOR EACH CRYSTAL SYSTEM

System	*Minimum symmetry elements*
Cubic	Four 3-fold rotation axes
Tetragonal	One 4-fold rotation or rotoinversion axis
Orthorhombic	Three perpendicular 2-fold rotation axes
Trigonal	One 3-fold rotation or rotoinversion axis
Hexagonal	One 6-fold rotation or rotoinversion axis
Monoclinic	One 2-fold rotation or rotoinversion axis
Triclinic	None

class in each system possesses the minimum symmetry for the system. The other classes in each system possess symmetry intermediate between those two extremes.

Another way of considering the crystal systems is in terms of the coordinate axes to which the translations of unit cells may be referred (Fig. 3-28). The cubic system is referred to three orthogonal axes of equal length. By convention, $\boldsymbol{a}_1$ is shown out of the paper, $\boldsymbol{a}_2$ to the right, and $\boldsymbol{a}_3$ to the top. The dimension $\boldsymbol{a}$ is the unit length along any of the axes. The tetragonal system is referred to three orthogonal axes, two of which are of equal length and one of which may be longer or shorter than the other two (Figs. 3-28b and c). The different axis is labeled $\boldsymbol{c}$, and is set vertical. The hexagonal system is referred to four axes, three of which are coplanar at 120° and one of which is of different length and at right angles to the plane of the other three. By conventions, $\boldsymbol{a}_1$ is shown out of the paper, $\boldsymbol{a}_2$ to the right, $\boldsymbol{a}_3$ into the paper, and $\boldsymbol{c}$ to the top of the paper (Figs. 3-28d and e). As discussed in Section 3-3 the third $\boldsymbol{a}$ axis is redundant, but is used for the sake of clarity and for its better representation of the forms that hexagonal crystals take.

The orthorhombic system is referred to three orthogonal axes, which are of different lengths. The $\boldsymbol{a}$ axis is shown out of the paper, the $\boldsymbol{b}$ axis to the right, and the $\boldsymbol{c}$ axis to the top (Fig. 3-28f). By convention, orthorhombic space lattices have the $\boldsymbol{b}$ axis longer than the $\boldsymbol{a}$ axis. There is, however, some disagreement among mineralogists as to how to assign the $\boldsymbol{c}$ axis in the classes $2/m\,2/m\,2/m$ and 222 (in class $mm2$, the 2-fold axis determines the $\boldsymbol{c}$ direction in conformity with the convention applied to the tetragonal, hexagonal, and trigonal systems). Some mineralogists adopt the convention that $\boldsymbol{c}$ be the longest; others make $\boldsymbol{c}$ the shortest axis.

The monoclinic system is referred to three axes of different unit lengths. Two of these axes, the $\boldsymbol{b}$ and the $\boldsymbol{c}$, are at right angles, and the third ($\boldsymbol{a}$) is inclined at some angle other than 90° to the $\boldsymbol{c}$. As in the orthorhombic case, there are two competing conventions for assignment of monoclinic coordinate axes. The first, congruent with tetragonal, hexagonal, and trigonal assign-

Table 3-5

CRYSTAL CLASSES

Crystal system	*Class*	*Symmetry*
Isometric	1. Normal	$\frac{4}{m}\bar{3}\frac{2}{m}$
	2. Pyritohedral	$\frac{2}{m}\bar{3}$
	3. Tetrahedral	$\bar{4}3m$
	4. Plagiohedral	432
	5. Tetartohedral	23
Tetragonal	6. Normal	$\frac{4}{m}\frac{2}{m}\frac{2}{m}$
	7. Hemimorphic	$4mm$
	8. Tripyramidal	$\frac{4}{m}$
	9. Pyramidal–hemimorphic	4
	10. Sphenoidal	$\bar{4}2m$
	11. Trapezohedral	422
	12. Tetartohedral	$\bar{4}$
Hexagonal	13. Normal	$\frac{6}{m}\frac{2}{m}\frac{2}{m}$
	14. Hemimorphic	$6mm$
	15. Tripyramidal	$\frac{6}{m}$
	16. Pyramidal–hemimorphic	6
	17. Trapezohedral	622
	18. Trigonal	$\bar{6}m2$
	19. Trigonal tetartohedral	$\bar{6}$
Trigonal	20. Rhombohedral	$\bar{3}\frac{2}{m}$
	21. Rhombohedral–hemimorphic	$3m$
	22. Trirhombohedral	$\bar{3}$
	23. Trapezohedral	32
	24. Trigonal tetartohedral hemimorphic	3
Orthorhombic	25. Normal	$\frac{2}{m}\frac{2}{m}\frac{2}{m}$
	26. Hemimorphic	$mm2$
	27. Sphenoidal	222
Monoclinic	28. Normal	$\frac{2}{m}$
	29. Hemimorphic	2
	30. Clinohedral	m
Triclinic	31. Normal	$\bar{1}$
	32. Asymmetric	1

Table 3-6

COMMON MINERALS CLASSIFIED ACCORDING TO CRYSTAL CLASS

	Class symmetry	Example
Cubic	$\frac{4}{m}\bar{3}\frac{2}{m}$	Cu, Au, pentlandite, galena, halite, sylvite, fluorite, cuprite, periclase, spinel, magnetite, chromite, garnet
	$\frac{2}{m}\bar{3}$	Pyrite
	$\bar{4}3m$	Sphalerite, lazurite, tetrahedrite, diamond
	432	—
	23	—
Tetragonal	$\frac{4}{m}\frac{2}{m}\frac{2}{m}$	Rutile, cassiterite, pyrolusite, zircon
	$4mm$	—
	$\frac{4}{m}$	Leucite
	4	—
	$\bar{4}2m$	Bornite, chalcopyrite
	422	Cristobalite
	$\bar{4}$	—
Hexagonal	$\frac{6}{m}\frac{2}{m}\frac{2}{m}$	Graphite, pyrrhotite, covellite, molybdenite, beryl
	$6mm$	Wurtzite, zincite, ice
	$\frac{6}{m}$	Apatite
	6	Nepheline
	622	High quartz
	$\bar{6}m2$	—
	$\bar{6}$	—

ments, makes the 2-fold axis the ***c*** axis. More commonly, however, the 2-fold axis is made the ***b*** axis, with the ***a*** longer than the ***c*** axis. The angle β, between ***a*** and ***c***, must be specified to complete the description of a monoclinic lattice.

The triclinic system is the most general and the least symmetric of all. It is referred to three axes that are of different unit lengths, and which are not at right angles to one another. Thus, to specify the size and shape of a triclinic unit cell, the lengths of the cell edges and their internal angles must be specified. The angles are labeled α between ***b*** and ***c***, β between ***a*** and ***c***, and γ between ***a*** and ***b***. Thus, the Greek letter for the angle is the same as the English letter for the axis that is not involved. The internal angles in the triclinic system may be all acute, all obtuse, or some combination of the two, although the last possibility is not preferred. The ***c*** axis may be either the longest or the shortest axis, but, by convention, the coordinates are chosen such that the ***b*** axis is longer than the ***a*** axis. However, for a variety of reasons, including historical precedent and the desire to show crystallographic similarities among related minerals, these conventions are not infrequently violated when the convention is arbitrary in its assignments.

Table 3-6 (Cont.)

	Class symmetry	Example
Trigonal	$\bar{3}\frac{2}{m}$	Corundum, hematite, brucite, magnesite, siderite, smithsonite, calcite, rhodochrosite
	$3m$	Tourmaline
	$\bar{3}$	Ilmenite, dolomite
	32	Low quartz, cinnabar
	3	—
Orthorhombic	$\frac{2}{m}\frac{2}{m}\frac{2}{m}$	α-S, stibnite, marcastite, chrysoberyl, perovskite, columbite–tantalite, goethite, aragonite, strontianite, witherite, cerrusite, anhydrite, celestite, barite, anglesite, olivine, sillimanite, andalusite, topaz, staurolite, zoisite, orthopyroxenes, orthoamphiboles
	$mm2$	Chalcocite
	222	—
Monoclinic	$\frac{2}{m}$	Arsenopyrite, realgar, orpiment, psilomelane, azurite, malachite, borax, gypsum, monazite, lazulite, carnotite, sphene, clinozoisite, epidote, allanite, clinopyroxenes, clinoamphiboles, talc, muscovite, biotite, chlorites, kaolinite, antigorite, chrysotile, sanidine, orthoclase, tridymite
	2	—
	m	—
Triclinic	$\bar{1}$	Turquoise, kyanite, chlorites, kaolinite, microcline, plagioclase
	1	—

These coordinate axes may be thought of as having their origin at the center of a real crystal when Miller indices of crystal faces are determined.

Although classes with symmetries lower than that of the normal class in each system exist, these classes are underpopulated in terms of mineral examples (Table 3-6). Most minerals belong to the class with the highest symmetry in each system. The least-symmetrical class in each system has few mineral examples, and none of them is in the least bit common.

3-8. Crystal Forms

Ions at the surface of a crystal are exposed to a different environment in at least one direction. The effects of the external environment are such that the most stable surface is the one which is the most regular (see Section 4-6). The arrangement at the surface is generally a plane that has the densest packing of ions, as compared with other possible planes in the crystal.

The planes that offer the densest packing of ions are those which have a rational set of intercepts in the appropriate space lattice (Fig. 3-29). Crystal

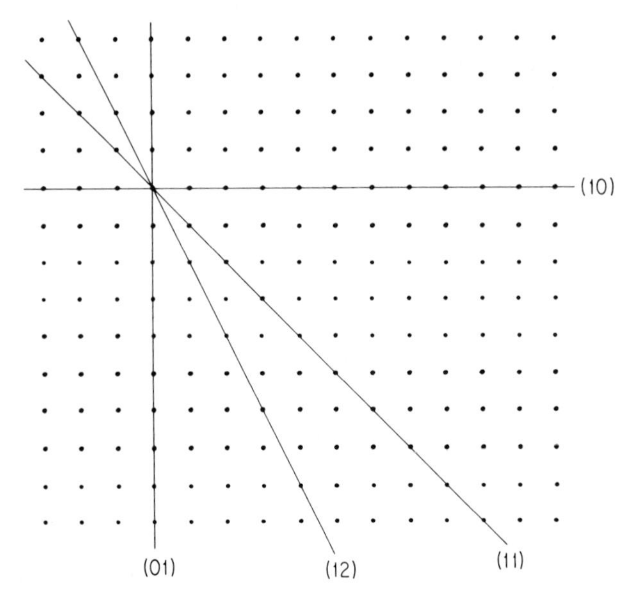

Figure 3-29 Lines with rational indices. Two-dimensional lattice with lines showing rational indices. These lines intersect more points than do lines drawn randomly.

faces, therefore, have rational intercepts with the coordinate axes. Indeed, Weiss intercepts and Miller indices were first derived from observation of crystal faces and crystal forms. It was a century later that the same planes of closest packing were observed to diffract X rays. Although not all mineral samples have crystal faces, potential for crystal faces can be observed from the X-ray diffraction pattern of any mineral.

A *crystal form* is a set of crystal faces that are symmetrically related. That is, in any crystal class one or more crystal faces may be equivalent by the symmetry operations which define the point group of that class. Conventionally, the Miller index of a crystal face is set off in parentheses, as (001), and the crystal form is designated by indices of one of the faces set off in braces, as {001} (Fig. 3-30a). Each crystal class has forms in common with one or more other classes, but there is usually at least one form in a class which is unique to that class. The more involved the symmetry of a class, the more faces there are in a given form.

The least-symmetrical forms are the open forms, some of which are illustrated in Fig. 3-30. The *pedion* is a form that consists of a single face. It is only possible in classes that lack a center of symmetry. A *pinacoid* (Fig. 3-30a) is a form consisting of two faces related by a center of symmetry, by a 2-fold axis of rotation, or by a mirror plane. A *basal pinacoid*, or a *base*, is the special pinacoid that intersects the ***c*** axis. Other pinacoids are referred to as ***a*** pinacoids or ***b*** pinacoids, depending upon their relationship to the crystallographic axes.

The stereogram for the basal pinacoid, which is shown in Fig. 3-30a, is given on the right-hand side of the figure. The symmetry shown in the stereogram is given to the right side of the page, in this case m. The indices of the poles of the faces are also given. [These stereograms, which illustrate the

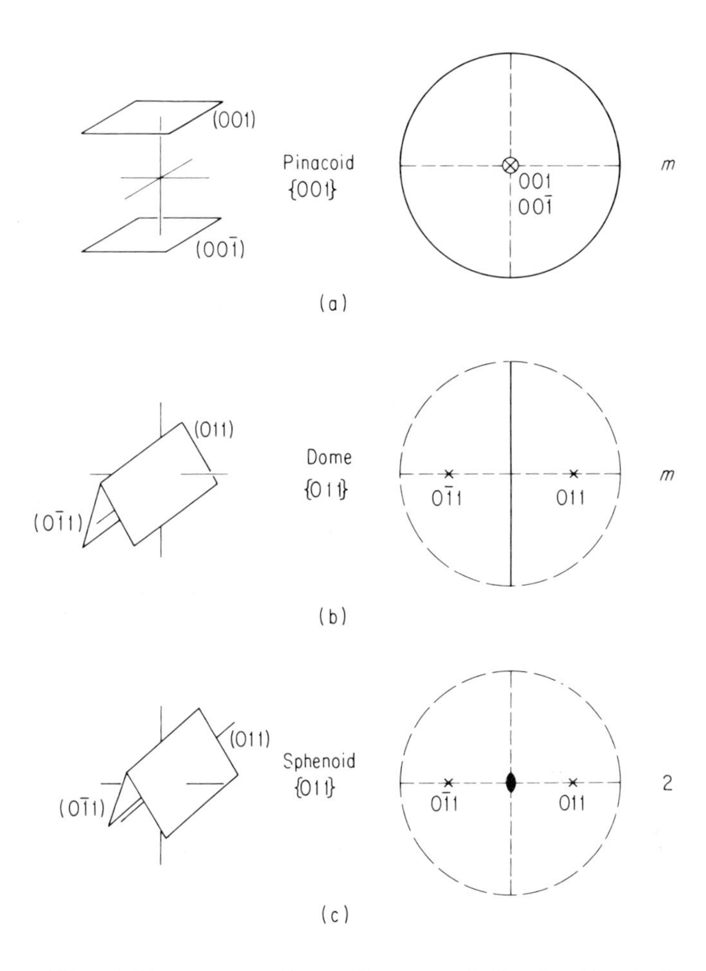

Figure 3-30 Open crystal forms. The symmetrically related faces and reference coordinates are shown to the left. The name and the form index are next to the right. On the right is the stereogram of the form with face poles shown as crosses in the upper hemisphere and circles in the lower. The symmetry elements relating the faces of the form are indicated on the stereogram. The point group for the symmetry elements is shown to the extreme right. (a) Pinacoid. (b) Dome. (c) Sphenoid.

forms, do not necessarily represent real point groups (Figs. 3-21 through 3-26).]

A *dome* (Fig. 3-30b) is an open form consisting of two faces that are related by a mirror operation. The dome, which intersects two crystallographic axes, is designated by the axis to which it is parallel, in this case an ***a*** dome. The stereogram to the right shows a mirror plane parallel to the ***a*** and ***c*** axes. A *sphenoid* is an open form of two faces, which is indistinguishable from a dome, but which is related by an axis of 2-fold rotational symmetry. The forms shown in Figs. 3-30b and c are distinguished only by the symmetry operation that relates them.

A *prism* is a form that has faces parallel to the maximum-fold axis of rotation. The faces may intersect one or more of the other crystallographic axes. If the faces intersect all of the ***a*** axes, the prism is of the first order. It is of the second order if the faces are parallel to one of the ***a*** axes. First- and second-order prisms do not differ geometrically, and either may be made into the other by reselection of the ***a*** axes. The correct choice of axial position is, by definition, the one with the ***a*** axes parallel to the unit cell edges, as determined by X-ray diffraction.

The prism shown in Fig. 3-32a is a second-order tetragonal prism because its four faces are parallel to the 4-fold rotation axis and they intersect but one of the ***a*** axes. The prism shown in Fig. 3-33a is a first-order trigonal prism, and that in Fig. 3-34a is a first-order hexagonal prism. The Miller indices of the faces are given in the diagram, along with the index of the form and the pole positions in the stereograms.

A *pyramid* is a form which has three or more faces that intersect the maximum-fold axis, as well as one or more of the other crystallographic axes. A tetragonal first-order pyramid is shown in Fig. 3-32b, with the Miller indices indicated for the faces as well as for the face poles on the accompanying stereogram. Trigonal and hexagonal pyramids are obtained where the maximum-fold axis is three and six, respectively.

With *open* forms, such as those shown in Fig. 3-30, two or more forms must be employed in combination in order for space to be completely enclosed. A prism capped by a pinacoid or by a pyramid would be an example.

A *closed* crystal form is one in which a single form completely encloses space. A *disphenoid* (Fig. 3-31a) may be thought of as two sphenoids that are symmetrically related. The addition of mirror planes to the stereogram shown in Fig. 3-31a gives the form shown in Fig. 3-31b, which is a *dipyramid.*

A *trapezohedron* is a closed crystal form with faces that are trapezia (Fig. 3-31c). (A *trapezium* is a plane figure having four sides, no two of which are parallel.) A *scalenohedron* is a closed crystal form with faces that are scalene triangles (Fig. 3-31d). (A *scalene* triangle is one with sides of unequal length.)

Some common tetragonal forms are illustrated in Fig. 3-32. A tetragonal prism is generated by the 4-fold rotation of any crystallographic face that is

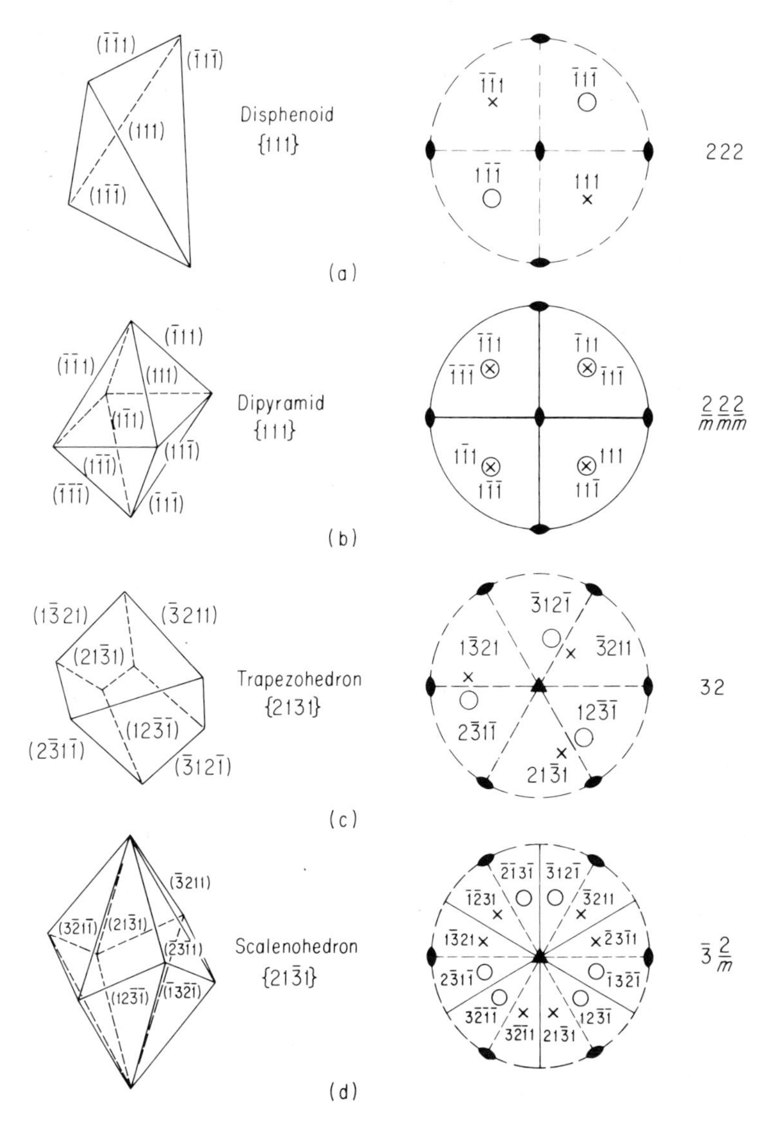

Figure 3-31 Closed crystal forms. Notation the same as in Fig. 3-30. (a) Disphenoid. (b) Dipyramid. (c) Trapezohedron. (d) Scalenohedron.

parallel to the ***c*** axis (Fig. 3-32a). A *ditetragonal* prism is an eight-sided open form, which is generated by a combination of a 4-fold rotation with 2-fold rotations (Fig. 3-32c). A similar relationship exists between the tetragonal pyramid (Fig. 3-32b) and the ditetragonal pyramid (Fig. 3-32d).

Some common trigonal forms are illustrated in Fig. 3-33. A trigonal prism

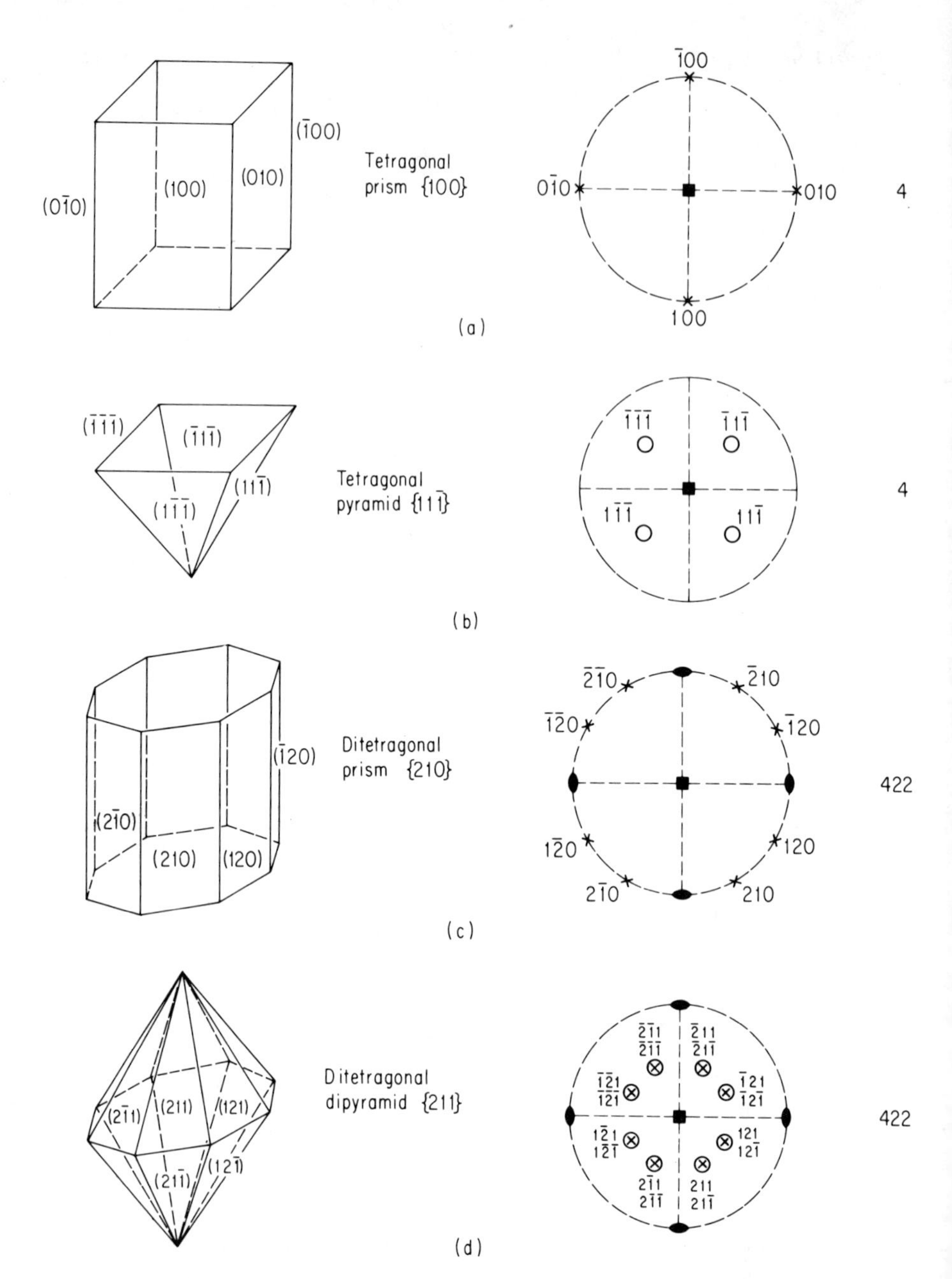

Figure 3-32 Tetragonal forms. Notation the same as in Fig. 3-30. (a) Second-order tetragonal prism. (b) First-order tetragonal pyramid. (c) Ditetragonal prism. (d) Ditetragonal dipyramid.

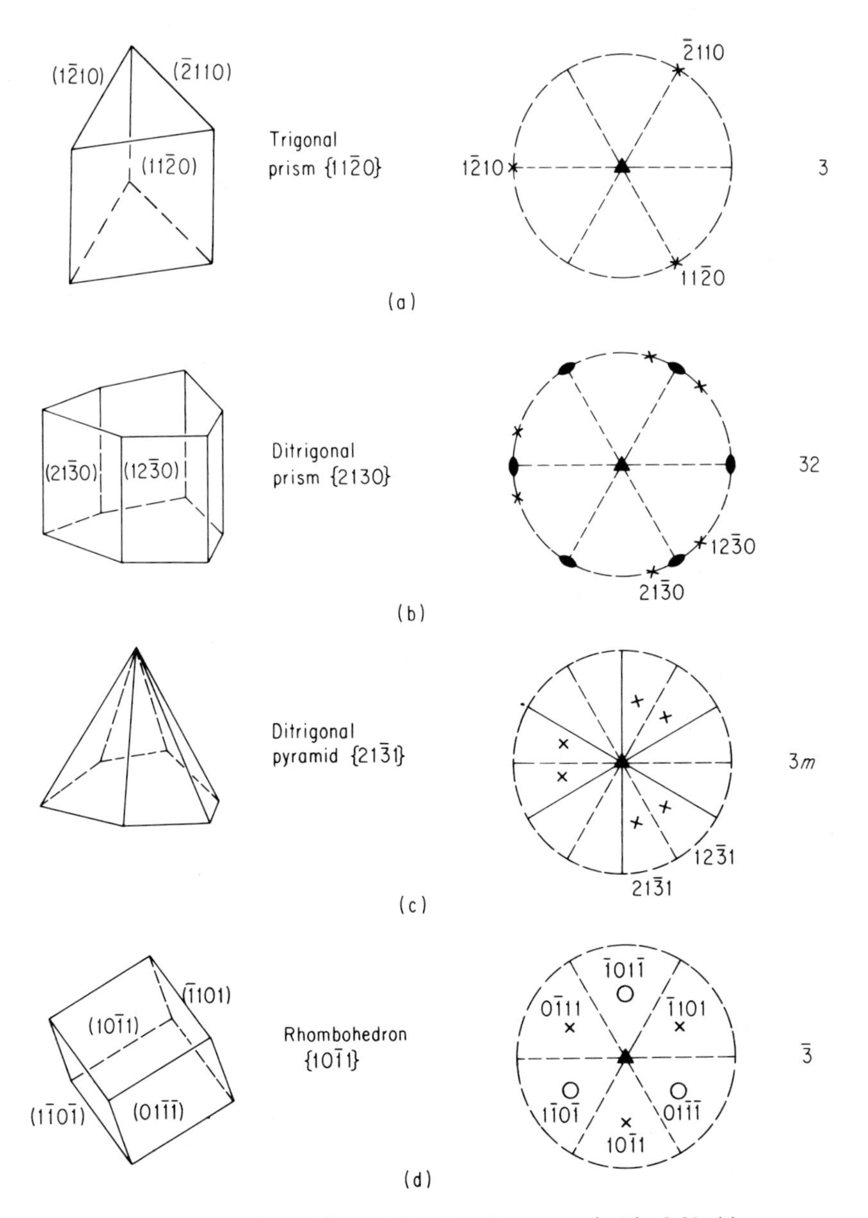

Figure 3-33 Trigonal forms. Notation the same as in Fig. 3-30. (a) Trigonal prism. (b) Ditrigonal prism. (c) Ditrigonal pyramid. (d) Rhombohedron.

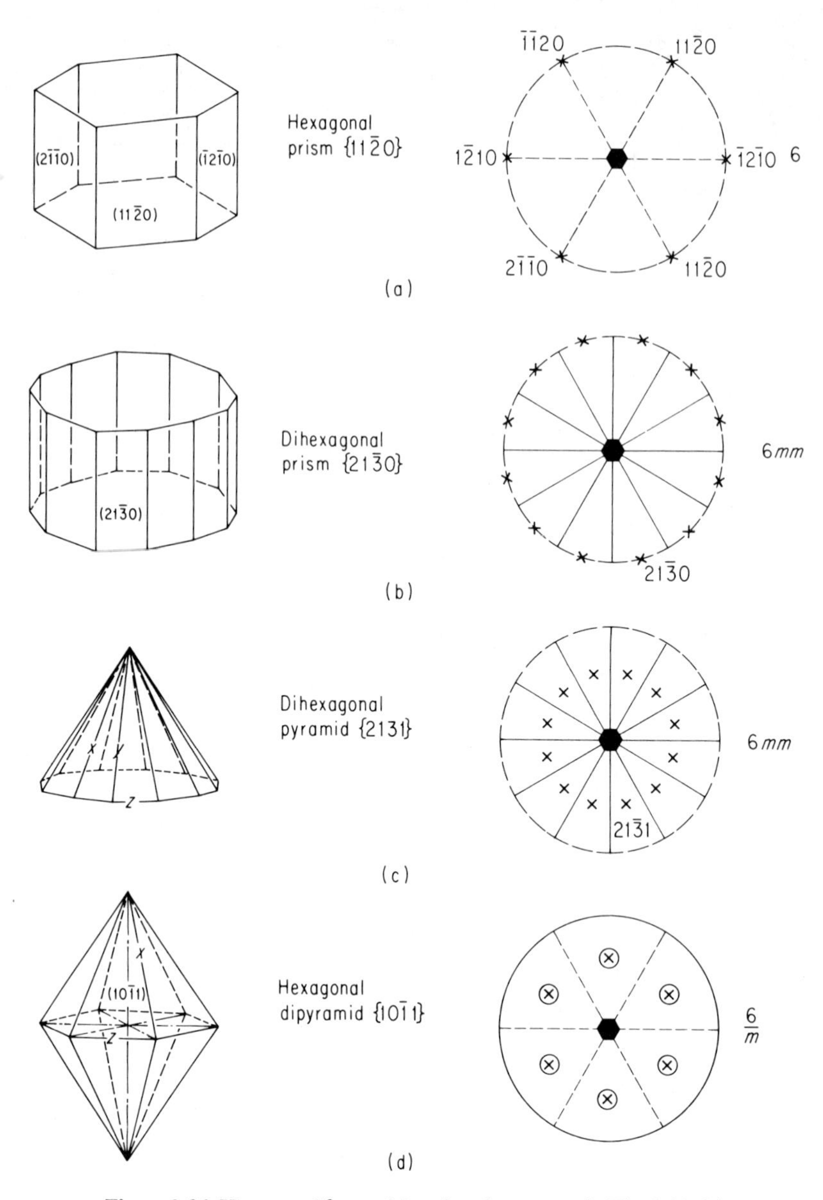

Figure 3-34 Hexagonal forms. Notation the same as in Fig. 3-30. (a) Hexagonal prism. (b) Dihexagonal prism. (c) Dihexagonal pyramid. (d) Hexagonal dipyramid.

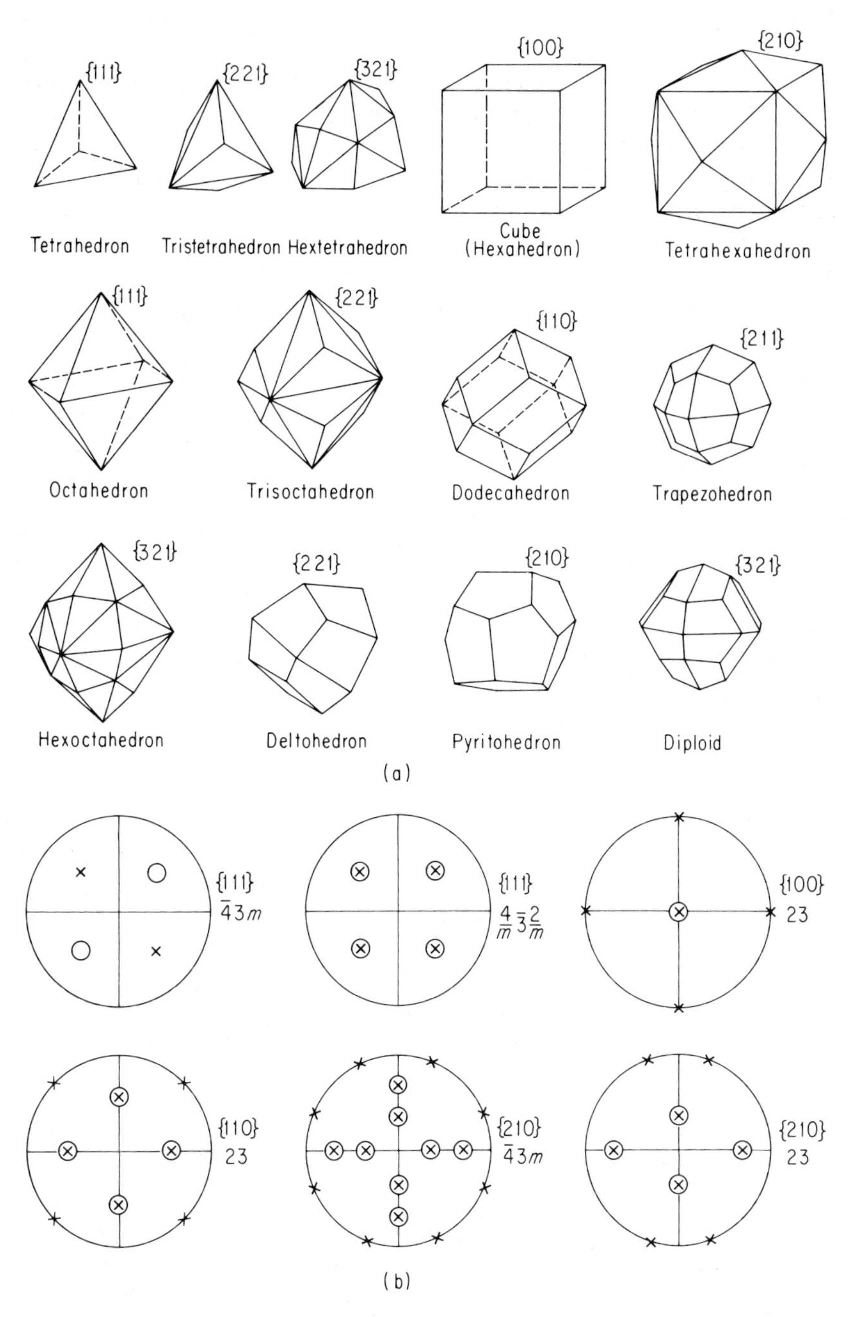

Figure 3-35 Cubic forms. (a) Sketches of 13 cubic forms with their names and indices. (b) Stereograms for six cubic forms with the form numbers and a point group that may produce them.

(Fig. 3-33a) has a cross section that is an equilateral triangle. A *ditrigonal* prism (Fig. 3-33b) is a six-sided open form with two different sets of internal angles; that is, not all the internal angles are equal, as they are in a hexagonal prism. A ditrigonal pyramid (Fig. 3-33c) is related to a trigonal pyramid in the same manner as a ditrigonal prism is related to a trigonal prism. A *rhombohedron* is a closed form of six faces (Fig. 3-33d), which may be thought of as consisting of the alternate faces of a hexagonal dipyramid (see Fig. 3-30g).

Some hexagonal forms are shown in Fig. 3-34. The hexagonal prism (Fig. 3-34a) differs from the ditrigonal prism (Fig. 3-33b) in the fact that all the internal angles of the former are equal, whereas those of the latter are not. The dihexagonal pyramid (Fig. 3-34c) is an open form of 12 faces, and the hexagonal dipyramid (Fig. 3-34d) is a closed form of 12 faces. Note the significance of the placement of the prefix *di*-.

All crystal forms in the cubic or isometric system are closed forms. Many such forms are related to others that have been discussed on the preceding pages, but they are given special names because of their regularity. This regularity of forms results from the high degree of symmetry in the system. Some of the more common forms of the cubic system are illustrated in Fig. 3-35, along with their names and form indices. A number of corresponding stereograms are also given.

3-9. Twin Crystals

Perfect, idealized crystals exist only in the mineralogist's mind (Sections 1-13 and 1-14). Different types of defect structures are illustrated in Fig. 1-16. *Crystal twinning* is a special type of defect wherein two parts of a crystalline material have different crystallographic orientations that are related by a symmetry operation. This operation may be reflection across a mirror plane, translation along a lattice plane, or rotation about an *n*-fold rotation axis. The twin operation cannot belong to the point group of the crystal class, but, in general, it is one that is very close to possible in the particular crystal structure.

The more nearly perfect a crystal is, the lower the internal energy of the crystal. Crystal twinning is a deviation from perfection, which increases the internal energy by a very small amount. That is, the internal energy of a twinned crystal is only slightly greater than the internal energy of an untwinned crystal, but is much less than that of a disordered structure.

Different kinds of twins develop by different means. If in a crystal growing from a solution or from a melt there is, on the accreting crystal face, a possible position that has an energy only slightly greater than the most stable position, an ion can occupy that position and cause the growth of the crystal to take on a new direction. It is a growth twin.

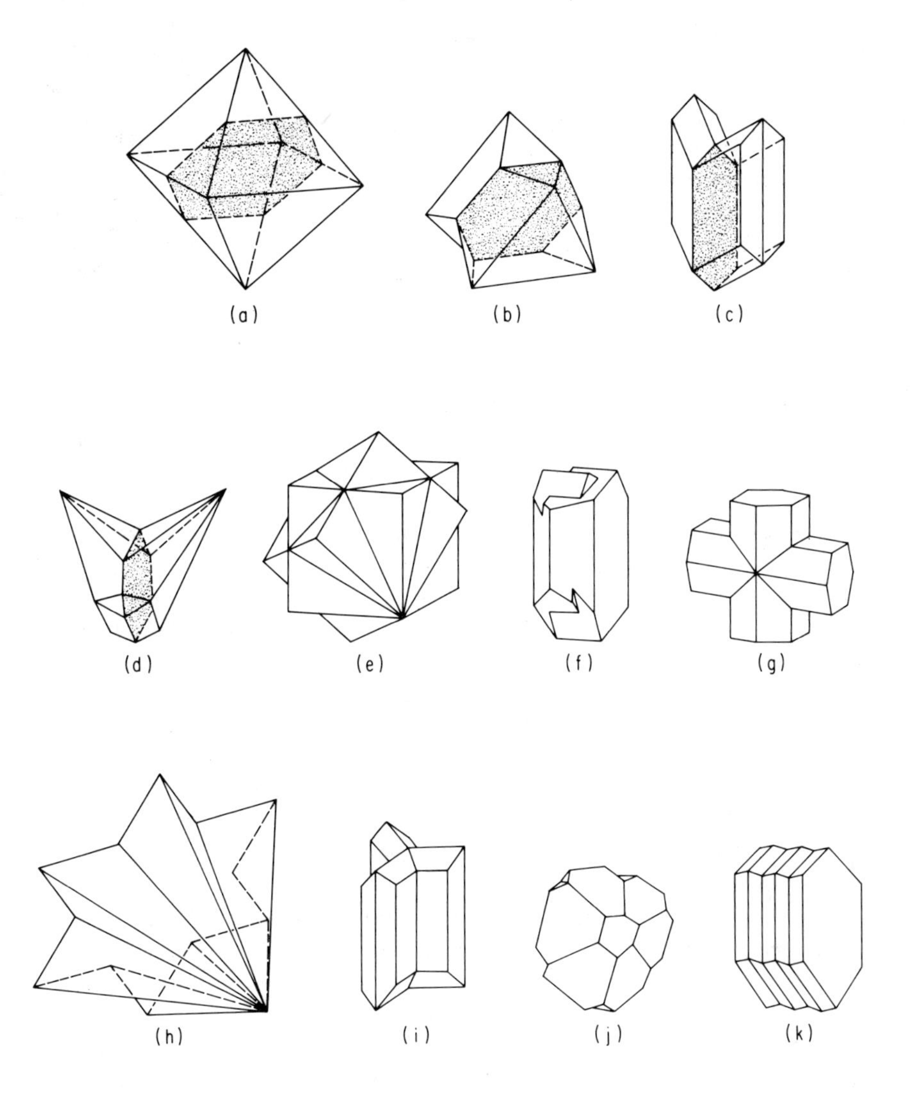

Figure 3-36 Crystal twinning. (a) Octahedron with octahedral twin plane. (b) Spinel twin formed by twinning across the octahedral twin plane. (c) Swallowtail twin (gypsum) formed by twinning axis being the ***c*** axis. (d) Calcite crystal with twin plane, one of the rhombohedral planes. (e) Iron cross (pyrite) formed by a dodecahedral plane being the twin plane. (f) Carlsbad twin (orthoclase) with ***c*** axis as twin axis. (g) Staurolite cross with (032) as twinning plane. (h) Multiple twin of tetrahedrite. (i) Multiple twin of aragonite with (110) as twin plane. (j) Multiple pyroxene twin crystal with ($\bar{1}22$) as twin plane. (k) Polysynthetic albite twinning with (010) as twin plane.

Transformation twinning occurs as a result of a polymorphic transformation, generally from a high-temperature form. The high-temperature polymorph has a higher symmetry than the low-temperature one. If there is more than one way in which the higher symmetry can be lost, then, as the inversion occurs, different parts of the crystal may start to invert in the different possible ways. The result is a low-temperature polymorph that contains both types of inversion—a *transformation twin* crystal.

If mechanical deformation causes a single crystal to yield by twinning, the result is a *glide twin*. The two parts of the twin are related in much the same way as the two parts of a growth twin.

Twin crystals joined by a common plane, known as a composition plane, are called *contact twins*. In those crystal classes that have a center of symmetry, the effect of reflection across the composition plane is the same as the effect of a 2-fold rotation about an axis normal to the composition plane. Such operations are not equivalent in crystal classes that lack a center of symmetry. The apparent twin axis is not necessarily a rational direction in classes in the monoclinic and triclinic systems. Therefore, some twin laws are defined by twin planes and some by twin axes.

The twinning of a crystal is *simple* if the two halves of a crystal are in twin relationship with each other. It is *multiple* if there are more than two twin-related parts. *Polysynthetic* twinning is a special case of multiple twinning in which three or more individual twins have the same twin plane. In plagioclase, the polysynthetic twin lamellae range in size from visible to the naked eye, appearing as striations on the {001} cleavages, to barely discernable with an optical microscope.

Twins in which a volume common to both parts is shared are known as *interpenetration* twins. The parts of the twin crystal follow the same laws that apply to contact twinning. That is, some common axis of rotation or common reflection plane exists. Interpenetration twins generally result from growth twinning or from transformation twinning.

Common twin forms for mineral species are described by the direction of the twin axis, or the Miller indices of the twin plane. The composition plane, if different from the twin plane, is also given. These data can be of immense aid in the identification of some mineral species. Several common types of twinning are illustrated in Fig. 3-36.

3-10. Mineral Habits

Recognition of crystal forms often leads to the identification of the crystal class to which the form belongs, and not infrequently leads to the identification of the mineral species, or at least the mineral family. Since most common minerals crystallize in the normal class of each crystal system, it is not necessary to cover in detail all their possible forms and combinations of forms in order to use them for purposes of mineral specification.

Minerals may nucleate and grow in a gaseous, liquid, or solid environment. Whether or not faces develop on a mineral grain depends on such factors as the difference between the chemical environment of ions in the grain and that in the growth medium, the rate at which crystal growth takes place, and the time and physicochemical conditions of any subsequent annealing or recrystallization. The structure and composition of the mineral itself are important. Some mineral species display crystal faces habitually, and others almost never do.

The more frequently encountered forms of the isometric system are illustrated in Fig. 3-35. Distribution of the cubic forms among the cubic classes is given in Table 3-7. Some common minerals that crystallize in this system are listed in Table 3-6. The stereogram of the point group for the normal class of the system ($4/m\,\bar{3}\,2/m$) is given in Fig. 3-25. Forms with simple indices, the cube {100}, the dodecahedron {110}, and the octahedron {111} dominate in any collection of isometric crystals (Fig. 3-37). Halite, sylvite, periclase, galena, and fluorite generally display cube faces, in some cases modified at their extremities by dodecahedral, octahedral, or hexoctahedral faces. Interpenetration twins of cubes with [111] twin axis are also known (Fig. 3-36b).

The spinel minerals typically crystallize in octahedra, some of which are

Table 3-7
CUBIC FORMS

Number of faces	*Form*	*Class*				
		2 3	4 3 2	$\frac{2}{m}\bar{3}$	$\bar{4}\,3\,m$	$\frac{4}{m}\bar{3}\frac{2}{m}$
4	Tetrahedron	×			×	
6	Cube (hexahedron)	×	×	×	×	×
8	Octahedron		×	×		×
12	Dodecahedron	×	×	×	×	×
12	Pyritohedron	×		×		
12	Tristetrahedron	×			×	
12	Deltohedron	×			×	
12	Tetratoid	×				
24	Tetrahexahedron		×		×	×
24	Trapezohedron		×	×		×
24	Trisoctahedron		×	×		×
24	Hextetrahedron				×	
24	Diploid			×		
24	Gyroid		×			
48	Hexoctahedron					×

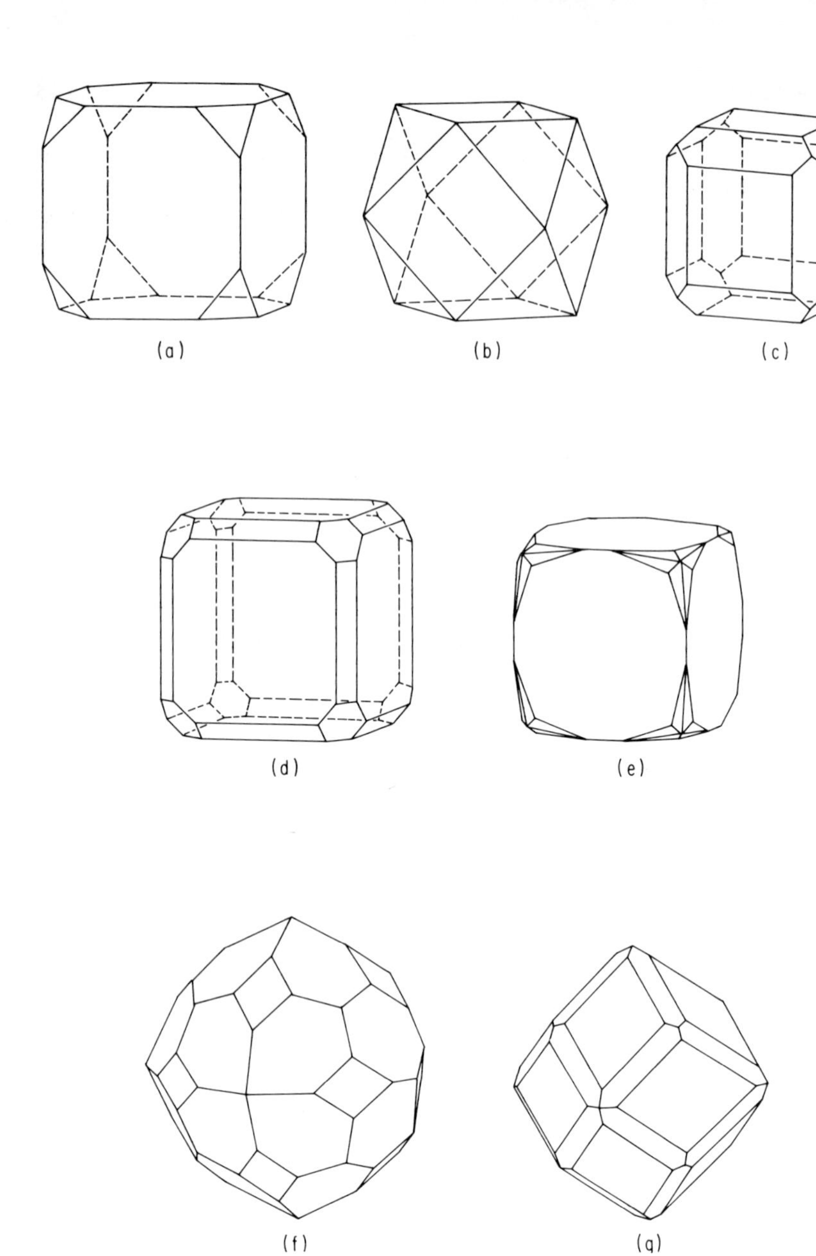

Figure 3-37 Cubic crystal habits. (a) Cube modified by octahedron. (b) Cube and octahedron equally developed. (c) Cube modified by dodecahedron. (d) Cube modified by dodecahedron and octahedron. (e) Cube modified by hexoctahedron (fluorite). (f) Trapezohedron modified by dodecahedron (garnet). (g) Dodecahedron modified by trapezohedron (garnet).

modified by dodecahedral faces. Contact twins of octahedra with (111) composition and twin plane are not rare. Cuprite, also, is found as octahedra, some highly modified.

The garnet minerals generally occur as very regular dodecahedral crystals, in some cases modified by trapezohedral faces {211}. Trapezohedra, with or without dodecahedral modification, are also to be found.

Pyrite is the only common mineral to crystallize in the pyritohedral class ($2/m\ \bar{3}$). Most pyrite crystals are cubes, although pyritohedra {210} are found in some deposits. The striations seen on most pyrite cube faces are due to oscillatory combination of pyritohedron faces.

The tetrahedral class ($\bar{4}\ 3\ m$) of the isometric system includes the minerals tetrahedrite, diamond, sphalerite, and lazurite. Tetrahedrite, as the name implies, crystallizes mostly in tetrahedra {111}, or in combinations of positive and negative tetrahedra, {111} and {$\bar{1}$11}. Sphalerite displays the same forms, although cube, dodecahedral, and tristetrahedral faces are present in some specimens. Lazurite displays cubes and dodecahedra when faces are present.

The most prevalent forms that occur in the tetragonal crystal system are illustrated in Fig. 3-32. Some common minerals of the system are listed in Table 3-6. The stereogram of the point group for the normal class ($4/m\ 2/m\ 2/m$) is given in Fig. 3-25. As in the cubic crystal system, forms with simple indices predominate. They include first- and second-order prisms, {110} and {100}, first- and second-order pyramids, {111} and {101}, and the basal pinacoid, {001}. Minutely alternating first- and second-order prism faces produce the striated prisms found on crystals of rutile and cassiterite.

Simple combinations of first- and second-order prisms with first- and second-order pyramids are to be found on cassiterite, rutile, zircon, and vesuvianite. In some cases, pyramids are modified by basal pinacoids. Octahedrite rarely displays prism faces, and most crystals of it show dipyramids with or without a basal pinacoid. Contact twinning, either simple or multiple, with (101) as the twin plane is not uncommon on crystals of zircon, rutile, and cassiterite. The most frequently encountered habits of tetragonal minerals are illustrated in Fig. 3-38.

Leucite is polymorphic, being cubic above and tetragonal ($4/m$) below about 625°C. (All temperatures cited in this book are in degrees Celsius.) However, crystals of leucite preserve their isometric forms, most commonly the trapezohedron {210}. The most common habit for chalcopyrite ($\bar{4}\ 2\ m$) is a disphenoid {112} or {11$\bar{2}$} closely approaching a tetrahedron in shape.

The point group for the normal class of the hexagonal system ($6/m\ 2/m\ 2/m$) is illustrated with a stereogram in Fig. 3-25. Forms of this class are sketched in Fig. 3-34. Some common minerals that crystallize in this class are given in Table 3-6. Of the minerals listed, only beryl develops crystal faces with any regularity—long hexagonal prisms that lack distinct termina-

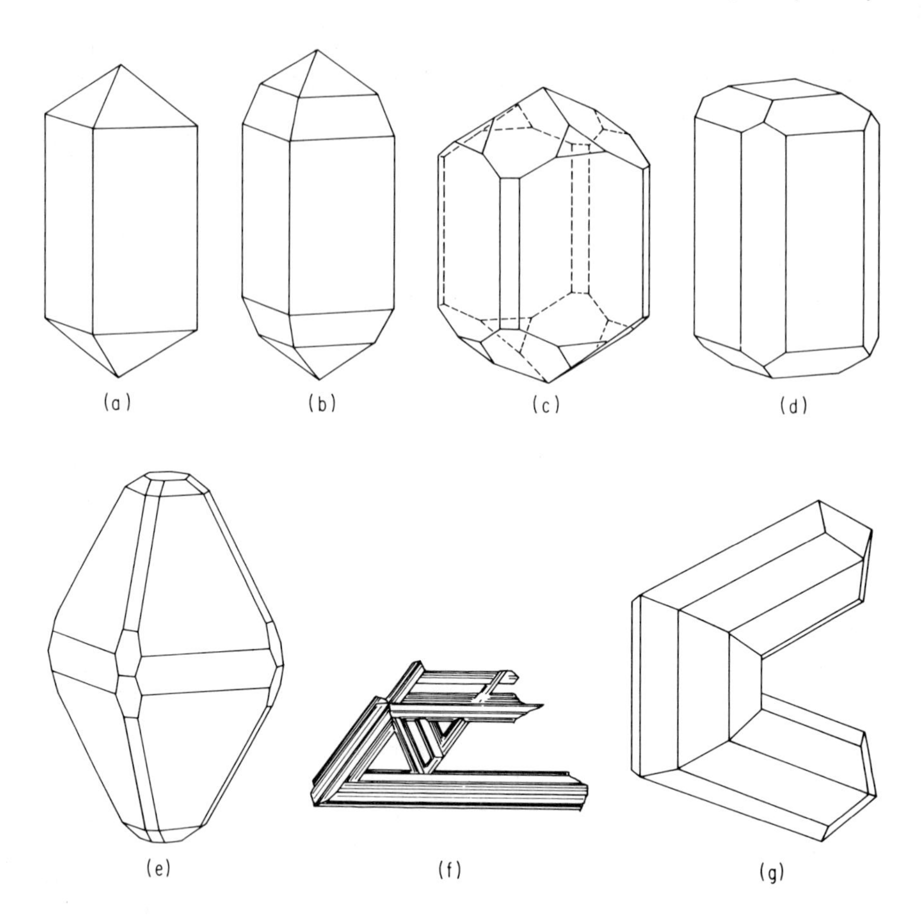

Figure 3-38 Tetragonal habits. (a) and (b) First-order prisms and pyramids (zircon). (c) First- and second-order prisms and pyramids (rutile). (d) First- and second-order prisms, first-order pyramid, and base (vesuvianite). (e) First-order prism, two first-order pyramids, three second-order pyramids, and base (octahedrite). (f) Twinned prisms (rutile). (g) Geniculated twinning with (101) as twin plane (rutile).

tions. Striations on these prism faces are due to alternating first- and second-order prisms. Crystals of molybdenite and graphite, when they do occur, are generally hexagonal plates.

Other common minerals, which crystallize in classes of lesser symmetry in the hexagonal system, are listed under the appropriate class in Table 3-6. Ice develops skeletal basal crystals (snow) and dendrites (frost). Apatite crystallizes in hexagonal prisms, with or without modification by second-order prisms. The prisms range from very long and acicular to quite thick and stubby. The prisms may be terminated by pyramids or by a basal pinacoid. Nepheline crystals are commonly short prisms. High-temperature quartz phenocrysts, such as found in some lavas, generally exhibit simple hexagonal dipyramids.

The normal class ($\bar{3}$ $2/m$) of the trigonal system (Fig. 3-26) contains the calcite group, hematite, corundum, and brucite. Calcite crystals exhibit a greater variety of forms and habits than any other mineral (Fig. 3-39). The more frequently encountered forms include rhombohedra $\{h0\bar{h}l\}$, which range from obtuse to acute; prisms, which range from tabular to long and have a variety of terminations; and scalenohedra $\{hk\bar{i}l\}$ of many types. Twinning, commonly polysynthetic, occurs with twin planes $(01\bar{1}2)$ and (0001). Siderite develops curved rhombohedra $\{10\bar{1}1\}$ or $\{11\bar{2}1\}$, commonly built up of subindividuals.

Hematite, if crystal faces are displayed, occurs in rhombohedra and in rhombohedra modified by basal pinacoids. Corundum generally occurs as barrel-shaped, hexagonal prisms. Twinning in the basal and rhombic directions causes parting in these directions in some crystals of hematite and corundum.

The rhombohedral–hemimorphic class ($3\,m$) of the trigonal system contains one important mineral—tourmaline. Tourmaline is prismatic in habit, the prisms being long and slender. The trigonal prism $\{10\bar{1}0\}$ and the hexagonal prism $\{11\bar{2}0\}$ may be combined in tourmaline crystals to give a nine-sided form. Rounded, three-sided prisms with striations result from minutely alternating prism faces of different forms. Terminations are generally combinations of rhombohedral and pyramidal forms.

The trirhombohedral class ($\bar{3}$) of the trigonal system contains the minerals dolomite and ilmenite. These may display rhombohedra, some of which are truncated by basal pinacoids and hexagonal prism faces. Dolomite crystals are commonly saddle-shaped composites of rhombohedral faces.

The trapezohedral class (3 2) of the trigonal system contains quartz, the mineral perhaps most frequently encountered with crystal faces in nature. After calcite, quartz crystals display the greatest variety of forms and habits among minerals (Fig. 3-39). Quartz crystals are generally prismatic $\{10\bar{1}0\}$, with the faces striated at right angles to the 3-fold axis by alternating positive and negative rhombohedral faces. Prisms are terminated by two rhombohedra, $\{10\bar{1}1\}$ and $\{01\bar{1}1\}$, which give the appearance of a hexagonal pyramid when equally developed. The angles between the rhombohedral faces are close enough to 90° to give a distorted cubic appearance to single forms. Twin crystals of the following types are found on many quartz crystals: twin axis parallel to the ***c*** axis; twin planes parallel to the ***c*** axis and one of the ***a*** axes; the twin plane $(11\bar{2}2)$; and twin planes parallel to the rhombohedral face.

Quartz is enantiomorphous, and simple crystals are either right or left handed, depending upon the apparent twist of the helices (Section 2-11). The

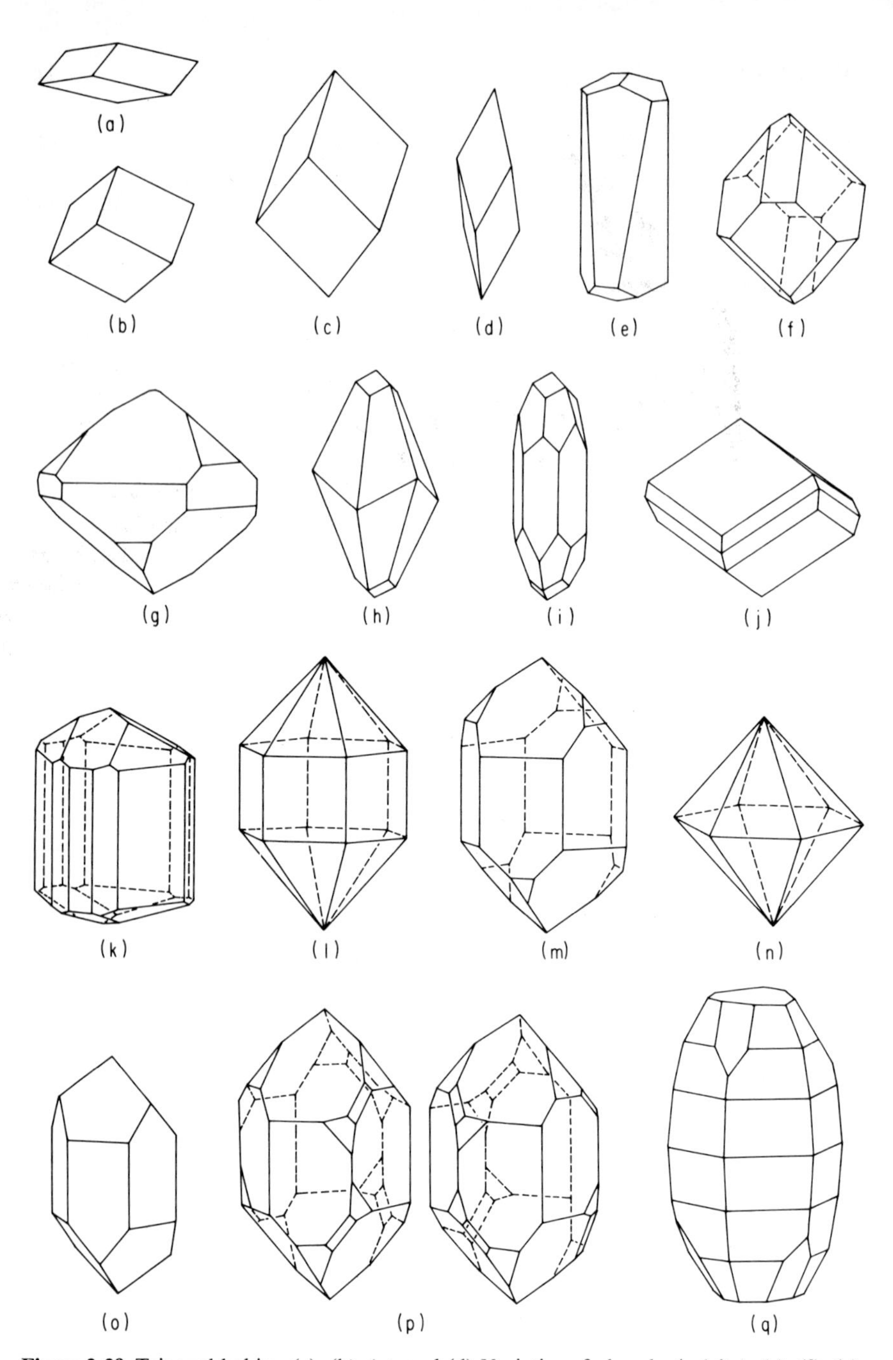

Figure 3-39 Trigonal habits. (a), (b), (c), and (d) Varieties of rhombs (calcite). (e), (f), (g) Combinations of varieties of rhombs (calcite). (h) Scalenohedron modified by rhomb (calcite). (i) Prism modified by scalenohedron and rhomb (calcite). (j) Rhomb modified by scalenohedron. (k) Prisms with rhomb and pyramid terminations (tourmaline). (l) Prism terminated with positive and negative rhombs (quartz). (m) Prism terminated with unequally developed positive and negative rhombs (quartz). (n) Positive and negative rhombs in pseudodipyramid (quartz). (o) Prism terminated by rhomb (quartz). (p) Right- and left-handed quartz. (q) Three pyramids terminated with rhomb and base (corundum).

difference may be recognized if either the right trigonal pyramid face $\{11\bar{2}1\}$ or the positive right trapezohedral face $\{51\bar{6}1\}$ is present. If the quartz is right handed, these faces are found to the right of the hexagonal prism face which is below the predominating positive rhombohedral face $\{10\bar{1}1\}$. Left-handed quartz crystals display these faces on the left (Fig. 3-39).

Most important orthorhombic minerals crystallize in the normal class ($2/m\ 2/m\ 2/m$) of the system. The forms of this class are all open (Fig. 3-30), and closed habits must be combinations of open forms. Most habits are various combinations of pinacoids, prisms, and domes (Fig. 3-40). The

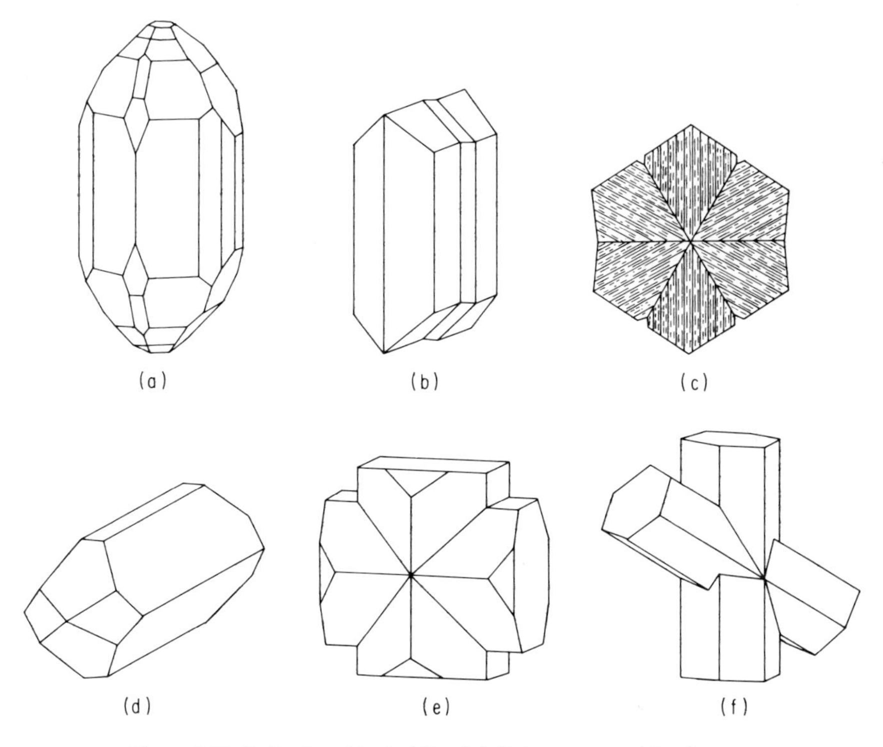

Figure 3-40 Orthorhombic habits. (a) Prisms, pyramids, domes, and pinacoids (topaz). (b) Multiple twinning with (110) twin plane (aragonite). (c) Multiple twinning across (110) to give pseudohexagonal basal section (aragonite). (d) Prism, domes, and base (barite). (e) Cruciform twin with (032) as twin plane (staurolite). (f) Cruciform twin with (232) as twin plane (staurolite).

prisms of sillimanite and andalusite are very nearly square, as is one of the prism forms of topaz {120}. Aragonite develops multiple twinning with the twin plane {110} and exhibits a pseudohexagonal habit. Crystals of staurolite are typified by interpenetration twins, with a cruciform shape of simple prisms and basal pinacoids. These twins are at right angles if the twin plane is (032), and at 60° if the twin plane is (232).

Many orthorhombic minerals are typically prismatic in habit (stibnite, columbite, goethite, aragonite, strontianite, cerrusite, anglesite, topaz, and zoisite); others are more often found to be distincitly tabular (marcasite, chrysoberyl, cerrusite, celestite, barite, and olivine); and some pyramidal (sulfur, cerrusite). Perovskite is pseudocubic, whereas witherite and aragonite develop pseudohexagonal shapes by multiple twinning.

The normal class ($2/m$) of the monoclinic system contains all the important minerals that crystallize in the system. The amphiboles, the clinopyroxenes (monoclinic pyroxenes), and orthoclase are the most important rock-forming minerals in this group. Crystals of clinopyroxenes are generally stubby prismatic shapes composed of the pinacoids {100} and {010} at right angles, or of the prism {110} at near right angles. Combinations of these three forms give eight-sided habits. Crystals are terminated with domes, some of which are modified by pyramid faces (Fig. 3-41).

Amphibole crystals are commonly prismatic in shape, with or without pinacoidal modification. Termination is mostly by the low dome {011}, or by the two domes {011} and $\{\bar{1}01\}$, which resemble a rhombohedral form. The prismatic habit of orthoclase derives from combinations of pinacoids, which are modified and terminated by domes. The most common twin habit found in orthoclase is the Carlsbad, either contact or interpenetration, with (100) as the twin plane.

Tabular gypsum crystals are pinacoids terminated by domes. Twinning across the twin plane (100) gives the familiar swallowtail twin. Sphene gets its name from its wedge-shaped habit, resulting from the combination of slightly acute and very highly acute domal forms. Epidote is generally prismatic parallel to the ***b*** axis, and has domal termination, usually at one end only. Spodumene crystals occur as nearly square prisms, commonly striated, which range up to 47 ft in length and up to 5 ft in diameter. (The larger crystals are obviously not to be found in the average mineralogy laboratory.)

Other monoclinic minerals display prismatic habits (arsenopyrite, monazite, realgar, and borax), some typically tabular (azurite, monazite, and the micas), some domal (azurite and lazulite). Muscovite and tridymite develop pseudohexagonal tabular crystals, whereas biotite and chlorite may be pseudorhombohedral.

Two of the most important rock-forming minerals, microcline and the plagioclases, crystallize in the normal class ($\bar{1}$) of the triclinic system. Because of the low symmetry, only one form, the pinacoid, occurs in this class. The habit of the triclinic feldspars, however, is quite similar to that of orthoclase. Indeed, microcline derives its name from the fact that the angle between the (010) and the (001) pinacoid faces is 89°30′, whereas the same angle in orthoclase is exactly 90°. This same angle in the plagioclase series varies from 86°24′ in albite to 85°50′ in anorthite. In addition to Carlsbad twinning, the triclinic feldspars habitually display polysynthetic twinning according to the albite and

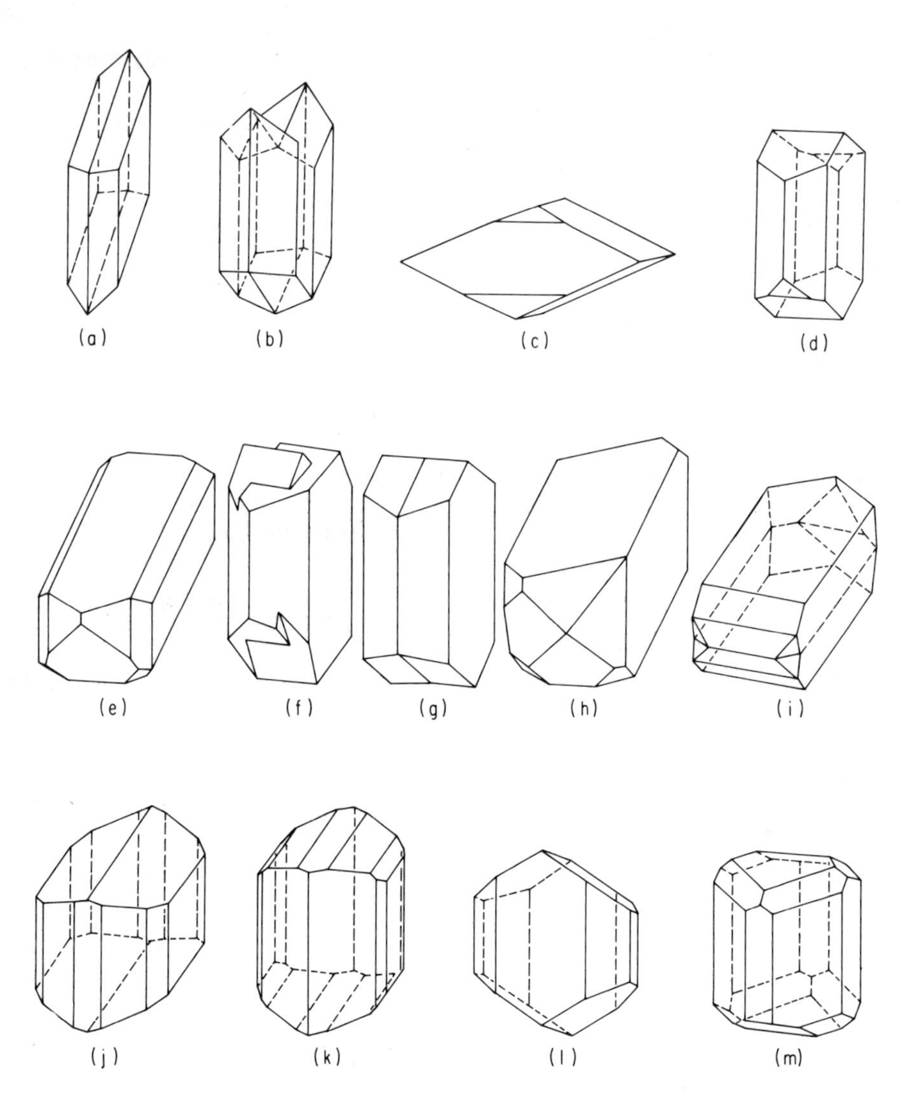

Figure 3-41 Monoclinic habits. (a) Prism, dome, and pinacoid (gypsum). (b) Swallowtail twin with (100) as twinning plane (gypsum). (c) Domes and pinacoid (sphene). (d) Prism, domes, and pinacoid (orthoclase). (e) Prisms, domes, and pinacoids (orthoclase). (f) Carlsbad interpenetration twin with (100) twinning plane (orthoclase). (g) Carlsbad contact twin with (010) as composition face (orthoclase). (h) Baveno contact twin with (021) as twinning plane (orthoclase). (i) Manebach contact twin with (001) as twinning plane (orthoclase). (j) Prisms, pinacoids, and domes (clinopyroxene). (k) Prisms, pinacoids, and domes (clinopyroxene). (l) Prism, pinacoids, and dome (amphibole). (m) Prism, domes, and pyramid (amphibole).

the pericline laws. The twin plane is (010) in the former, and the twin axis is [010] in the latter. The combination of these two forms at nearly right angles is diagnostic for microcline. They are not so frequently encountered together in plagioclase. It is the albite twinning of plagioclase that produces the familiar striations of the {010} cleavage faces of this mineral.

Kyanite most commonly occurs as long, bladed crystals composed of pinacoidal forms. These crystals rarely terminate in a form. Turquoise with crystal faces is quite rare.

Recognizable, and therefore identifiable, crystal faces are far more frequently encountered in mineralogical museums and laboratories than they are in the general run of rocks of the earth's crust. Nevertheless, recognition of distinctive forms is a powerful tool in the identification of mineral species.

FURTHER READINGS

BUERGER, M. J., *Elementary Crystallography*. New York: John Wiley & Sons, Inc., 1963.

DANA, E. S., and W. E. FORD, *Textbook of Mineralogy*, 4th ed. New York: John Wiley & Sons, Inc., 1932.

PHILLIPS, F. C., *An Introduction to Crystallography*, 3rd ed. Essex, England: Longman Group Ltd., 1963.

4

Physical Properties of Minerals

Physics may be defined as that branch of science which deals with the inanimate part of the universe, and mineralogy as that branch which deals with the crystalline part. Clearly, there is a large part of the universe which is both inanimate and noncrystalline, that is, which belongs exclusively to the physicist, and, therefore, does not concern the mineralogist. The area of overlap between physics and mineralogy—the interaction of energy and crystalline matter—is covered in this chapter and in the next one.

Two kinds of energy of concern to the mineralogist are mechanical energy (Chapter 4) and radiant energy (Chapter 5). The way that a mineral reacts to the application of either form of energy depends upon its chemical composition and upon its crystal structure. The physical properties of minerals, as they are observed in the laboratory or in the field, are related directly to the chemical and structural characteristics, as set forth in the first three chapters of this book.

One of the first things that an individual is likely to notice about any object that he may happen to encounter is its weight, and, especially, its weight as compared with other objects of similar size, shape, and appearance. Consider, for a moment, three objects the size and shape of a brick. One is silvery and metallic, one is dull and reddish, and the third is brown and grainy. Most people would decide, on the basis of appearance alone, that the metallic object is heaviest and the woody one is lightest. Hefting the blocks usually verifies this preconceived idea. If the wooden block were teak and the metallic bar were aluminum, most people would consider the wooden one to be exceptionally heavy and the metallic one to be exceptionally light, even though the relative weights remain unchanged. And most people would feel tricked somewhat if the reddish block were not the brick that they thought it to be, but instead a bar of iron disguised by rust.

This thought experiment should demonstrate to the reader that the correlation of the appearance of solid objects with other physical properties is really common knowledge. Density is associated with appearance, and both are related to the chemistry and the structure of the material.

4-1. Density and Specific Gravity

The amount of mass that can be accommodated in a given volume depends upon the atomic masses of the elements present, the radii of the atoms or ions, the way in which the atoms or ions are packed together, the temperature, and the pressure. How much mass a given volume contains is measured as either the quantity *density* or the quantity *specific gravity*. Density is defined as the mass per unit volume. The units of measurement must be stated, usually grams per cubic centimeter. Specific gravity is defined as the ratio of the weight of a material to the weight of an equal volume of water at a temperature of 4°. Specific gravity is, therefore, dimensionless.

The numerical value of density, when stated in grams per cubic centimeter, is identical to the specific gravity of a material, because the density of pure water is 1 g/cm^3 at 4°. An unfortunate result of this numerical identity is the common failure to distinguish between these two different quantities, with the secondary result that frequently the terms are used interchangeably. Density and specific gravity are not the same, and the methods for measuring each are different, although conversion between the two is simple and direct.

The dependence of density upon atomic masses should be obvious from the discussion in Chapter 1. The electronic structure of an atom or ion determines its size, and ionic radii vary by a factor of 2 or 3. The atomic mass of an element depends upon the number of protons and neutrons in the nucleus, and this number varies by a factor greater than 200. The degree of dependence upon atomic masses can be derived from the data in Table 4-1, where the densities of several examples of isostructural matter are given.

The density of an individual ion depends on the size of the ion, which is decreased by oxidation and increased by reduction. This obvious effect is difficult to illustrate simply, because oxidation and reduction change both stoichiometry and the coordination number of the cation. The oxidation of magnetite [$Fe^{2+}Fe_2^{3+}O_4$] to maghemite [$Fe_2^{3+}O_3$ with spinel structure] does not lead to an increase in density because the Fe to O ratio is decreased. The dependence of density on ionic radius, however, may be illustrated with minerals having the halite structure. The shorter the interionic distance, the greater the density (Table 4-2).

Table 4-1

DEPENDENCE OF DENSITY ON ATOMIC MASS

	Composition	Atomic mass	Density (g/cm^3)
Olivine			
Forsterite	Mg_2SiO_4	Mg 24.31	3.22
Fayalite	Fe_2SiO_4	Fe 55.85	4.41
Carbonate			
Calcite	$CaCO_3$	Ca 40.08	2.71
Magnestite	$MgCO_3$	Mg 24.31	3.12
Rhodochrosite	$MnCO_3$	Mn 54.94	3.60
Siderite	$FeCO_3$	Fe 55.85	3.88
Smithsonite	$ZnCO_3$	Zn 65.37	4.45
Feldspar			
Orthoclase	$KAlSi_3O_8$	K 39.10, Si 28.09 } 67.19	2.56
Celsian	$BaAl_2Si_2O_8$	Ba 137.34, Al 26.98 } 164.32	3.37
Spinel			
Spinel	$MgOAl_2O_3$	Mg + 2Al = 78.27	3.55
Chromite	$FeOCr_2O_3$	Fe + 2Cr = 159.85	5.09
Magnetite	$FeOFe_2O_3$	3Fe = 167.55	5.20
Franklinite	$ZnOFe_2O_3$	Zn + 2Fe = 177.07	5.34

Table 4-2

DEPENDENCE OF DENSITY ON BOND LENGTH IN MINERALS HAVING HALITE UNIT CELL

Mineral	*Formula weight*	*Bond length* (Å)	*Density* (g/cm^3)
Sylvite	74.6	3.14	2.0
Halite	58.5	2.75	2.2
Periclase	40.3	2.07	3.6

The dependence of density upon physical environment can be visualized with reference to the dependence of effective ionic size upon the same environment. When a mineral is heated, the energy is taken up by increased lattice vibration, which means that the ions themselves require more space for oscillation. The ions become effectively larger without changing their mass, and the mineral becomes less dense as a result. Pressure, on the other hand, squeezes the electron clouds closer together (see Fig. 1-5), literally cramming more mass into the same volume, with a concomitant increase in density.

Any change in the structure of isochemical matter also changes the density of it. The more efficient the packing, the more ions there are in a given volume, and the higher the density of the mineral. The degree to which packing

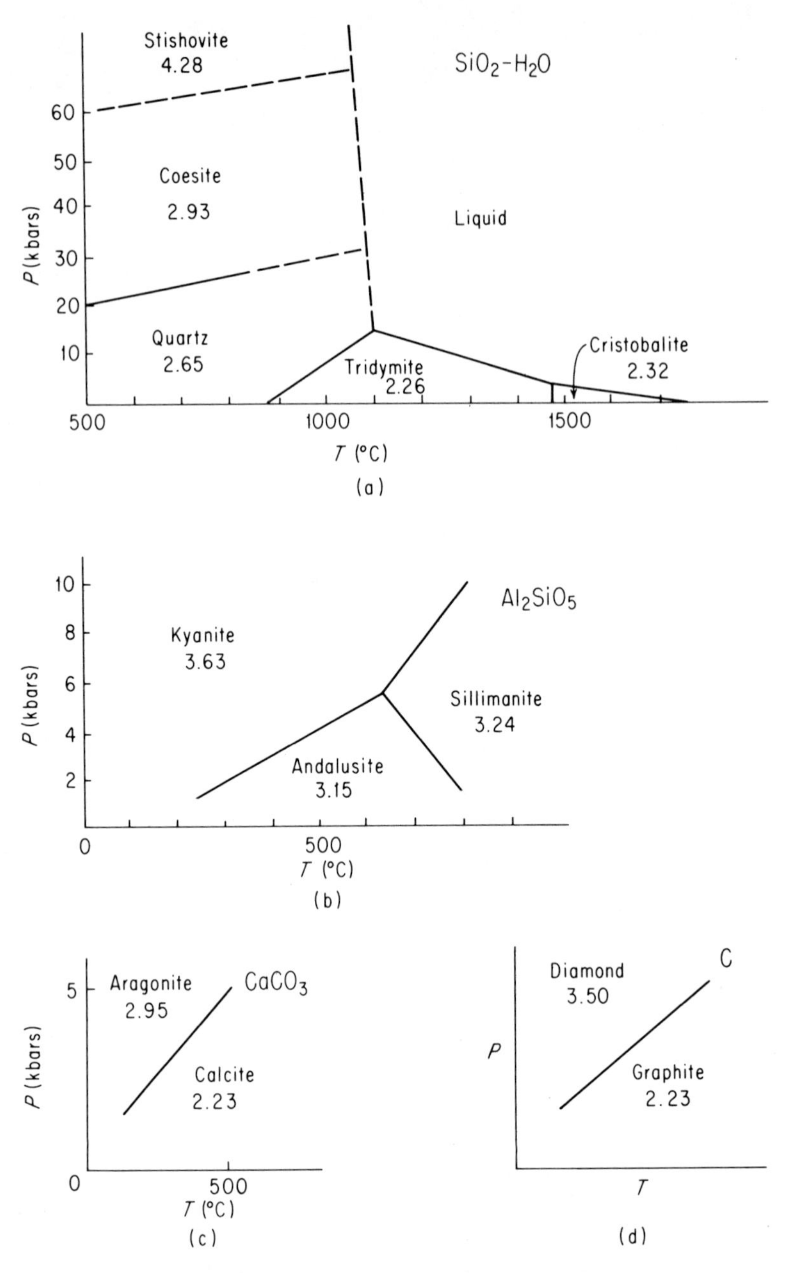

Figure 4-1 Density differences in isochemical matter. Temperatures and pressures of equilibrium crystallization of various polymorphs, and their densities (g/cm³) under laboratory conditions are shown. High-pressure forms tend to be denser and high-temperature forms tend to be less dense than the low-temperature and low-pressure polymorphs.

influences density can be derived by inspection of the data in Fig. 4-1, where the densities at standard pressure and temperature of several different polymorphs are plotted on their phase diagrams. As expected from the discussion in the preceding paragraph, polymorphs formed at higher pressures are denser than those formed at lower pressures. Similarly, the low-temperature polymorphs are denser than the high-temperature polymorphs.

Density may be measured directly on a mineral sample, or it may be calculated from the results of chemical analysis and of X-ray diffraction. The volume of the unit cell can be calculated from diffraction data, and the mass of the unit cell from chemical data. These two quantities yield a theoretical density, which can be compared with measured densities.

That density of an object is related to its appearance is demonstrated at the beginning of this chapter. This observation can be traced to the kinds of chemical bonds present in a mineral. Most metals crystallize in one of the two close-packed arrays (Section 1-9). This packing gives the maximum number of atoms per unit volume, and the atoms are not diluted by the presence of anions having low atomic mass. Both factors contribute to making metallic crystals relatively dense. Metallic luster of metallically bonded crystals is explained in Section 5-3. Both properties are related to the kinds of bonding and to the kinds of structure present.

The directional nature of covalent bonds (Section 1-4) militates against efficient packing in crystal structures where these bonds are present. Organic compounds, such as wood, generally have the lowest densities of common solid materials. The directional character resulting from partial covalency of the Si—O bond in silicate minerals contributes to their low densities, which are intermediate between organic structures and metallic structures.

Coordination polyhedra (Section 1-11) represent the most efficient packing of ions of different sizes and charges into a unit-cell volume. The intermediate densities of ionic compounds result from a balance of the effects of efficient packing, a contributor to high density, and the presence of anions with low atomic mass, which contributes to low density. Except for a few minerals that contain ions of high atomic mass, densities of ionic compounds range between 2 and 3 g/cm^3. Density and appearance obviously are associated, as most people learn at a very early age.

4-2. Principles of Mechanics

Stress is produced in a mineral when force is applied on it. On the atomic scale, this means that forces other than those of the chemical bonds are present. Ions are removed from their equilibrium positions, that is, *strained*, if the stresses are great enough. Although stress and strain are used popularly as though they were synonymous, they refer to very different kinds of things.

Stress is a resolution of force, and is measured in force units, which are those of mass times acceleration. *Strain* is deformation, and is measured in dimensional units, which are those of ordinary linear measurement.

The manner in which a mineral reacts to stress and the way in which it is strained can be characteristic, or even diagnostic, of it. The mechanical properties of a mineral depend on the strength and distribution of chemical bonds and on the degree of perfection of crystallinity in the lattice. The theoretical strength of a mineral may be calculated from the strength of the chemical bonds between the ions. Such a calculation, however, yields strengths that are commonly more than 100 times greater than strengths obtained from direct measurement. The great difference between theoretical strength and measured strength can be attributed to the presence of defects in the crystal lattice (Section 1-14).

Crystal imperfections are illustrated schematically in Fig. 1-16 and described in Table 1-9. Wherever an ion is missing from a lattice site, some chemical bonds are also missing, and there is local electrostatic repulsion. Obviously, a crystal is weaker at the point of a Schottky defect than at points where perfect order is to be found. Ions are displaced slightly from equilibrium positions wherever an interstitial ion exists in a lattice, and there is local, residual strain about the interstitial. Further weakening of the crystal occurs if defects are mobile and strain causes them to be concentrated.

Zones of strain and of irregular bonding extend along line defects in the crystal lattice. An edge defect or a screw dislocation defines a line of ions in an irregular and, therefore, structurally weak environment. The result of a lineage defect is a series of edge dislocations spaced at regular intervals along a plane in the lattice. Grain boundaries exist in polycrystalline materials; ions at these boundaries have an irregular environment on at least one side. The net result in all these cases is local strain and fewer chemical bonds than in the well-ordered part of a crystal lattice.

Stacking faults and twinning (Section 3-9) are characterized by planes of ions in perfect crystal order with respect to their nearest neighbors, but out of order with regard to the next most distant ions. These second-nearest neighbors contribute a repulsive force across the stacking fault or the twin plane, which is not present in unfaulted and untwinned parts of the lattice.

All crystalline materials, with the exception of flawless single-crystal whiskers, are weaker than would be expected from calculations of chemical bond strengths, because of the ubiquitous presence of one or many kinds of crystal imperfections. How, then, can the mechanical properties of minerals be related to the kind and distribution of chemical bonds? The answer lies, in part, with the fact that material weakening due to defect structures is systematic. The relative strengths of materials are mostly unchanged because the defects are universal. Another factor that permits relating strength to structure is due to the inequality of chemical bonds. The strength of a mineral is dependent upon the weakest bonds present. If the weakest bonds are systematically oriented with respect to the crystallographic axes, the mechanical properties of a mineral are also related to crystal direction. This effect can be considerably larger than the combined effects of the defects.

It is not always possble to be completely precise in relating mechanical properties to chemical bond types, however, because the kinds and distributions of defects are a function of the past history of a mineral specimen, in addition to depending upon the bond types present. In other words, because of structural defects in crystal lattices, it is not possible to explain all mechanical properties at the atomic scale.

4-3. Mechanical Behavior without Rupture

The mechanical behavior of crystalline materials may be conveniently and logically divided into two parts, that which involves rupture and that which does not. It may be noted, in passing, that the physical principles of the latter are much better understood than those of the former.

One of the most important definitive properties of minerals is their *hardness*. The hardness of a substance is related both to the strength of the chemical bonding and to the method of measurement. For materials which deform plastically, that is, which may flow without rupture, relative hardness of a sample may be determined by indentation with a stylus. The shape of the stylus is determined by convention, and the hardness of a sample is related to the amount of force necessary to cause a standarized indentation (Brinell, Knoop, and Rockwell hardnesses).

For brittle materials, hardness may be related to small-scale rupture. One method of measurement involves relating the hardness inversely with the amount of material removed from the sample by abrasion (Rosiwall hardness). Although very quantitative, this method of hardness determination requires that conditions of measurement must be standardized completely for all comparisons.

Friedrich Mohs proposed, in 1822, a series of 10 minerals, numbered in order of increasing resistance to abrasion, to define a scale of hardness. The minerals are

1. Talc
2. Gypsum
3. Calcite
4. Fluorite
5. Apatite
6. Orthoclase
7. Quartz
8. Topaz
9. Corundum
10. Diamond

Each mineral can scratch all those lower in number in the scale, and is scratched by all that are higher. The intervals on the scale are roughly equal, except for that between 9 and 10, which is considerably greater than the other intervals. The scale has the advantage that all the minerals are commonly available in most mineralogy laboratories, and the relative hardness of unknown brittle materials may be compared directly to them.

When trying to relate mineral properties, such as hardness, it should be remembered that, although structure depends upon the strongest bonds present in the crystal structure, physical properties depend upon the weakest bonds in structures having more than one kind of bond. This is most obvious among the silicate minerals that have Si^{4+} in tetrahedral coordination. The strength of the Si—O bond is approximately the same in all of them, but the hardness of silicates ranges from 1 in talc to 7 in quartz. The hardness is controlled by bonds other than the Si—O in the structure wherever they are present. Even in materials having chemical bonds all of one kind, hardness depends upon the number of bonds per unit area (*bond density*) and, thus, varies with crystallographic direction.

The strength of ionic bonds between divalent ions should be expected to be stronger than bonds between monovalent ions, if they occur in the same kind of structure. This relationship is clearly demonstrated by halite [NaCl], which has monovalent ions and a hardness of $2\frac{1}{2}$, and periclase [MgO], which has divalent ions and a hardness of 6. These minerals are isostructural (Section 2-2). There is the added effect of the ions in halite being larger than those in periclase. Since electrostatic attraction falls off in proportion to the square of the distance, the attraction between Na^+ and Cl^- ions in halite is less than half the attraction between Mg^{2+} and O^{2-} in periclase. Still, the difference in the hardness of these two isostructural minerals is dramatic.

As already indicated, ionic size should affect the hardness of isostructural matter, because electrostatic attraction falls off with the square of the distance (Table 4-3). The greater the cation-to-anion distance is, the lower is

Table 4-3

DEPENDENCE OF HARDNESS ON IONIC RADIUS

	Hardness	*Ion*	*CN*	*Radius* (Å)
Olivine structures				
Forsterite, Mg_2SiO_4	7	Mg	VI	0.80
Fayalite, Fe_2SiO_4	$6\frac{1}{2}$	Fe^{2+}	VI	0.86
Monticellite, $CaMgSiO_4$	$5\frac{1}{2}$	Ca	VI	1.08
Corundum structures				
Corundum, Al_2O_3	9	Al	VI	0.61
Hematite, Fe_2O_3	6	Fe^{3+}	VI	0.73
Halite structures				
Halite, NaCl	$2\frac{1}{2}$	Na	VI	1.10
Sylvite, KCl	2	K	VI	1.46

the hardness. The difference in hardness between corundum (9) and hematite (6) is a reflection of the difference in the ionic radius of Al^{3+} and Fe^{3+}, and, therefore, the longer bonds in the latter.

The more chemical bonds there are in a given volume, the harder is the mineral. Most high-pressure polymorphs are measurably harder than their low-pressure form. Calcite, the low-pressure polymorph of $CaCO_3$, has a hardness of 3; aragonite, the high-pressure polymorph, has a hardness of 4. The high-pressure polymorph of C, diamond, is substantially harder than the low-pressure polymorph, graphite. This generalization is not always valid—andalusite, which is the low-pressure aluminosilicate, is harder than kyanite, the high-pressure polymorph.

The reason for the general validity of the correlation of hardness with pressure of mineral formation lies in the fact that the higher-pressure polymorphs are generally denser than the lower-pressure ones. This means that there are more ions per unit volume and, therefore, more and shorter chemical bonds per unit volume. The relationship fails in those cases, such as that of the aluminosilicates, where the bonds of the high-pressure polymorph are less equally distributed than those of the low-pressure one.

The tacit assumption has been made in the foregoing argument that the harder mineral is stressed, but not strained, during the abrasive testing. Stated another way, it is assumed that the abrading minerals are *brittle*. Brittleness is a characteristic of materials with purely ionic bonds. However, the purely ionic bond is a concept, not a reality. Real chemical bonds in most minerals are partially covalent, partially metallic, or partially van der Waals (Fig. 1-6). In general, the more ionic a bond is, the more brittle a mineral it makes. Brittle materials cannot store energy in the form of strain, and, as a consequence, react to stress by rupturing (Section 4-4).

One characteristic of purely metallic bonds is that they yield to stress by deforming plastically. This means that ions can be rearranged without losing cohesion. In ionically bonded structures, there is only one place for the cation, and that is surrounded by anions in its coordination polyhedron. Movement of a cation by one ionic diameter places it next to another cation, and electrostatic repulsion occurs. In the metallic-bond model, the cations are conceived as existing in a cloud of mobile electrons. Thus, a metallic cation can slide past another without being repulsed electrostatically. As a result, metallic compounds are *malleable* in that they can be hammered flat without losing cohesion.

Covalent bonds involve the sharing of electrons, and, therefore, are spacially directed. Covalent bonding results in a structure which is more open than that of ionic structures, where radius ratio determines the packing. Directional bonds can be bent from their equilibrium position by stress without juxtaposing ions of like charge. The Si—O bond is about half covalent in character. Bending without rupture is quite obvious in the layer silicates.

Sheets of mica are quite *elastic* and will snap back to their original plane after considerable bending. Chlorite sheets are *flexible* in that they can be bent, but do not snap back. In both cases, chemical bonds are bent.

The difference between these two characteristics may be interpreted from consideration of their structures. The layers in mica have a net negative electrostatic charge that is balanced ionically by K^+ ions in the interlayer position. The K^+ ion occupies a rather large basket formed by the two opposing six-member tetrahedral rings. These rings can be distorted and translated somewhat without destroying the basket. The K^+ ion pulls the structure back to equilibrium as soon as stress is removed (Fig. 2-21).

In the chlorite structure there is a net negative charge on the talc-like layer and a net positive charge on the brucite layer. When these layers are slid past each other, new bonds are formed, and there is no restoring force stored in the mineral. Thus, both mica and chlorite are flexible; but the deformation is permanent in chlorite, whereas mica can store and release energy elastically.

4-4. Rupture: Fracture and Cleavage

If the strain in a crystalline substance exceeds the elastic limit of the material, it breaks. The actual mechanism of breakage, or rupture, is quite complicated and poorly understood at best. It involves concentration of stress at ends of microfractures and in the vicinity of structural defects. Fortunately for the study of mineralogy, the actual mechanism of rupture is less important than is the resulting breakage pattern. These patterns of rupture depend on the orientation and distribution of chemical bonds in the crystal structure.

In some crystal structures, such as those of quartz, garnet, or olivine, the strength of the bonds is approximately the same in all crystallographic directions. Rupture follows no particular crystallographic direction in such cases, and it is called *fracture*. The difference in the fracture pattern of minerals depends upon the way in which a fracture is propagated in the mineral, which in turn depends upon the abundance, kind, and distribution of microfractures and other defects in perfect crystalline order.

Without attempting to relate fracture characteristics to specifics of chemical bonding, it is still possible to note that some minerals have very distinctive fracture patterns which can be used in their identification. The *conchoidal fracture* of quartz is a case in point.

The kind and distribution of chemical bonds depend upon crystallographic direction in some minerals. These minerals are systematically weaker in certain crystallographic directions than in others. Wherever this occurs, minerals tend to break preferentially along the weaker directions, and the mineral is said to have *cleavage*.

Layered silicates (Figs. 2-19, 2-20, and 2-21) all have strong Si—O chemical bonds in the plane of the layers. However, the chemical bonds between the layers are much weaker than those within the layers. Minerals with the 7-Å structure, kaolinite and septechlorite, and minerals with the 9-Å structure, talc and pyrophyllite, are composed of layers that are electrostatically neutral and held together only by van der Waals forces. The mica-group minerals have layers that are electrostatically negative and held together by interlayer cations. In both cases the bonding in the direction of the ***c*** axis is considerably weaker than in the other crystallographic directions, and these minerals have *basal cleavage* (Fig. 4-2).

If there are two sets of intersecting planes which have chemical bonds across them that are weaker than the bonds in any other direction, two cleavage directions exist. Such cleavage is *prismatic* (Fig. 4-2). The chain silicates (Figs. 2-17 and 2-18) have strong Si—O bonds in the ***c*** direction, but weaker cation-to-oxygen bonds linking the chains together. Pyroxene has systematic bond weakness across planes of the form {110} and, therefore, cleavage along those planes. The angle between the (110) plane and the $(1\bar{1}0)$ plane is approximately 87°. Amphiboles have systematic weakness across the planes of the form {110}, but have cleavage planes at approximately 56°. The difference in cleavage angle between pyroxenes and amphiboles derives from the fact that the ***b*** dimension of the unit cell of pyroxenes is about equal to that of the ***a*** dimension, whereas in amphiboles the ***b*** dimension is about twice as large as the ***a*** dimension.

Inspection of Fig. 2-29 should reveal that there are fewer chemical bonds across the (001) and the (010) planes of feldspars than there are in other directions. This results in perfect prismatic cleavage in the feldspar minerals (Fig. 4-2) and illustrates a second cause for cleavage. In the sheet and chain silicates, cleavage results from the fact that weaker chemical bonds are located along certain crystallographic directions.

Cleavage is also possible in minerals having only one kind of bond if the bond density is less across certain crystallographic planes than across others. Halite has cleavage in the form {100} (Fig. 4-2). Inspection of Fig. 2-2 should reveal that there are fewer Na^+—Cl^- bonds across planes which are parallel to the cube faces than across other crystallographic planes. The rhombohedral $\{10\bar{1}1\}$ cleavage of calcite bears the same relation to a face-centered rhomb as does the cubic cleavage of halite to a face-centered cube (Fig. 2-9).

Minerals with the fluorite structure have octahedral {111} cleavage. This structure has cations in a FCC array. The (111) planes are those with closest-packed layers (Fig. 1-10), and also those with the fewest Ca^{2+}—F^- bonds across them.

Dodecahedral cleavage occurs in the mineral sphalerite, which is made up of two interpenetrating lattices of FCC ions. The {110} form has fewer bonds across its planes than any other.

Crystallographic planes with fewer or longer, thus weaker, chemical

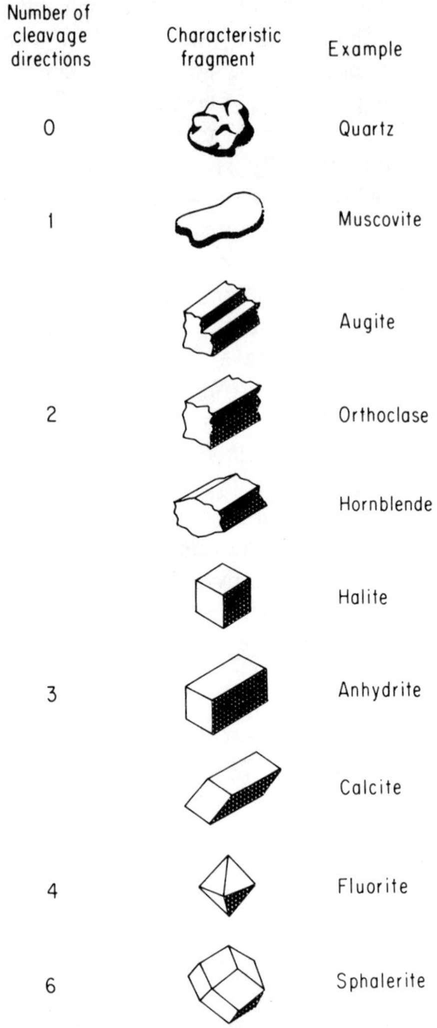

Number of cleavage directions	Characteristic fragment	Example
0		Quartz
1		Muscovite
		Augite
2		Orthoclase
		Hornblende
		Halite
3		Anhydrite
		Calcite
4		Fluorite
6		Sphalerite

Figure 4-2 Cleavage patterns. The characteristic shapes of broken fragments depend upon the number and the quality of the cleavage directions.

bonds across them account for the phenomenon of cleavage. Many minerals, however, display something akin to cleavage in some specimens, whereas other specimens of the same species do not. Such development of planar fractures, which are crystallographically controlled, is called *parting*. It results from systematic weakness across certain crystallographic planes, which are present or absent, depending upon the history of a particular sample. Contact twinning occurs in structures where there is an alternative orientation of a crystal plane, which is energetically only slightly less stable than the perfect orientation (Section 3-9). Coordination of the nearest-neighbor ions is unaffected, but second-nearest neighbors are systematically out of order. The contact plane is thus weaker than other planes in the same direction that do not have twinning across them. Corundum may display a basal {0001} or rhombohedral {1011} twin plane, and some crystals part parallel to these planes.

If a crystal solution series is complete at high temperatures but restricted at low temperatures, exsolution may occur preferentially at certain crystallographic planes. Thus, grain boundaries are formed along these planes, permitting parting parallel to the nascent crystal faces.

Two major distinctions between cleavage and parting permit their discrimination. First, parting occurs on some specimens of a mineral species and not on others. Second, where parting does occur in a mineral, the planes of parting are at discrete intervals, which are commonly visible to the unaided eye. Thus, a cleavage fragment can be further cleaved along a very large number of parallel planes, until, conceptually, only one layer remains. A parting fragment, on the other hand, cannot be parted further.

4-5. Radioactivity

Some atomic nuclei are not stable, but alter spontaneously to a different kind of nucleus. The causes for this instability are poorly understood, and quite beyond the scope of mineralogy. But certain properties of *radioactive decay* (spontaneous nuclear change) are important to any thorough study of mineralogy—properties that can be stated rather briefly.

There is no way to predict when a particular unstable nucleus of an isotope will decay or how it will decay if more than one mode is possible. However, given a sufficiently large number of such nuclei (any quantity that can be measured chemically is sufficiently large), the rate at which they decay is a quantity that is proportional to the number of nuclei present. The constant of proportionality is unique for each radioactive isotope. This decay constant may be used to calculate the length of time necessary for half the nuclei present to decay, the *half-life* of the isotope.

Radioactive isotopes change their chemical identity by one or more of four processes. The nucleus may emit an alpha particle, which is He^{2+} with a charge of $+2$ and mass of 4. Alpha decay thus changes the nucleus to a daughter nucleus, which is two elements back in the periodic chart of the elements. Thus, a U ion or atom, atomic number 92, transmutes by alpha-particle decay to a Th ion or atom, atomic number 90.

A second decay scheme involves the emission of a beta particle (an electron) from the nucleus. There is no effective nuclear mass change with beta-particle decay, but the daughter element is one atomic number higher in the periodic chart of the elements. The Rb isotope with atomic mass 87 decays by beta-particle emission to the Sr isotope with the same mass. In effect, a neutron becomes a proton, thereby increasing the atomic number of the atom by 1.

The third mode of transmutation involves the nuclear capture of one of the *K* electrons. This, in effect, converts one of the protons to a neutron, so that the daughter isotope is one element back in the periodic chart. Capture of a *K* electron by the isotope of *K* (potassium), with a mass of 40, transmutes it to the noble gas Ar with the same mass.

The fourth mode of atomic transmutation is spontaneous nuclear fission. In this process, the unstable isotope splits apart into two or more atomic nuclei, which may or may not be stable isotopes of the elements they represent. In addition, there may be release of alpha, beta, or gamma particles accompanying fission.

Decay of a radioactive nucleus of an ion in a crystal structure can affect the structure profoundly. The first effect, which was noted in Section 1-18, results in complete destruction of the crystal lattice by destroying the coordination polyhedra. The complete decay scheme for U^{238} is illustrated in Fig. 4-3. The direct transmutation of U^{4+}, with a radius of 1.08 Å, to Pb^{4+}, with a radius of 0.86 Å, would not be expected to destroy a coordination polyhedron of O^{2-} ions; but the intermediate daughter ions have not only a much greater range of radii, but have different valences as well. Thus, the crystal structure in the immediate vicinity of the transmuting ion is severely strained, and the entire crystal lattice is broken up if sufficient numbers of radioactive nuclei are present.

Most minerals that have essential U or Th are *metamict*. That is, whatever crystal structure they had originally is now broken up. Minerals that have U or Th substituted in their structure may become partially metamict over long periods of geologic time.

The radioactive decay of K^{40} to Ar^{40} (K-electron capture) or to Ca^{40} (beta emission) involves a substantial change in size and valence. However, K^{40} makes up only about 0.01 percent of natural K, and structural damage in K minerals is not noticeable. Radioactive Rb^{87}, which makes up more than one third of natural Rb, does little structural damage to the mineral in which it is found, because of its very long half-life and the fact that it is a dispersed element (Section 7-1).

Alpha particles, being identical with He^{2+} once they are liberated from the nucleus, are very strong oxidizing agents; that is, they have a strong affinity for electrons. Inspection of Table 1-4 reveals that the second ionization potential for He, and, therefore, the power of He^{2+} to attract electrons, is substantially greater than the third and fourth ionization potentials of

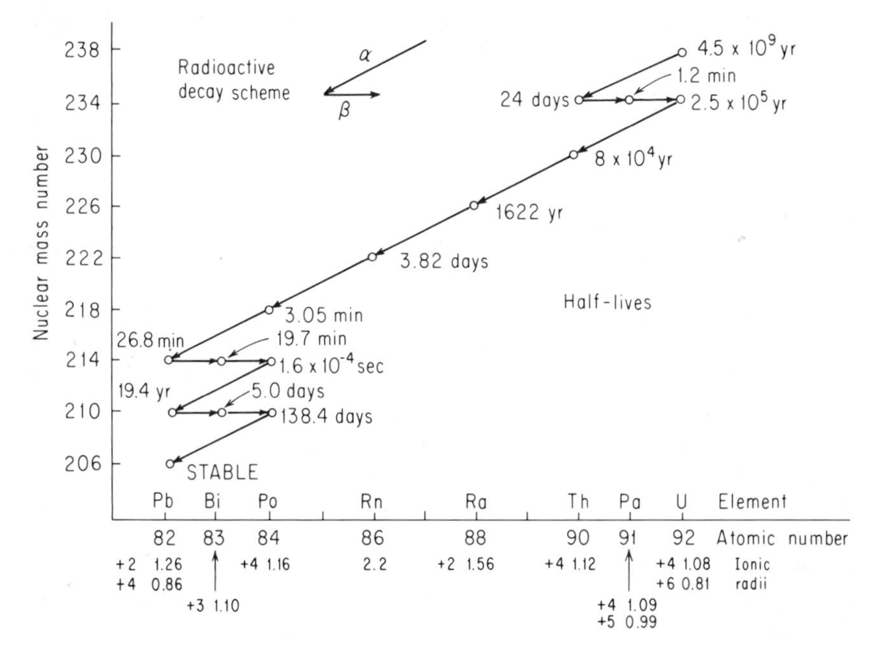

Figure 4-3 Radioactive decay of U^{238}. Uranium decays to daughter elements, the ions of which have sizes and charges different from U^{4+} or U^{6+}. The change in size and charge of the cation breaks up the coordination polyhedron about it, and causes metamictization. The decay scheme of Th^{232} differs in detail, but the changes in charge and radius are just as drastic.

many transition metal ions, which may be in a partially reduced state. The first ionization potential of He would oxidize many such metals. The result of alpha-particle decay is the oxidation (loss of valence electrons) of partially reduced transition metal ions, with concomitant changes in radius and valence leading to the breakup of lattice structures.

Finally, it should be noted that alpha particles are emitted from radioactive nuclei with sufficient energy to knock ions out of their positions in the crystal lattice, which amounts to the same thing as the mechanical breakup of a mineral.

4-6. Surface Properties

In all preceding discussions, it has been assumed explicitly that the ions under discussion were in a crystalline environment and, furthermore, that the crystal lattice extended infinitely in all directions. The possibility of a slightly irregular crystalline environment was noted in the discussion of defect structures (Section 1-14). In this context, the surface of a crystalline solid might be considered to be the most drastic defect of all, because on one side there is present no crystalline environment whatsoever.

An ion in the interior of a crystal, with the absence of defects, is in a uniform crystalline environment. An ion at the surface is subject to a net attraction toward the bulk of the crystal, which is not compensated by an equal attraction from the liquid, vapor, or other crystal across the boundary. This imbalance tends to strain the surface ions into positions somewhat different from those which they would occupy in the interior.

A second effect, which tends to alter the arrangement of ions at the surface, derives from cation shielding. Pauling's rules, which are stated in terms of individual ions, have been interpreted in terms of interior ions and the manner in which they behave in a crystal structure (Section 1-12). These same principles apply to ions at a crystal surface, since coordination polyhedra are formed about cations, but they are either deformed or different from the ones in the interior. This means that cations, especially highly charged cations, tend not to occur at the surface. Rather, they distort the structure so that they are shielded, and anions mostly make up the surface. This distortion may extend several ionic radii into the interior of the crystal so strong is the tendency for cations to shield themselves. Electrostatic neutrality is maintained over the shortest possible distance, but not in the same pattern as in the interior. The cations still separate themselves as much as possible, but they may be closer at the surface than in the interior because of the shielding requirement.

Although crystal surfaces, whether generated by crystal growth, cleavage, or fracture, are structurally different from crystal interiors, the surface structure is just as characteristic of a mineral species as is the internal structure. Differences in surface properties are of considerable importance in commercial mineral-separation procedures.

In general, the higher the charge on the cation and, hence, the stronger its screening requirement, the greater the polarity of the surface of the crystal. Also, the more ionic the character of the bond, the more polar the surface. Different surface polarities result in different surface *wettability*, which is a measure of the ease with which a surface is coated with water or with some other fluid. Since water is a polar molecule, polar surfaces are more easily wetted with it than nonpolar ones. Oil, being nonpolar, wets surfaces of low polarity more easily than does water.

Diamonds, being covalently bonded, can be separated easily on the basis of wettability from the silicate minerals with which they are found. The crushed diamond ore is flushed as an aqueous slurry down a grease table. The well-wetted silicates pass on, while the poorly wetted diamonds are entrapped in the grease.

Sulfide ores are separated from silicate and carbonate gangue minerals on the basis of the same properties, in a technique known as *flotation*. The crushed ore is mixed with water to which has been added a small amount of oil and a foaming agent. The sulfides, with their low-polarity surfaces, are wetted with oil, while the gangue minerals, with their high-polarity surfaces, are wetted with water. When air is blown up through the slurry, the oily sulfide particles are carried up with the foam; the gangue minerals sink.

Surface polarity can be used to explain why it is possible to make mudpies. The water molecules next to the silicate grains are oriented with their positive, proton sides away from the surface. This means that there is an electrostatic attraction between the negative, proton-deficient water next to the mineral surface and the positive, proton-rich water in the central part of the interstitial space, which binds the particles together (Fig. 4-4). The reader may compare the quality of castle he can build with wet sand with one built of wet Al metal filings of the same size.

Colloidal suspensions result when the force of mutual repulsion between polarized surfaces exceeds the force of gravity. The water between the col-

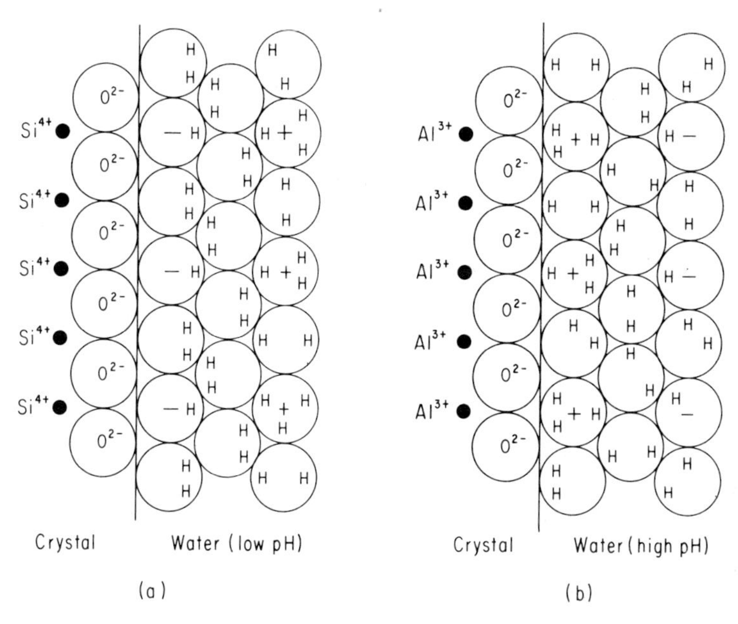

Figure 4-4 Polar surfaces. Highly charged cations distort their coordination polyhedra near the crystal surface in such a manner that the surface is composed wholly of anions. (a) Interstitial water with low pH caused by repulsion of H_3O^+ by Si^{4+} cations. (b) Interstitial water with high pH caused by attraction of H_3O^+ by anionic surface.

loidal particles is forced by the polar surfaces to assume an almost crystalline arrangement.

A slurry of freshly crushed mineral in distilled water develops an acidity or basicity, known as *abrasion pH*, which is characteristic of the mineral species. At the one extreme, for example quartz, protons are concentrated in the centers of interstices, thereby lowering the pH of the interstitial water (Fig. 4-4). Corundum and hematite tend to concentrate protons at their surface, rendering the interstitial water more basic. Silicates containing alumina or iron have intermediate abrasion pH's, which depend upon the relative proportions of the oxides. It should be obvious that other factors, such as impurities and particle size, would overwhelm this effect, and, therefore, abrasion pH is not reliably diagnostic for mineral identification.

FURTHER READINGS

AZAROFF, L. V., *Introduction to Solids.* New York: McGraw-Hill Book Company, 1960.

CLARK, S. P., JR. (ed.), *Handbook of Physical Constants*, Memoir 97. New York: The Geological Society of America, 1966.

KITTEL, C., *Elementary Solid State Physics.* New York: John Wiley & Sons, Inc., 1962.

5

Radiant Energy and Crystalline Matter

The nature of light was a topic of scientific and philosophic controversy from the seventh decade of the seventeenth century until the third decade of the twentieth century, a period of two and one half centuries. Newton (1675) held that light was corpuscular, on the basis of its rectilinear propagation. Huygens (1678) proposed that light traveled in waves, analogous to waves on water. The issue was settled in favor of wave theory by Fresnel and Young (between 1800 and 1820), who observed that under certain conditions light is bent around sharp edges and interferes with itself.

The wave theory of light was given a strong theoretical basis by Maxwell (1865) when he demonstrated that light had properties in congruence with all other electromagnetic radiation. Thus, light waves were postulated to be perturbations of a luminiferous ether that pervaded all the universe. However, certain experimental results—the photoemission of electrons from metal surfaces, the long wavelength shift in Compton scattering, and even the exposure of photographic emulsions—remained completely inexplicable by the electromagnetic radiation theory.

Einstein (1905) explained these effects with his quantum theory and, thereby, reopened the eighteenth century controversy over the nature of light. The question of the nature of electromagnetic radiation became even more disturbing when Davisson and Germer (between 1923 and 1927) demonstrated that electrons were diffracted by crystal structures, and, therefore, behaved as waves. Before long, diffraction of He atoms and H_2 molecules had been accomplished. The question had become how to explain electromagnetic radiation that behaved as particles, and particulate matter that behaved as though it were a wave.

5-1. Nature of Electromagnetic Radiation

The resolution to the question of the dual nature of light is to be found in a speculative argument set forth by L. de Broglie in 1922. He observed that determination of the position of electrons in atoms involved integers (Section 1-1, Fig. 1-1). At that time, the only other known phenomena in physics that involved integers were those of light interference and harmonic vibrations. From these observations, de Broglie reasoned that periodicity, that is, wave behavior, must be assigned to the electrons in orbitals about atoms.

The hypothetical *de Broglie wavelength* of electrons comes from the relation

$$\frac{h}{\lambda} = mv$$

where h is Planck's constant, λ is the hypothetical wavelength, and mv is the momentum, mass times velocity. This equation is analogous to the one for the energy of a photon

$$\frac{hc}{\lambda} = E$$

where h is Planck's constant, c the velocity of light, λ the wavelength of light, and E the energy. The first equation describes the diffraction of particulate matter, and the second the quantum behavior of light.

The conclusion to be drawn is that the old question, "Is light wavelike or corpuscular?", is meaningless. A billiards cue is a finely designed piece of sporting equipment in the hands of a gentleman, whereas in the hands of a rioter it is a lethal weapon. It is meaningless to demand which is its real nature. The same is true of light; how it is visualized depends on what is being done with it.

What, then, is the significance of using waves to describe the motion of quanta or particles? Max Born answered that question in 1926. Certain experimental facts cannot be interpreted if it is assumed that quanta and particles follow the laws of Newtonian mechanics. Therefore, another fundamental law, which describes the behavior of these quanta and particles, must be found. The hypothesis is made that the square of the amplitude of the wave, computed for any spot in a diffraction pattern, is proportional to the probability of finding electrons or photons traveling through that spot. Thus, de Broglie waves are not identical with matter, and electromagnetic waves are not identical with light. Rather, the waves describe the behavior of photons and other non-Newtonian particles in a probabilistic manner. A useful fiction is to think of light waves *guiding* photons and de Broglie waves *guiding* particles under conditions of diffraction.

That Newtonian mechanics does not always apply to particle interactions on this scale should not be too distressing. Planetary orbits about the sun cannot be completely described by purely Newtonian mechanics, either. Relativity must be brought in to complete the description. However, in the range between photons and planets—the familiar world of direct experience—not the slightest deviation from Newton's laws of physics has yet been observed.

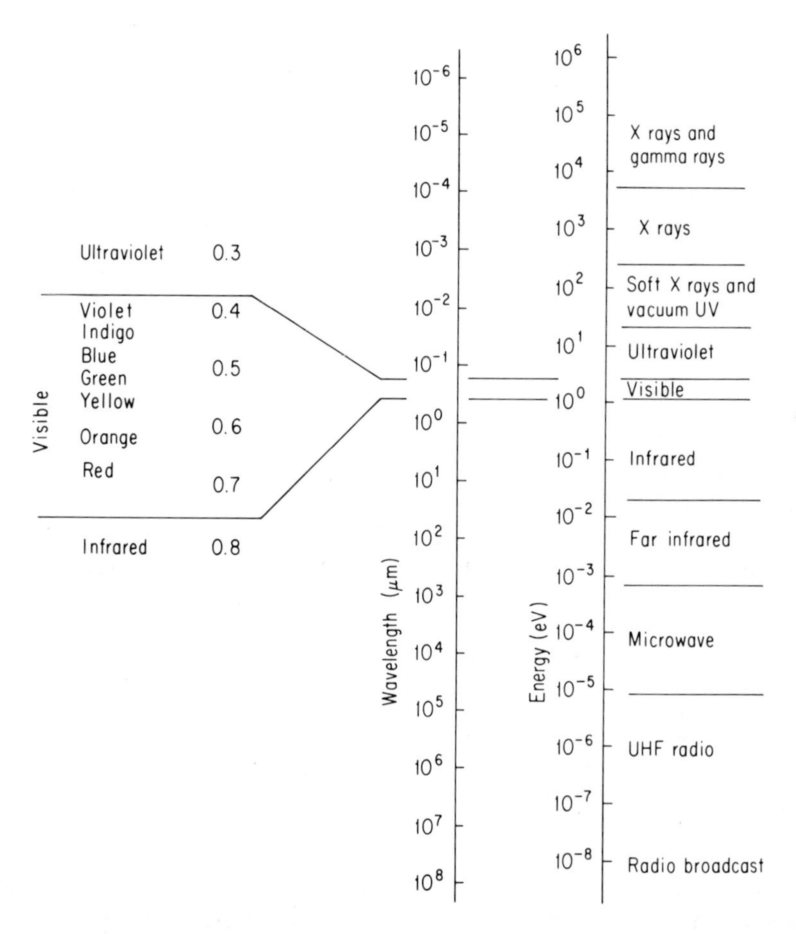

Figure 5-1 Electromagnetic spectrum. The different regions of the electromagnetic radiant energy spectrum are labeled, with the associated wavelength given in micrometers and the quantum energy given in electron volts (1 eV = 1.24 μm).

The foregoing discussion should not lead to the inference that light is a special kind of matter. Remember that light always travels at 3×10^{10} cm/s in vacuo, and matter never does.

Many of the interactions between radiant energy and matter, which are dealt with in this chapter, are such that the flood of energy involved completely masks the individual quanta, and such that no energy transfer is involved. For these interactions, wave theory provides a completely adequate explanation. Many other interactions involve individual quanta transferring energy to individual ions, and quantum theory is needed for these. The beauty of an ice-blue diamond requires wave theory to explain the fire and quantum theory to explain the color. The focus of a camera cannot be explained without reference to light waves, and the exposure of the photographic emulsion cannot be explained without reference to photons.

The complete electromagnetic spectrum is illustrated in Fig. 5-1. It ranges from radio frequencies at the long-wavelength end to gamma radiation at the short-wavelength end. The main divisions in the spectrum are based on methods of generation and detection, so that the boundaries within the spectrum overlap. The visible range (390–760 nm) is that limited part of the electromagnetic spectrum which is detected by the normal human eye.

5-2. Interaction of Quanta with Matter

Light, when incident upon a mineral, may be transmitted, refracted and scattered, reflected, or absorbed. Absorption and emission spectral phenomena of minerals require the quantum description of light for their explanation. The color and luster of minerals result from absorption in the visible range of radiant energy, but instruments have been developed that extend the range of spectral studies far beyond the visible range. The *absorption spectrum* of a mineral (a diagram relating absorption and transmission of radiation as a function of energy or wavelength) is related to the kinds of ions and the kinds of chemical bonds present in the mineral (Fig. 5-2).

For the present discussion, the important facts pertaining to the electronic structure of atoms and ions are that the electrons in atoms and ions exist in discrete orbitals, and that these orbitals are separated by finite differences in energy (Sections 1-1 and 1-2). The Pauli exclusion principle limits the number of electrons in a given orbital, and all the inner orbitals of an ion are filled under equilibrium conditions. There are unfilled orbitals, which have energies greater than this *ground state* of an atom or ion. The difference in energy between an electron in a ground state and one in an excited state is referred to as an *energy gap* or as a *forbidden zone.*

When ever a quantum of radiant energy, one that has the same energy as an energy gap, strikes an excitable electron, the energy of the quantum is transferred to the electron. The excited electron then occupies the orbital on the high-energy side of the forbidden zone, and the quantum is, thus, absorbed. In a crystal lattice with many such excitable electrons, all quanta with that particular energy, that is, with that particular wavelength, are absorbed. If the quantum has less than sufficient energy to excite the electron across the forbidden zone, energy cannot be transferred, and the quantum passes on. If the energy of the quantum is greater than the energy of the gap, energy transferral would excite the electron into a higher forbidden zone, which is also impossible by wave theory (Fig. 1-1). Thus, the presence of ions with excitable electrons operates as a filter, removing from a stream of white radiation all quanta of a certain energy and wavelength.

In actuality, owing to various defects in crystal structures (Section 1-14) and the fact that the width of an energy gap depends in large part on local ionic environment, energy absorbed by crystal lattices in this fashion has a finite and appreciable width in the electromagnetic spectrum (Fig. 5-2). The

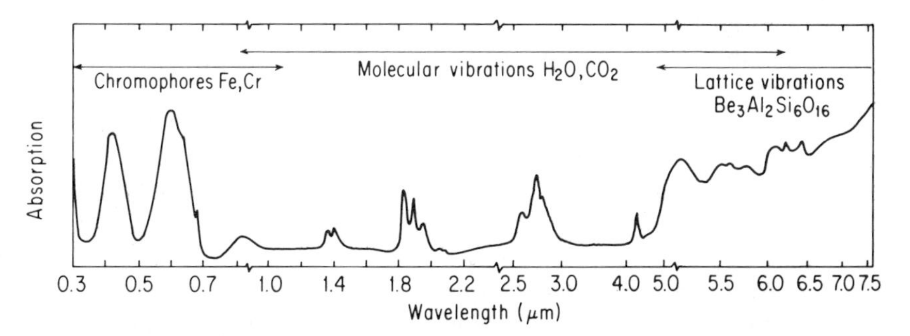

Figure 5-2 Absorption spectrum of beryl. The energy absorption of a crystal varies with the wavelength of the radiant energy incident upon it. Maximum transparency (minimum absorption) in the visible range is at 0.5 μm and maximum absorption (minimum transparency) is at 0.4 and at 0.7 μm. Thus, both red and violet light are absorbed and the mineral appears green to the eye. (After D. L. Wood and K. Nassau, "The Characterization of Beryl and Emerald by Visible and Infrared Absorption Spectroscopy," *Amer. Mineral. 53*, p. 778.)

presence of more than one kind of ion with excitable electrons and the existence of different excited states above the ground state lead to the possibility of many different bands of radiant energy being absorbed by a crystal structure.

There are ways, other than the simple excitation of an electron into an orbital of higher energy, as in the foregoing discussion, by which a crystal lattice can absorb radiant energy. All are quantized. Without going into the specific detail of the mechanics of the absorptions, which are beyond the scope of an introductory text in mineralogy, they include harmonic lattice vibrations, lattice vacancies that can trap excited electrons, impurities in the crystal structure, and electronic charge transfer of multivalent ions. Each of these modes of energy absorption adds to the complexity of the absorption and transmission of quanta as a function of the wavelength of the radiation impinging upon a crystal lattice.

5-3. Color and Metallic Luster

Mineral coloration may be interpreted from the preceding discussion of absorption spectra. White light is a mixture of all the wavelengths of light in the visible part of the electromagnetic spectrum (Fig. 5-1). That is, what the normal human eye sees as white is the result of a mixture of photons of all different energies between 0.32 and 0.61 eV (wavelengths between 390 and 760 nm). Colors are narrow bands of light or mixtures of narrow bands of light within the visible range.

Photons that have energies which correspond to an energy gap are absorbed by a crystal lattice. Photons that do not may be transmitted. If part of the visible spectrum is absorbed by a mineral, the color of the mineral is a mixture of the wavelengths of light that are transmitted or reflected. Thus, a crystal which contains ions that have any kind of energy gap between 0.5 and 0.6 eV (wavelength between 600 and 700 nm) absorbs strongly in the red end of the visible spectrum and appears blue in color. Similarly, absorption in the blue end of the spectrum leads to a red mineral.

Some general statements about mineral color may be made without going into great detail on the specifics of optical absorption. Ionic compounds with ions having noble gas configurations (Section 1-4) are generally colorless. The energy gap between an occupied *p* orbital and the closest available unoccupied orbital is considerably greater than the energy of photons of visible light.

As is discussed in Section 1-1 and shown in Fig. 1-2, *d* orbitals are directional in spacial orientation. In a free ion, the five *d* orbitals all have the same energy, and are fivefold degenerate. However, if an ion is in a crystal lattice, the electron clouds of the *ligands*— anions of the coordination polyhedron— interact with the *d* electrons. The energy levels of the *d* orbitals are split by electron–electron repulsion between the *d* electrons and the electron cloud of the ligands.

The d_{xy}, d_{xz}, and d_{yz} orbitals are oriented away from ligands in octahedral coordination, and the $d_{x^2-y^2}$ and d_{z^2} orbitals are oriented toward them (Figs. 1-2 and 5-3a). Electrostatic repulsion, thus, raises the energy levels of the $d_{x^2-y^2}$ and d_{z^2} orbitals, while it lowers the energy levels of the d_{xy}, d_{xz}, and d_{yz} orbitals. This energy separation of *d* orbitals is called *crystal field splitting* and

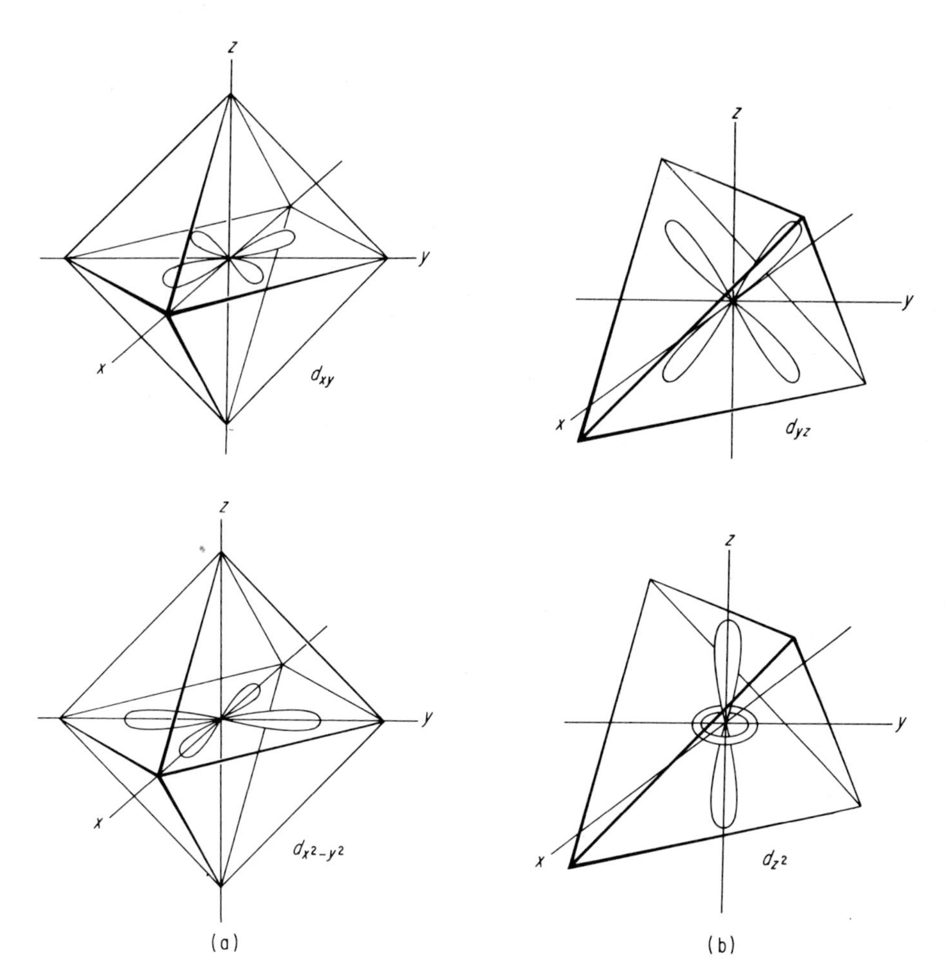

Figure 5-3 Coordination polyhedra and *d* orbitals. Electron–electron repulsions between *d* electrons and those of their ligands split these orbitals into separate states. (a) Octahedral coordination lowers the energy of the d_{xy} orbital, but raises that of the $d_{x^2-y^2}$, because $d_{x^2-y^2}$ orbitals are directed toward the ligands and d_{xy} orbitals are directed between ligands. (b) Tetrahedral coordination stabilizes d_{z^2} and destabilizes d_{yz}. (After Burns, 1970.)

is designated Δ_o (delta subscript "oh," for octahedral). Each electron in one of the lower three *d* orbitals stabilizes a transition metal ion by $\frac{2}{5}\Delta_o$, and each electron in one of the upper two orbitals reduces stability by $\frac{3}{5}\Delta_o$ (Fig. 5-4). Summation of the Δ_o terms gives the *crystal field stabilization energy* (CFSE) for a particular ion (Table 5-1).

Ligands in tetrahedral and cubic coordination favor $d_{x^2-y^2}$ and d_{z^2} over

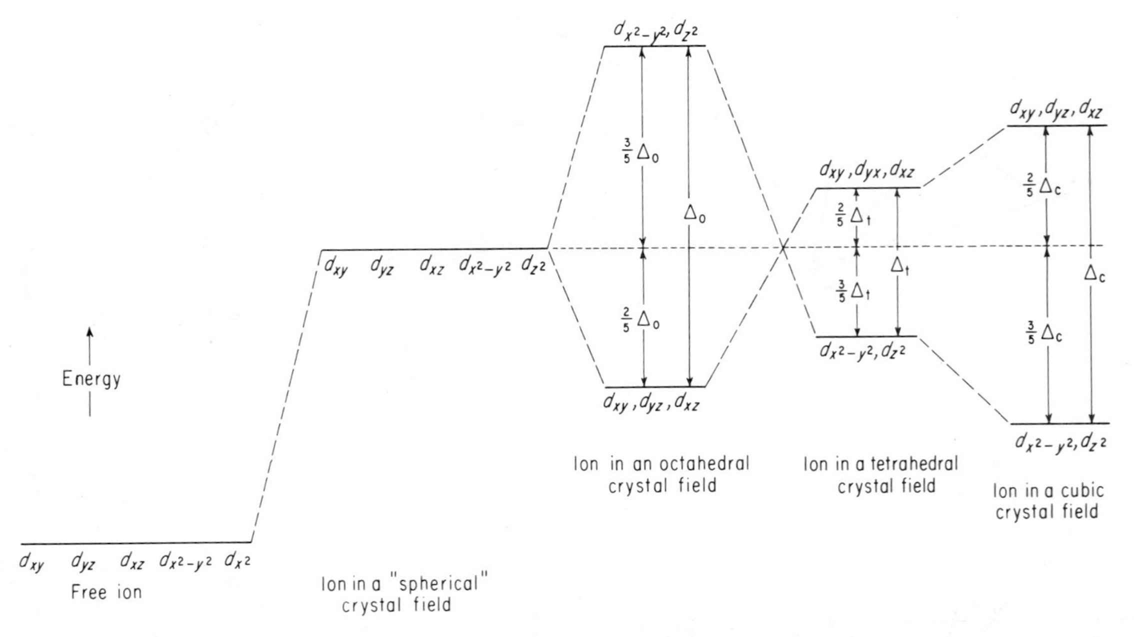

Figure 5-4 Crystal field splitting. A hypothetical "spherical" crystal field raises the energy state of *d* orbitals above the free ion state. Octahedral, tetrahedral, and cubic crystal fields split the *d* orbitals by an amount **Δ**. Mineral coloration occurs, in part, because the values of **Δ** lie in the visible and near-visible spectrum.

Table 5-1

ELECTRON DISTRIBUTION IN *d* ORBITALS AND CRYSTAL FIELD STABILIZATION ENERGIES (CFSE) OF IONS OF THE FIRST TRANSITION SERIES AND IN OCTAHEDRAL COORDINATION (HIGH-SPIN CONFIGURATION)[a]

Number of d electrons	*Ions*	d_{xy}	d_{xz}	d_{yz}	$d_{x^2-y^2}$	d_{z^2}	*Number of unpaired electrons*	CFSE
0	Ca^{2+}, Sc^{3+}, Ti^{4+}						0	0
1	Ti^{3+}	↑					1	$\frac{2}{5}\Delta_o$
2	Ti^{2+}, V^{3+}	↑	↑				2	$\frac{4}{5}\Delta_o$
3	V^{2+}, Cr^{3+}, Mn^{4+}	↑	↑	↑			3	$\frac{6}{5}\Delta_o$
4	Cr^{2+}, $Mn^{3+(b)}$	↑	↑	↑	↑		4	$\frac{3}{5}\Delta_o$
5	Mn^{2+}, $Fe^{3+(b)}$	↑	↑	↑	↑	↑	5	0
6	Fe^{2+}, Co^{3+}, $Ni^{4+(b)}$	↑↓	↑	↑	↑	↑	4	$\frac{2}{5}\Delta_o$
7	Co^{2+}, $Ni^{3+(b)}$	↑↓	↑↓	↑	↑	↑	3	$\frac{4}{5}\Delta_o$
8	Ni^{2+}	↑↓	↑↓	↑↓	↑	↑	2	$\frac{6}{5}\Delta_o$
9	Cu^{2+}	↑↓	↑↓	↑↓	↑↓	↑	1	$\frac{3}{5}\Delta_o$
10	Zn^{2+}, Ga^{3+}, Ge^{4+}	↑↓	↑↓	↑↓	↑↓	↑↓	0	0

[a]After Burns, 1970.
[b]Also has a low-spin configuration.

d_{xy}, d_{xz}, and d_{yz} orbitals, and CFSE's of Δ_t and Δ_c, respectively (Fig. 5-3b). The relative energies of the crystal field splitting are $\Delta_t = -\frac{4}{9}\Delta_o$ and $\Delta_c = -\frac{8}{9}\Delta_o$, partitioned as indicated in Fig. 5-4.

In filling *d* orbitals, the first electrons clearly go into the lower energy states. Because of electron spin interactions, electrons tend to space themselves among the orbitals and to align their spins in a parallel configuration. The first three electrons in octahedrally coordinated ions clearly go into the lower energy orbitals indicated in Fig. 5-4 and Table 5-1. The next electron may either pair up with one of the other electrons and enter one of the lower orbitals, or it may maintain its spin parallel to that of the previous entrants and enter a higher-energy orbital. The former is a *low-spin configuration* and the latter a *high-spin configuration*. The energy of spin coupling is greater than the CFSE both in oxide and in silicate structures; therefore, only radii for high-spin configurations are given in Table 1-5.

Reduction in the symmetry of the ligands by distortion of the octahedra further reduces the degeneracy of the *d* orbitals, as shown in Fig. 5-5. Depending upon the number of *d* electrons, a transition metal ion will seek out crystal lattice sites that give it the lowest possible CFSE, even to the extent of distorting a coordination polyhedron. The energy differences, Δ, between *d* orbitals of transition metal ions in crystal lattice sites are influenced, then, by the number of *d* electrons, the coordination number of the cation, the length of the cation–anion bond, the charge on the cation, and the symmetry of the coordinated ligands.

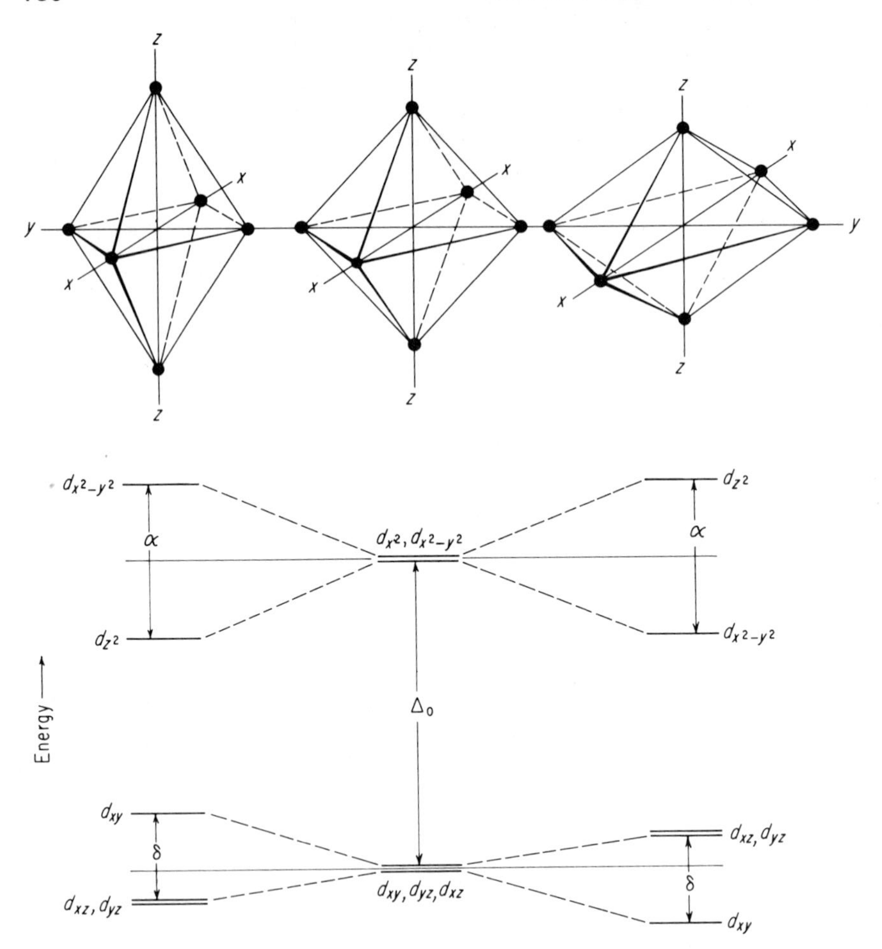

Figure 5-5 Distorted octahedral sites cause further splitting of *d* orbitals.

Many of these **Δ** values for metals of the first transition series are the same as the energy of photons in the visible and near-visible range of the electromagnetic spectrum (Fig. 5-2). For this reason, minerals containing transition metal ions tend to be colored. The typical green color of olivine derives from Fe^{2+}, which has, in the near infrared, a strong absorption band that has a tail extending into the red portion of the visible spectrum.

Many transition metal ions have more than one oxidation state, for example, Fe^{2+} and Fe^{3+}. If a mineral contains ions in both oxidized and reduced oxidation states, an electron may be transferred between two ions, oxidizing the one ($Fe^{2+} \longrightarrow Fe^{3+} + e^-$) and reducing the other ($Fe^{3+} + e^- \longrightarrow Fe^{2+}$). This process, known as *charge transfer*, is also quantized with energy values in the visible spectrum. The very dark coloration of ferromagnesian silicates results from such charge transfer.

Mineralogists distinguish two classes of colored compounds, idiochromatic and allochromatic. An *idiochromatic* mineral is one in which the *chromophore* (color-absorbing ion) is an essential part of the mineral composition. Olivine is an idiochromatic mineral, as are many of the pyroxenes and amphiboles. An *allochromatic* mineral is one in which the chromophore is not an essential ion, but is a substituent or impurity that may or may not be present, depending upon the genesis of the specific mineral sample. Pure quartz is colorless, and the colored varieties depend upon impurities and structural defects for their absorption. This is true for most pale pastel-colored minerals. A partial list of chromophores is given in Table 5-2.

Table 5-2

CHROMOPHORES IN ALLOCHROMATIC MINERALS

Chromophore	*Color*	*Examples*
Fe^{2+}	Green	Microcline
Fe^{3+}	Pink to red	Calcite, quartz, microcline, kyanite
	Greenish yellow	Corundum
Cr^{3+}	Red	Corundum (synthetic ruby), alexandrite
	Green	Beryl (emerald), alexandrite
Ti^{3+}	Blue	Corundum (synthetic sapphire), kyanite
	Pink	Synthetic corundum
Ni^{2+}	Yellow	Synthetic corundum
V^{3+}	Green	Synthetic corundum
$Fe^{3+} + Fe^{2+}$	Blue	Corundum (sapphire)
	Yellowish green	Corundum
$Ti^{3+} + Fe^{2+}$	Blue	Corundum (synthetic sapphire)

The familiar division of minerals on the basis of luster into metallic and nonmetallic groups also has its foundation in the nature of the energy gaps in the electronic structure. Electrons are localized at specific lattice sites if the bonding is primarily ionic, covalent, or van der Waals in character (Section 1-4). The ground states and the excited states are separated by specific energy gaps, most of which involve energies considerably greater than those of photons of visible light. In minerals with metallic bonds, however, the energy gaps are very much smaller than the energy of visible photons, and there exist a great number of excited states having excitation energies covering the entire range of the visible spectrum. Therefore, any photon of visible light that strikes the surface of a mineral with metallic or with partially metallic bonding is absorbed immediately. Some of this energy is dissipated as heat, but most of it is reemitted immediately as light. Surfaces of objects with metallic luster, therefore, reflect light completely (Fig. 5-6).

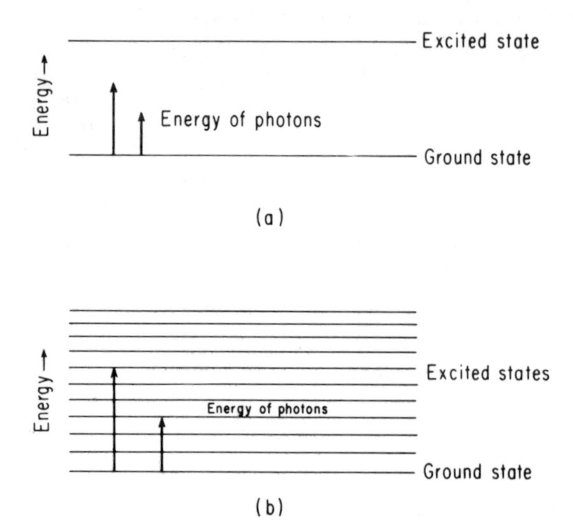

Figure 5-6 Metallic and nonmetallic lusters. (a) For nonmetallic luster, the energy of photons in the visible range is substantially less than the energy gap between the ground state and the next excited state for outer electrons. (b) With metallic luster, there are many excited states that have energies covering the entire visible spectrum. Thus, all photons are absorbed at the surface and none are transmitted.

The other kinds of nonmetallic luster depend upon phenomena that require explanation in terms of a wave description of light (Section 5-6).

5-4. Quantum Emission

Some crystalline substances can be caused to emit light in the visible range (quite apart from heating materials to red or white heat). *Luminescence* is the result of energy being absorbed by a crystal lattice and then reemitted with a lower energy, that is, with a longer wavelength. Energy is not destroyed in the process of luminescence, but is reemitted in two stages, one of which is in the visible range. This occurs when an electron is excited to a state that has other excited states between it and the ground state (Fig. 5-7).

Quanta in the blue end of the visible, in the ultraviolet, and in the X-ray spectra provide the excitation energies to the higher energy state. The electron then drops back to a lower excited state, emitting radiation corresponding to that energy gap before dropping to the ground state. One of these changes produces a photon in the visible range. Thus, luminescence is colored rather than white.

Luminescence that occurs immediately is known as *fluorescence*. Fluo-

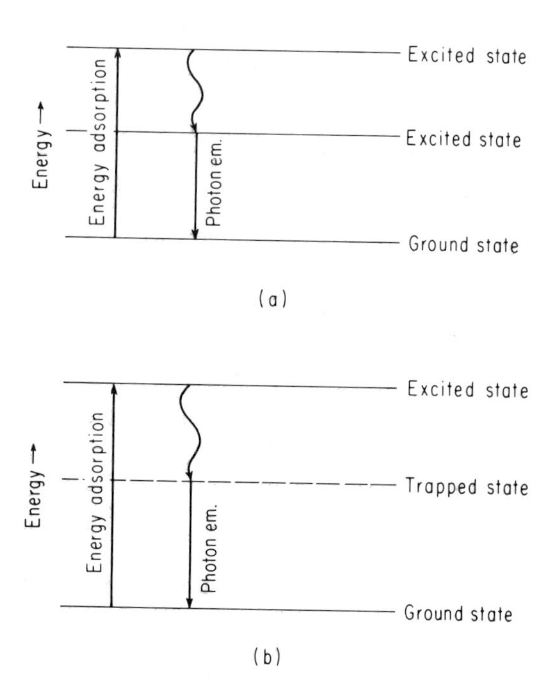

Figure 5-7 Luminescence in minerals. (a) Fluorescence. A photon in the violet, ultraviolet, or X-ray range of the electromagnetic spectrum excites an outer electron to a high energy state. The electron then falls from the high excited state to a lower excited state before falling further to the ground state. Fluorescence occurs when one of these energy gaps lies in the visible range. (b) Phosphorescence. The electron is raised to an excited state as in fluorescence, but falls into a trapped state, where it remains for a finite time before falling to the ground state. Thus, phosphorescence continues after the energy source has been cut off.

rescent minerals emit visible photons only so long as the activating energy is supplied. In some crystal structures the excited electron becomes trapped—for instance, in an omission defect or in a forbidden zone. If the entrapment is temporary, the excited electrons fall back into their ground state at some finite rate. However, the luminescence continues after the excitation energy has been stopped. Slowly decaying luminescence is called *phosphorescence*.

If an energy barrier prevents the excited electron from falling into its ground state, it is entrapped until that threshold energy is provided by some outside means. If the trapped electrons can be activated over the energy barrier by heating, the effect is known as *thermoluminescence*. The essence of the workings of a laser involves *pumping* electrons into an excited state in a forbidden zone until they are discharged back to ground state in an intense, monochromatic emission of coherent light.

Chemically, luminescence is generally associated with the presence of rare-earth elements that have unfilled *f* electron orbitals. Fluorite and calcite, which have some lanthanide-element substitution for Ca^{2+} in the crystal structure, generally fluoresce. Phosphates with the same substitution phosphoresce. This type of photoemission is analogous to allochromatic coloration in minerals. The characteristic luminescence of lanthanide-, uranium-, and thorium-bearing minerals is analogous to idiochromatic coloration. The phosphores of color television tubes are rare-earth element phosphate compounds.

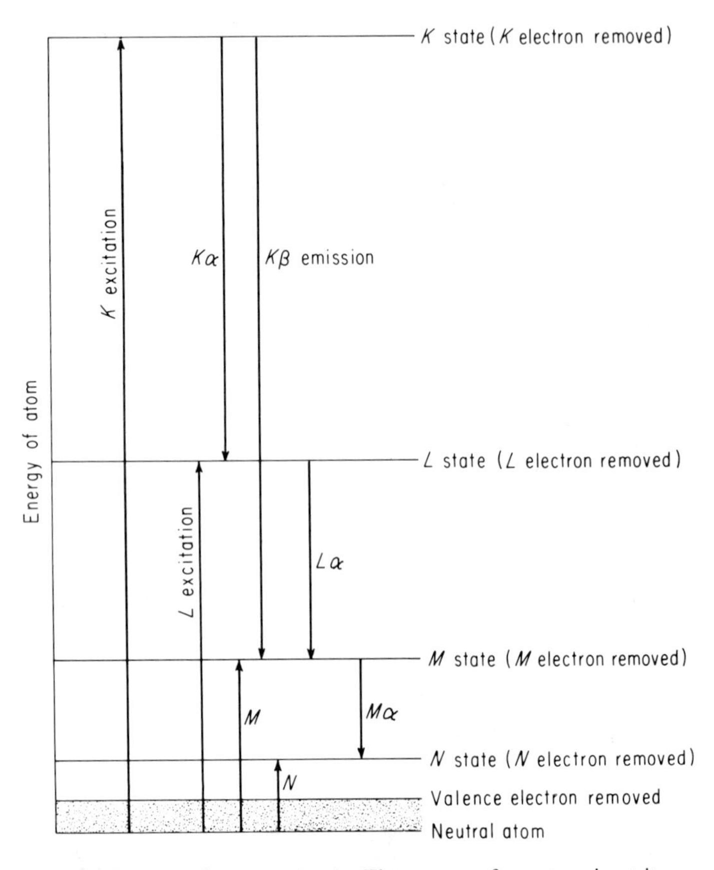

Figure 5-8 Atomic energy levels. The energy of an atom is at its lowest when all electrons are in their ground states. A small amount of energy is added to the atom when a valence electron is removed; progressively more energy is added as *N*, *M*, *L*, or *K* electrons are removed. $K\alpha$ emission of energy occurs when an *L* electron falls into a vacant *K* orbital. $K\beta$ emission occurs when an *M* electron falls into a vacant *K* orbital. *L* and *M* emissions occur when *L* and *M* vacancies are filled from overlying shells.

Fluorescence is not limited to the visible region of the electromagnetic spectrum. In other words, energy may be released by orbital electrons in other parts of the spectrum as well. A common method of nondestructive chemical analysis makes use of fluorescence in the X-ray part of the electromagnetic spectrum.

If an atom, either in a crystal structure or in some other physical state, is bombarded with a high-energy X-ray beam, some of the incident quanta have sufficient energy to eject a *K*- or *L*-shell electron from the vicinity of the atom. An atom in this excited state has an inner electron missing. The atom returns to ground state when one of the outer electrons falls into the vacancy, thereby emitting a quantum of energy in the X-ray part of the spectrum. Since the energy gaps between the orbitals are unique, the exact energy and wavelength of the emitted quantum may be used to identify the atom that was excited.

The possible excitation states and the fluorescence spectra are illustrated schematically in Fig. 5-8. When a K electron is ejected from its orbit, the atom is raised to a K excitation state. When an outer electron falls into the vacancy in the K orbital, K emission takes place. An electron from the L orbital gives $K\alpha$ emission, and an electron from the M orbital gives $K\beta$ emission. Similarly, ejection of an L electron gives rise to an L excitation state, with L emission quanta from outer electrons falling into the L vacancy. Removal of an M electron results in M fluorescence.

Fluorescence occurs in other parts of the electromagnetic spectrum, as well as in the visible and X-ray ranges. However, X rays longer than 10 Å and ultraviolet light shorter than 500 Å are absorbed very strongly by air. Many materials that are transparent to visible light are opaque to ultraviolet light and to large parts of the far infrared spectrum. Thus, problems of instrumentation limit the usefulness of fluorescence in spectral ranges other than the visible, the X ray, and parts of the infrared. Furthermore, much of the longer-wavelength fluorescence is allochromatic, and, therefore, not useful for determining essential constituents.

5-5. Atomic Absorption and Emission

The Bohr model of the atom is set forth, almost without justification, in Section 1-1, and has been appealed to as fact in several of the foregoing discussions. The lack of justification is rectified in this section, with the added benefit of providing theoretical background for some methods of instrumental chemical analysis (flame coloration, emission spectrography, atomic absorption, infrared absorption, and X-ray fluorescence).

A single de Broglie wave for an electron in orbit about an atomic nucleus is illustrated schematically in Fig. 1-1, and the relative energies for electron shells are shown in Fig. 1-3. The exact spacing for the energy levels, however, is unique for each element (Fig. 5-9). Furthermore, the amount of energy necessary to remove outer electrons from an atom, that is, its ionization potential, is unique (Table 1-4).

The explanation of the electronic structure, set forth in Chapter 1, is required to explain the absorption and emission of energy by the outer electrons of ions and atoms. Although emission and absorption spectra had been associated with specific atoms before, the first analysis of spectral lines was presented by Balmer in 1885. He determined the wavelengths of specific

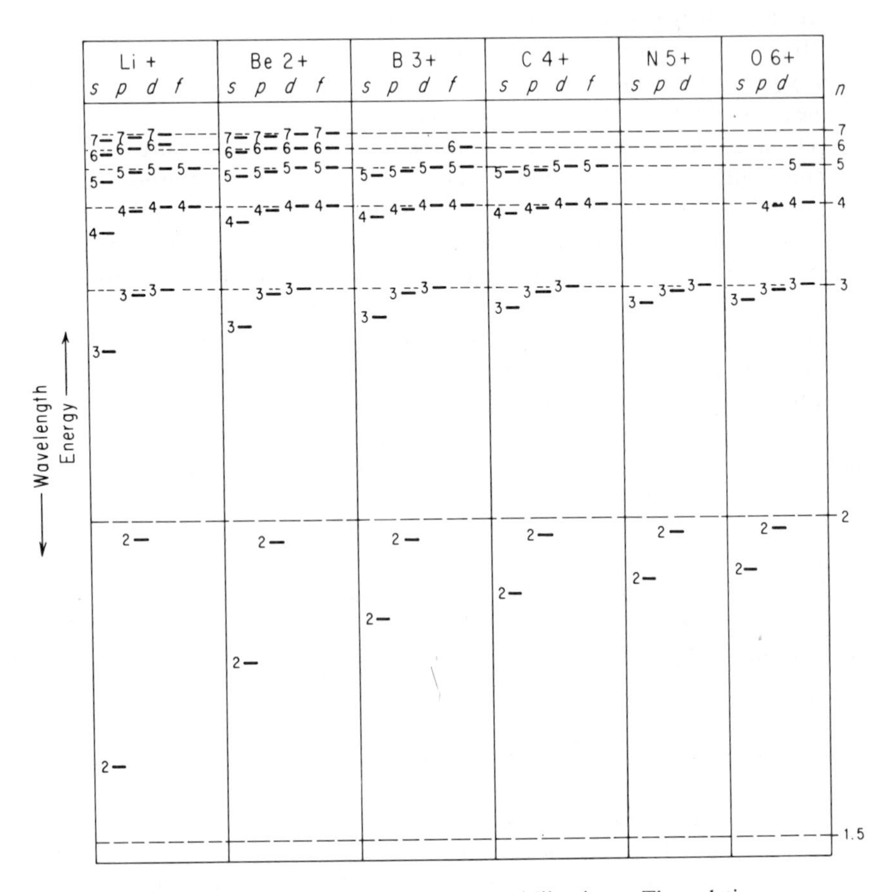

Figure 5-9 Energy levels of Li and Li-like ions. The relative energies of the electron orbitals change with atomic number. The letters *s*, *p*, *d*, and *f* refer to the azimuthal quantum number *l* (Section 1-1), and the numbers 1–7 are the principal quantum numbers *n* (Table 1-1). Transfer of electrons between orbitals, represented by horizontal bars, is quantized. Differences between orbital energies determine which quanta are absorbed or emitted, and each element has a unique spectrum of absorptions and emissions. (After Herzberg, 1944.)

emissions from an electric discharge through H_2, and represented the energy spacing of these emission lines in the visible spectrum with an equation containing integers. From this and other spectra not in the visible range, Niels Bohr (1913) proposed the model of the atom set forth in Section 1-1. He reasoned that if only discrete orbitals existed about the H atom, the radiation that was observed by Balmer and others was the result of a transition of an electron from one orbital to another by a quantum jump. Further study of the emission spectra of other elements led to the building up of the electronic structure of the atom and the periodicity of the chart of the elements that is set forth in Section 1-2.

Now the situation may be turned around, and the spectra, which were first used to build up the model for the electronic structure of the atom, can be used to identify the chemical elements in a substance. For *emission spectrographic analysis*, matter is vaporized by an electric arc or laser beam. The atoms are thermally excited and each emits its characteristic radiation. The emission energies, which represent transitions between possible orbitals, are unique for each element. The intensity of the emission of a particular energy is proportional to the amount of the element present, to a first approximation. An analysis of the particular energies of emission in the visible range can give both the kinds of elements present in an unknown sample and the relative proportions of each element.

Atomic absorption spectroscopy makes use of the same electronic transitions in a slightly different way. Light of a particular energy and wavelength is passed through a flame, which contains the vapor of an unknown sample. If an element which has a possible electronic transition that corresponds to the energy of the activating monochromatic light is present in the vapor, that light will be absorbed to a degree proportional to the amount of the element present, again to a first approximation. A different monochromatic light source must be used for the detection and analysis of each element.

Quantized lattice vibrations have excitation energies in the near infrared part of the electromagnetic spectrum. *Infrared spectrography* long has been used in the analysis of hydrocarbon molecules, but more recently has been employed in mineralogy. Most chemical bonds in minerals are only partially ionic, and the precise nature of specific bonds can be determined from interpretation of absorption spectra in the infrared range.

5-6. Interaction of Light Waves and Isotropic Matter

Light that is transmitted and refracted by matter requires description based on a wave theory of electromagnetic radiation. Such waves do not exist, but the probability of finding a photon at a given point is proportional to the square of the amplitude of the wave at that point (Section 5-1). Since transmission and refraction of light involve such a great flood of photons, treatment of light as a wave phenomenon is perfectly adequate to describe its behavior under these conditions. An analogous situation arises when a geologist describes a beach without reference to the individual sand grains that make it up.

Electromagnetic radiation, of which visible light is a small portion, is characterized by electric (**E**) and magnetic (**H**) vectors (read: *E* vector and *H* vector), which oscillate sinusoidally at right angles to each other and to the direction of propagation (Fig. 5-10a). Inasmuch as the **H** is always to be found

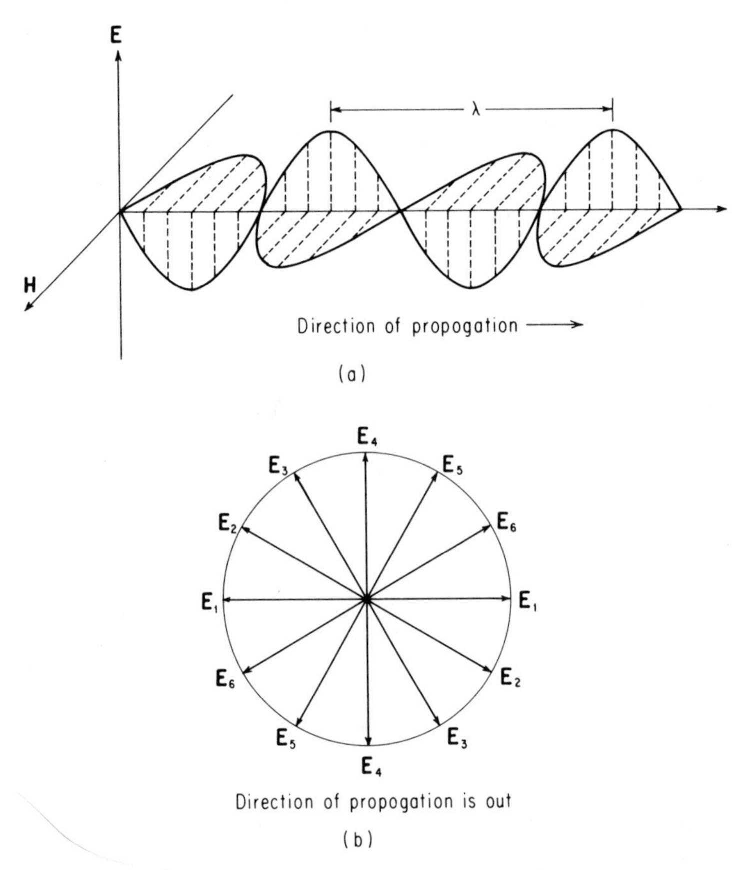

Figure 5-10 Electromagnetic waves. (a) The electric vector **E** and the magnetic vector **H** are at right angles to each other and to the direction of propagation. Both oscillate sinusoidally. (b) The **E**'s of unpolarized light are randomly distributed about the direction of propagation.

at right angles to the **E** and to the direction of propagation, specification of **H** adds no new information about a light wave, and its direction and phase may be assumed.

Ordinary light sources, such as the sun or an incandescent wire, produce light with **E**'s randomly distributed about the direction of propagation (Fig. 5-10b). *Plane-polarized light* has an **E** restricted to one plane only. Inasmuch as microscopic studies of the optical properties of minerals are generally carried out with polarized light, only plane-polarized light will be considered in this section. This restriction simplifies discussion without affecting the general validity of statements about light.

A second, simplifying restriction on the wavelength of light must be placed on subsequent discussion. White light is a composite of all wavelengths between 390 and 760 nm. Such a mixture is impossible to illustrate; therefore, monochromatic radiation is assumed for further discussions.

If **E**'s of two waves of plane polarized and monochromatic light are superimposed, the two waves interfere with each other as shown in Fig. 5-11. If the two waves, a and b, are exactly in phase with each other, interference results in the vector addition of the amplitudes to form the wave $a+b$ (Fig. 5-11a). Addition of **E**'s of two waves that are out of phase by one half-wavelength results in the cancellation of the waves by destructive interference $a+b$ (Fig. 5-11b). If two waves are out of phase by an arbitrary amount other than one half-wavelength, interference produces a third wave $a+b$ of different phase and amplitude than the original two (Fig. 5-11c).

In outer space, light has a velocity of 3×10^{10} cm/s. The velocity of light is less than this value for all transparent materials. (The velocity of light in air is so slightly less than in vacuo that the difference may be ignored.) The reason for the diminished velocity of light in transparent media is illustrated in Fig. 5-12. Consider that each ion in a crystal lattice is made up of a nucleus surrounded by electron waves forming an electron cloud (Fig. 1-1). The **E** of the light wave perturbs the electron cloud by causing it to oscillate about the nucleus (Fig. 5-12a). The nucleus remains effectively immobile because it contains more than 99.9997 percent of the mass of the ion.

An oscillating electric field, such as an electron cloud vibrating about a nucleus, generates its own electromagnetic radiation. The radiation generated by the electron cloud lags behind—is out of phase with—the primary electromagnetic radiation because of the small, but finite, inertia of the electron cloud itself. Therefore, the secondary radiation interferes with the primary radiation (Fig. 5-11), with the result that the combined wave also lags behind a parallel wave which is traveling in vacuo.

Slowing down of light in a crystal lattice is illustrated schematically in Fig. 5-12b. A light wave arrives at lattice plane *1*, where part of its energy is used to oscillate the ions in that plane. The amplitude is decreased, but its wavelength remains unchanged as it goes from lattice plane *1* toward lattice plane *2*.

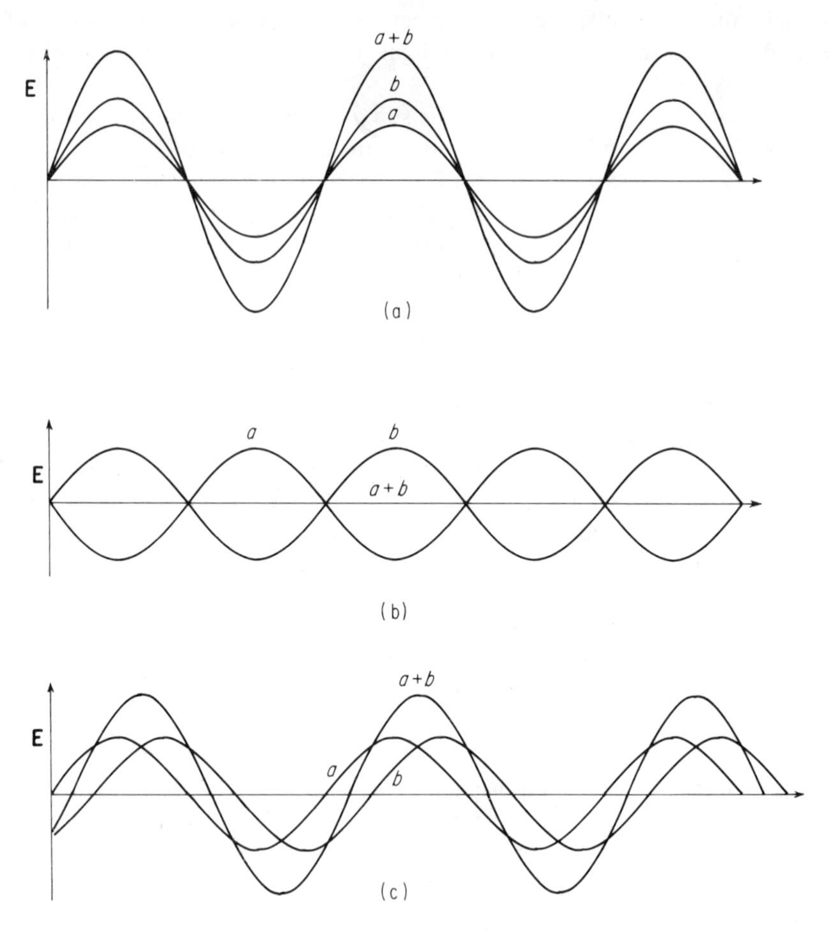

Figure 5-11 Wave interference. (a) Constructive interference. Wave *a* interferes constructively with wave *b* so that the resultant wave $a+b$ has an amplitude equal to the sum of the interfering waves. (b) Wave *a* interferes destructively with wave *b* so that the resultant wave $a+b$ has no amplitude and, therefore, does not exist. (c) Wave *a* partially interferes with wave *b* such that a new wave of different amplitude and position results.

However, each of the ions in lattice plane *1* also sends out radiation as a result of oscillation of its electron cloud. Since each ion in the lattice plane is radiating light as a point source, only those traveling in the direction of the original wave are not destroyed by interference. The secondary wave has the same wavelength as the primary, but is out of phase with it and, therefore, interferes with it. The net result is that the light wave is slowed and arrives

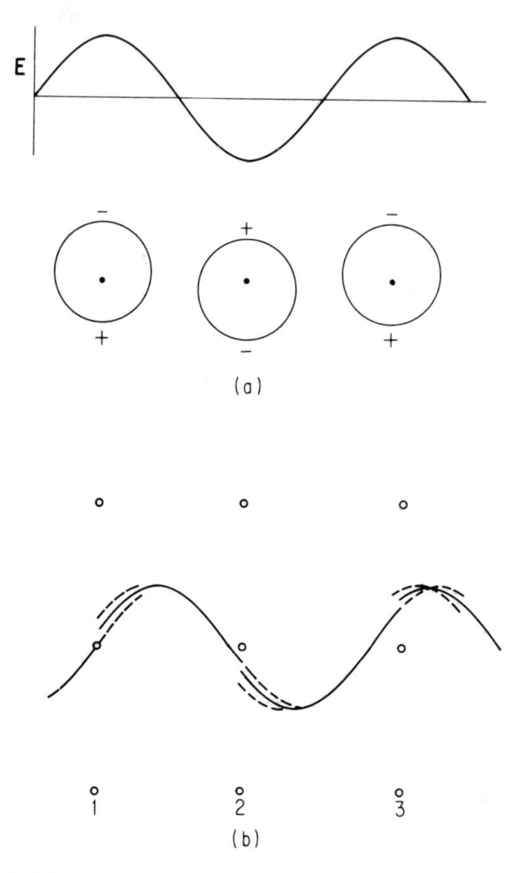

Figure 5-12 Retardation of light in crystals. (a) The **E** of the light wave interacts with the electrons of the ion causing them to oscillate. Because of the mass of the electrons, their response lags behind the original wave. The oscillation of the electron clouds sets off secondary radiation, which lags behind the primary and interferes with it. (b) Light waves arriving at layer *1* of ions set their electrons to vibrating. The secondary waves partially interfere with the primary so that the resulting waves lag behind the primary waves. The resultant wave arrives late at row *2* and the process is repeated, with further retardation as the new resultant wave arrives late at row *3*.

late at lattice plane *2*. This resultant wave then acts as a primary wave on lattice plane *2*, and the process is repeated.

A stick of wood, stuck half in the water at an angle, appears to be bent as a direct result of the diminished velocity of light as it travels through a transparent medium. The ratio of the velocity of light in vacuo to the velocity of light in a transparent medium is a number larger than 1, which is known as the *index of refraction n*. The larger the index of refraction, the slower the velocity of light and the more a light wave is bent on passing obliquely into a transparent medium. The index of refraction n is related to the angle of incidence i and the angle of refraction r by the equation, $n = \sin i/\sin r$, known as *Snell's law* (Fig. 5-13).

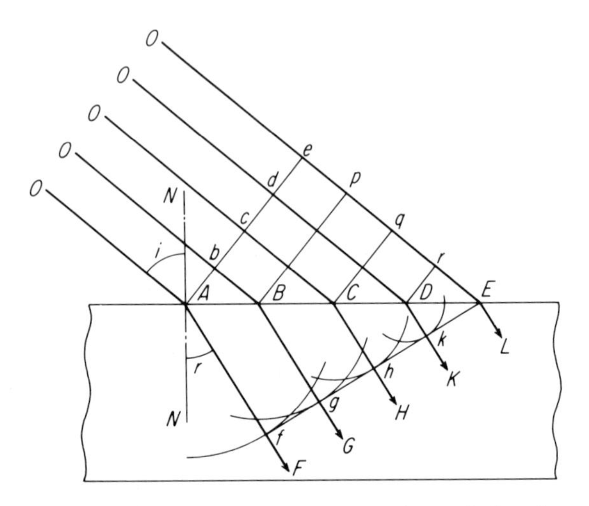

Figure 5-13 Snell's law. Light rays with wave front *Abcde* arrive at a crystal surface. Ray *OA* travels the distance *Af* through the crystal, while ray *OE* travels the distance *eE* through air. The new wave front for light traveling through the crystal is parallel to *fE*, and the new ray direction is at right angles to it. See text for derivation of Snell's law.

Consider a beam of light, made up of waves *OA*, *OB*, . . . , *OE*, striking the surface of a transparent crystal. When the wave front has reached the position *Abcde*, wave *OA* enters the crystal. Point *A* may be thought of as a new point source of light, which is spreading out through the crystal. Waves *OB*, *OC*, *OD*, and *OE* strike the surface sequentially. However, in the time that *OE* has traveled from *e* to *E*, *OA* has traveled from *A* to somewhere on the surface of a sphere represented at *f*. The distance *Af* is less than the distance *eE* because of the reduced velocity of light in the crystal. Similarly, while *OE* has traveled from *p* to *E*, *OB* has traveled the lesser distance *Bg*. The same argument can be applied to waves *OC* and *OD*. To find the direction in which a wave is traveling, it is only necessary to connect up the wave fronts of several parallel waves with a common tangent plane at an arbitrary time. This is the plane represented by line *fghkE* in Fig. 5-13. All waves traveling in directions other than at right angles to this plane are destroyed by interference. Thus, the waves take on a new direction upon entering a crystal obliquely.

Snell's law may be derived with reference to Fig. 5-13. The angle of incidence i is equal to the angle eEa, by the laws of plane geometry. Sin i is, therefore, eE/EA. The angle of refraction r is equal to the angle fEA, and sin r equals Af/EA. Since EA is common to both equations, they may be rearranged to

$$EA = \frac{eE}{\sin i} = \frac{Af}{\sin r} \quad \text{and} \quad \frac{eE}{Af} = \frac{\sin i}{\sin r}$$

But the construction was so devised that eE is proportional to the velocity of light in vacuo, and that Af is proportional to the velocity of light in the crystal. Therefore, $eE/Af = n$, and $n = \sin i/\sin r$, which is Snell's law.

The actual velocity of light in a mineral is not generally determined directly, but measurement of the index of refraction is a simple operation.

If the angle i equals zero, then sin i also equals zero. Since the index of refraction has a finite value, sin r and the angle r must also equal zero. This means that if the incident wave of light is normal to the surface, it proceeds through the interface with diminished velocity and no refraction (Fig. 5-12b).

While swimming under water, anyone who has rolled over on his back and tried to look up out of the water has noticed that he can see through the air–water interface only through a cone directly above him. At angles greater than 48.5° from the vertical, the interface has a metallic luster and reflects light totally. This phenomenon can be explained by reference to Snell's law.

As the angle of incidence approaches 90°, sin i approaches 1, and Snell's law becomes $n = 1/\sin r$, or $\sin r = 1/n$. The value $1/n$ for sin r defines the critical angle. In physical terms, the critical angle means that a light wave which is just skimming the surface of a transparent medium will be refracted into it at an angle, the sine of which is $1/n$, and that no greater angle of refraction is possible for the medium.

Snell's law is equally valid for light traveling from a transparent medium into air, so the direction of wave propagation may be reversed in Fig. 5-13. However, to avoid confusion, r may be retained for the angle in the denser medium and i for the angle in the rarer medium. At the critical angle, then, light traveling from the medium into air is refracted parallel to the interface. At all angles greater than the critical angle, refraction is impossible, and the wave is reflected totally (Fig. 5-14). When viewed from below the surface of water, or from the inside of a crystal, the interface beyond the critical angle appears metallic. Clearly, from Snell's law, the greater the index of refraction, the smaller the critical angle.

Use is made of the critical angle in cutting gem stones. The facets of a gem are cut so that the internal angles between facets are total reflecting surfaces. Light, which enters the gem through one of the upper facets, is reflected internally and reemitted through one of the upper facets.

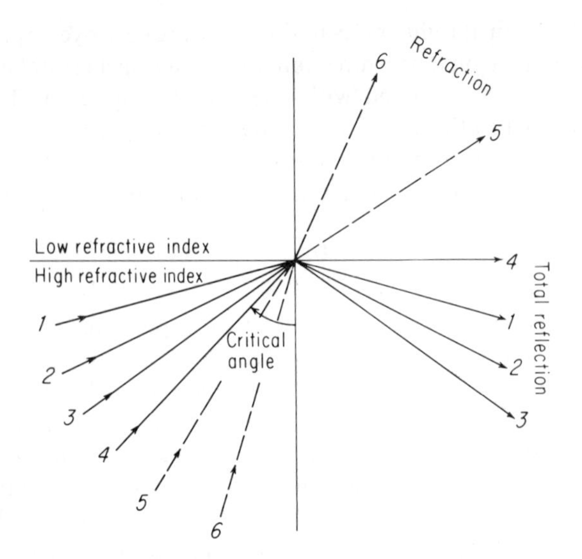

Figure 5-14 Critical angle of reflection. Some of the light traveling from material of high refractive index to material of low refractive index is reflected totally at the interface between the two materials. Ray *4* strikes the interface at the critical angle and is refracted parallel to the interface. Ray *5* strikes the interface at an angle less than the critical angle, and is, therefore, refracted into the low index material. Rays *1*, *2*, and *3* strike the interface at angles greater than the critical angle and are totally reflected by the interface.

Light has been assumed to be monochromatic in the foregoing discussions. Light from all sources, other than special monochromatic sources, is, of course, a mixture of all wavelengths in the visible range. The velocity of light in transparent media varies with wavelength (Fig. 5-15). It follows that the index of refraction varies with wavelength, as do the angle of refraction and the critical angle of reflection. This is called *dispersion.* As is stated in Section 5-1, the color and the fire of an ice-blue diamond require different kinds of explanation. Color is described in terms of photons of a particular energy being absorbed by excitation in the electronic structure of an ion (Section 5-3). Fire in a diamond may be described in terms of dispersion of light and the critical angle of reflection. The facets of a diamond are cut so that part of the spectrum, which is split apart by dispersion, strikes a facet from the inside at an angle less than the critical angle, and is refracted out at that place. Part of the spectrum is reflected totally by that facet and comes out by an upper facet as "fire."

All minerals disperse light, few more than diamond, however. Newton was able to display the spectrum of visible light by use of a trigonal prism cut from transparent quartz.

The luster of nonmetallic substances is related to the way in which light is

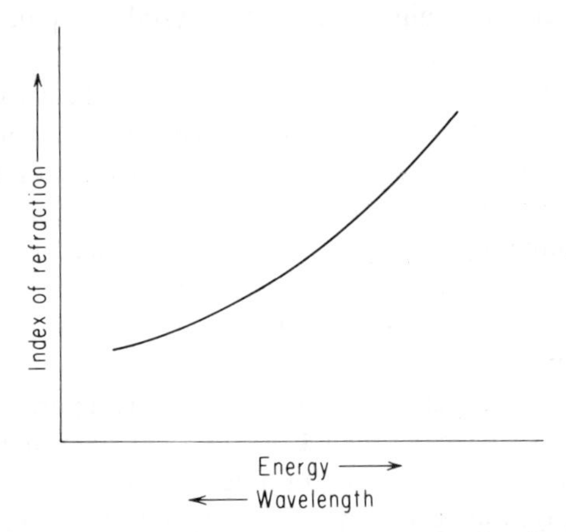

Figure 5-15 Dispersion of light. The index of refraction of light varies with the wavelength of the light. In most materials the index increases with increasing energy and decreasing wavelength of the incident light.

reflected from the surface of a mineral. This, in turn, is partially related to the index of refraction. To a first approximation, the greater the index of refraction, the brighter the luster. Minerals with indices of refraction greater than 3 generally display *metallic luster*. Indices of refraction between 2.6 and 3 show a *submetallic luster*. *Adamantine luster*, that typical of diamond, is associated with minerals having indices of refraction between 1.9 and 2.6. Approximately three fourths of all non-metallic minerals have indices of refraction between 1.3 and 1.9, and give a *vitreous luster*, that of broken glass.

Other qualities of luster, which are recognized by mineralogists, have to do with special mechanical or chemical properties of the mineral surfaces. For instance, halite, which is vitreous when freshly cleaved, develops a *greasy luster* due to a surface hydration layer having a different index of refraction. Very fine grained and porous minerals, such as the clays, scatter light completely to produce a *dull* or *earthy luster*. Minerals with the luster of pearls, such as talc, mica, and some varieties of calcite and gypsum, owe their *pearly luster* to the existence of many successive reflecting surfaces, which are internal to the mineral specimen.

5-7. Interaction of Light Waves and Anisotropic Matter

Throughout Section 5-6, the velocity of light was assumed tacitly to be independent of the direction of the light wave. This is true of light passing through vapors, liquids, glasses, and minerals crystallizing in the cubic system, all of which are *isotropic*. All other crystalline materials are *anisotropic*.

That is, the velocity of light varies with crystallographic direction in the crystal.

The source of the variability of the velocity of light with differing crystallographic directions is to be found in the distribution of ions and chemical bonds in a crystal lattice. In cubic crystals this distribution is the same along all three crystallographic axes. It is, clearly, not so in noncubic crystals. The structural environment in hexagonal and in tetragonal crystals is different in the ***c*** direction than in the ***a*** directions. In orthorhombic, monoclinic, and triclinic crystals, the structural environment in all three crystallographic directions is different.

When light enters any anisotropic mineral, it is separated into two plane-polarized light rays that have their **E**'s at mutual right angles. This is called *double refraction.* Since the velocity of a light ray depends upon the direction of its **E**, the two rays then travel at different speeds in different directions. The effect of double refraction is best seen with a cleavage fragment of clear calcite (Fig. 5-16). Here the double refraction is sufficient to cause double

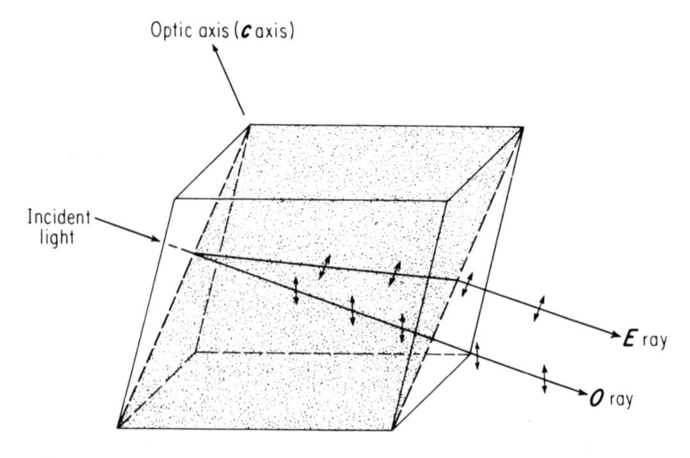

Figure 5-16 Double refraction in calcite. Light, which is incident upon a cleavage fragment of calcite, is split into two rays having their **E**'s at mutual right angles (shown by arrows). These two rays diverge while passing through the crystal and reconverge into parallel directions upon emergence.

images to be seen through the cleavage fragment. The great difference in the velocities of the two rays shows up in the different apparent depth of the two images, since apparent thickness is inversely related to light velocity. That the two waves are, in fact, polarized may be demonstrated by donning polaroid sunglasses and rotating the cleavage fragment of calcite. All anisotropic minerals refract doubly, but few have greater double refraction than calcite, and most have considerably less.

As illustrated in the previous section, the velocity of light through any transparent medium is dependent upon an interaction between the **E** of the light ray and the electron clouds of the ions in the medium. The greater the interaction, the more the ray is slowed in passing through the medium.

The spacial relationships among the ray direction, its **E**, and the index of refraction are shown conveniently with a solid geometric construction known as an *indicatrix*, first proposed by L. Fletcher in 1891. A light wave traveling through isotropic matter is shown schematically in Fig. 5-18a, and the associated indicatrix in Fig. 5-17. The ray direction is at right angles to the **E**.

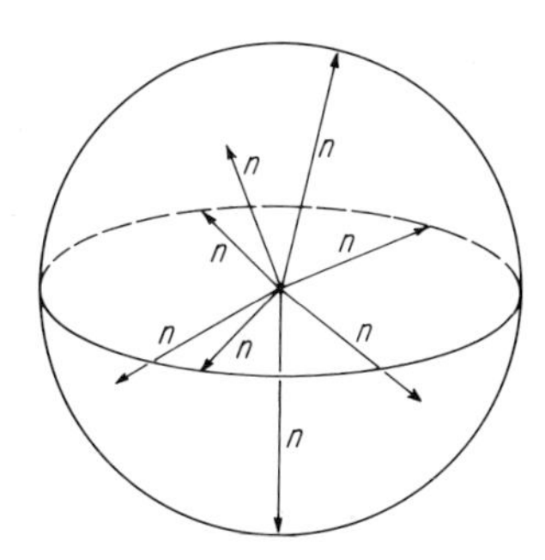

Figure 5-17 Isotropic indicatrix. An indicatrix is a geometric figure, the radius of which is proportional to the refractive index of a light ray having its **E** in that direction. The isotropic indicatrix is the surface of a sphere having radius n.

The indicatrix is constructed such that its radius is proportional to the index of refraction n in the **E** direction. Since light traveling through isotropic substances, by definition, has the same index, regardless of **E** direction, the isotropic indicatrix is a sphere with its radius proportional to n.

As seen in the example of calcite (Fig. 5-16), anisotropic minerals refract doubly, the two light rays being plane polarized with **E**'s at right angles and having different velocities. Thus, it is clear that the velocity and, therefore, the index of refraction of a light ray will vary with the direction of its **E** in anisotropic substances. It follows, then, that the indicatrix will not be spherical.

Two classes of anisotropic crystals are recognized, *uniaxial* and *biaxial* (the significance of these names will become apparent as this discussion develops). All minerals crystallizing in the tetragonal and hexagonal systems are uniaxial, and all minerals crystallizing in the orthorhombic, monoclinic, and triclinic systems are biaxial.

When an anisotropic mineral refracts doubly, two mutually orthogonal **E**'s are said to be *permitted*. These are also called *allowed vibration directions*. In isotropic substances, and at right angles to certain special ray directions in

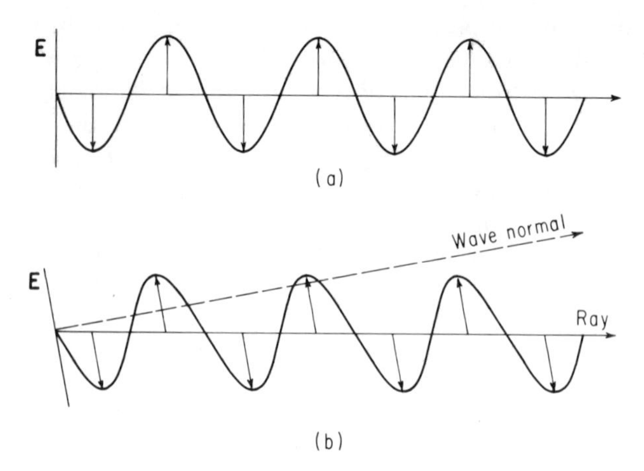

Figure 5-18 Ordinary and extraordinary wave and ray directions. (a) The ordinary ray has its **E** at right angles to the ray direction, the wave and ray direction being parallel. (b) The extraordinary ray has its **E** not at right angles to the ray direction. Hence, the ray and wave directions are not parallel.

anisotropic minerals, all **E**'s are permitted. The index of refraction depends upon the velocity of a ray with a particular permitted **E**.

In uniaxial minerals, double refraction produces two rays, known as the *ordinary* (*o* ray) and the *extraordinary* (*e* ray) rays. The ordinary ray has its permitted **E** at right angles to the ***c*** axis and to the ray direction. Upon entering and leaving a uniaxial crystal, this ray follows the ordinary laws of refraction (that is, Snell's law).

The distribution of electron clouds in isotropic minerals is the same along all three crystallographic axes. In tetragonal and in hexagonal minerals, the distribution of electron clouds is clearly not the same along the ***c*** axis as it is along the ***a*** axes. For this reason, **E**'s suffer different interference in the ***c*** direction than in the ***a*** directions. The velocity and index of a ray with **E** not in the ***a*** plane, therefore, are different from the velocity and index of a ray with **E** in the ***a*** plane.

A second effect of anisotropy is that for rays with **E**'s not in the ***a*** plane, the **E** is not at right angles to the ray direction (Fig. 5-18b). When a light wave impinges on ions that are asymmetrically constrained, the electron clouds do not oscillate parallel to the **E** of the wave. Averaged over a finite part of a crystal, their secondary radiation is not only out of phase with the primary wave, but has an **E** that is not parallel as well. Interference produces a wave that has its **E** not at right angles to the ray direction (Fig. 5-18b).

The direction, which is at right angles to **E** and in the plane that contains both **E** and the ray, is called the *wave normal*. Since the wave normal is not parallel to the ray direction, these rays do not follow the ordinary laws of refraction and are, thus, extraordinary. The **E** of the *e* ray, being at right angles to that of the *o* ray, always lies in the plane that contains the ***c*** axis, the ray direction, and the wave normal.

The foregoing observations may now be assembled schematically into the uniaxial indicatrix (Fig. 5-19). The index of refraction of the ordinary ray is

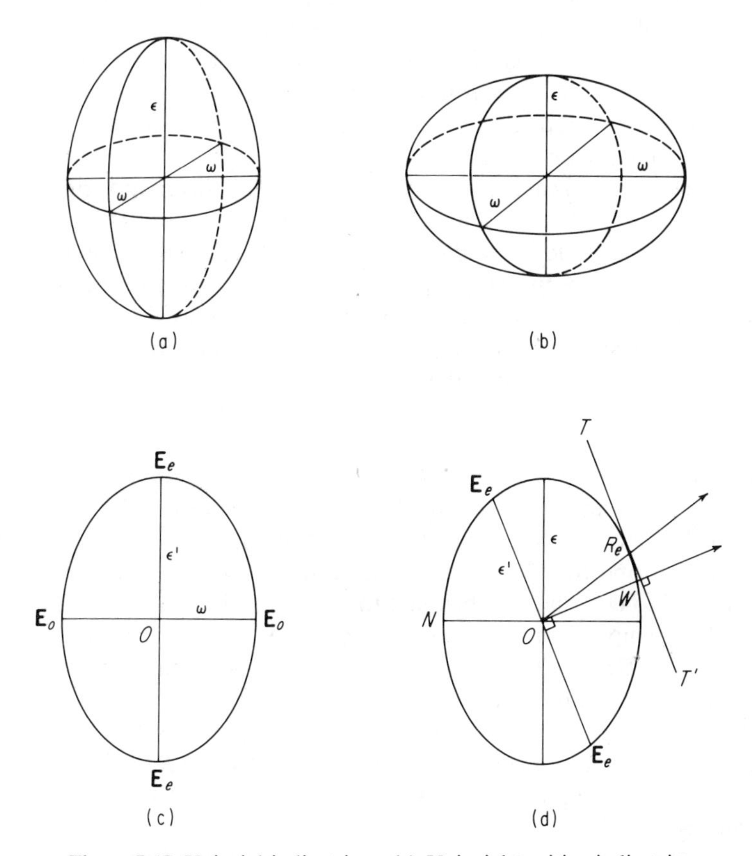

Figure 5-19 Uniaxial indicatrices. (a) Uniaxial positive indicatrix. (b) Uniaxial negative indicatrix. (c) Wave-front section through a uniaxial positive indicatrix. (d) Wave and ray directions in the principal section. OW = wave direction. OR_o = ordinary ray direction. OR_e = extraordinary ray direction. TT' is normal to OW and tangential to the ellipse at R_e. $\mathbf{E}_e O\mathbf{E}_e'$ = wave front. ϵ' = index of refraction of extraordinary ray. ON = ordinary index of refraction.

labeled ω and that of the extraordinary ray ϵ'. The maximum or minimum index, that which is associated with a ray having its **E** parallel to the ***c*** axis, is labeled ϵ. The *optic sign* of uniaxial minerals is *positive* if ϵ is greater than ω (Fig. 5-19a), and *negative* if ϵ is less than ω (Fig. 5-19b). The index ϵ' may take on all values between ϵ and ω. The uniaxial indicatrix is an ellipsoid of revolution, prolate if positive and oblate if negative.

Before discussing the general relationships of the uniaxial indicatrix, two special cases may be considered. First, all rays traveling in the **c**-axis direction have their **E**'s in the ***a***-axis plane, which has identical electron-cloud distributions in each of the ***a*** directions. Since rays with **E**'s in this plane all have the same index of refraction, ω, the section containing the ***a*** axes is circular. All **E**'s are permitted in the *circular section*, and a ray traveling in the direction of the **c** axis behaves as though it were isotropic with index of refraction ω. This direction is called the *optic axis*.

The second special case to consider is that of a ray traveling at right angles to the optic axis. Such a ray is broken into two rays. One has its **E** parallel to the ***c*** axis and has index ϵ, and the other is in the circular section and has index ω. Since the **E** of the *e* ray is parallel to the optic axis and at right angles to the ray direction, the wave normal is parallel to the ray direction, and the ray follows the ordinary laws of refraction in this special direction.

Consider a section through the indicatrix that contains the wave normals for both rays, the ray directions for both rays, the **E** of the *e* ray, and the ***c*** axis (Fig. 5-19d). The **E** of the *o* ray is at right angles to the section. The semimajor axis (positive) of the section is the optic axis and is proportional to ϵ. The semiminor axis is proportional to ω. The wave normal for both rays is OW, which is also the *o*-ray direction. The **E** of the *o* ray is normal to the section and proportional to ω and, therefore, equal to ON. The **E** of the *e* ray is normal to OW and proportional to ϵ' (OE_e). The *e*-ray direction is the radius of the ellipse that is conjugate to OE_e. (A radius OR is conjugate to another radius OE_e if the tangent to the ellipse at R_e is parallel to OE_e.)

Given the *e*-ray direction OR, it is possible to locate the wave normal and to determine the value of ϵ'. First construct the tangent to the ellipse TRT'. Next construct the radius of the ellipse OE_e parallel to TRT'. The length of the radius OE_e is proportional to ϵ'. The wave normal OW is the radius at right angles to OE_e.

Optical properties of minerals may be deduced by imagining the appropriate section of the indicatrix to be located wherever convenient within a mineral grain. The case in which the mineral slice is oriented neither parallel to nor normal to the ***c*** axis of the mineral is illustrated in Fig. 5-20. The indicatrix is transposed such that the ray of light is normal to the surface of the slice at the center of the indicatrix. Upon entering the mineral, the light is broken into two rays with **E**'s in the permitted directions. The *o* ray with **E** normal to the section travels at reduced velocity but without change in direction. The *e* ray with **E** parallel to the surface of the slice and in the plane of the paper, travels toward R, which is where the wave front and the surface of

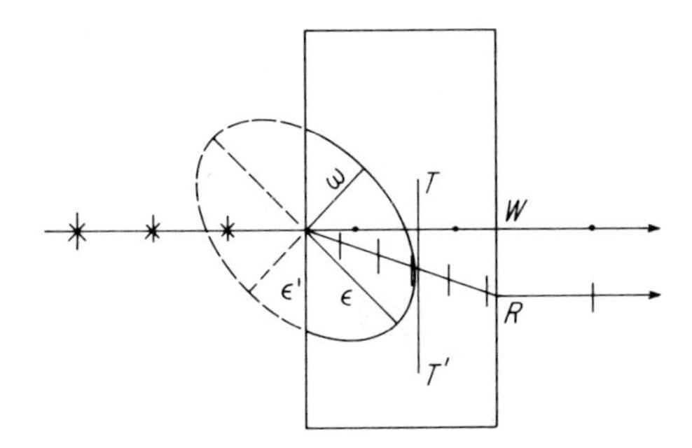

Figure 5-20 Light transmission through uniaxial crystals. Unpolarized light ray is resolved into two rays, which are polarized with **E**'s normal to one another. OW = wave direction and ordinary ray direction. OR = extraordinary ray direction.

the slice are tangential to the ellipsoid. It continues in this direction until it emerges from the far side of the slice, where the ray direction resumes a path at right angles to the **E** of the ray.

As with uniaxial minerals, the index of refraction in biaxial minerals varies with the **E** of the light ray, from a maximum value of γ to a minimum value of α, which are at right angles to one another. There is, however, at right angles to both γ and α, a unique value β, which is intermediate in value. These three permitted directions in a biaxial crystal are labeled $\mathbf{E}_z$ for γ, $\mathbf{E}_x$ for α, and $\mathbf{E}_y$ for β, or more commonly X, Y, and Z. For **E**'s that do not coincide with the principal axes X, Y, and Z, the indices of refraction will be intermediate between γ and β, or between β and α. These indices are labeled γ' and α', respectively.

The distribution in space of these **E**'s and their indices of refraction are illustrated in Fig. 5-21a. The shape of the indicatrix is a triaxial ellipsoid. Since β has a value between γ and α, there must exist in each quadrant of the XZ section through the indicatrix an index that is equal to β. Furthermore, in each octant there must exist a whole family of radii equal to β and extending from the β value in the XZ plane toward Y. These indices, being all equal, define the two circular sections of the biaxial indicatrix (Fig. 5-21b). The two optic axes, hence the name *biaxial*, are normal to each of the two circular sections. The optic axes are in the XZ plane since they are normal to Y.

All sections through the ellipsoid, except the circular sections, are ellipses. The principal sections XZ, XY, and YZ have semimajor and semiminor axes equal, respectively, to γ and α, β and α, and γ and β. All other sections through the indicatrix are ellipses with semimajor axes γ' and semiminor axes α'.

A light ray traveling normal to a circular section, as in the uniaxial case, has all **E**'s permitted and, therefore, behaves as though the crystal were

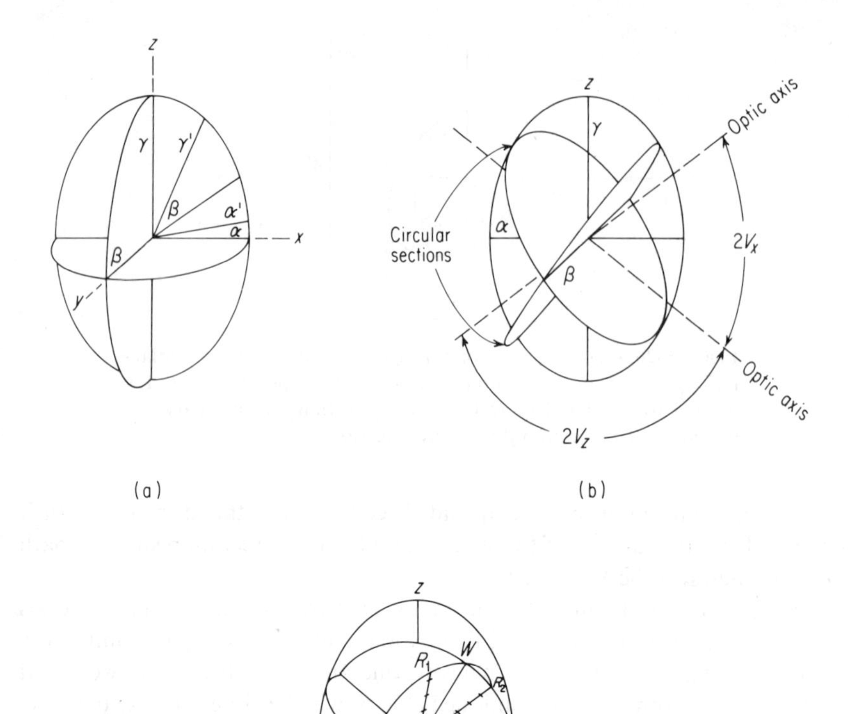

Figure 5-21 Biaxial indicatrix and sections. (a) Biaxial indicatrix with principal planes shown. (b) Circular sections and optic axes. (c) Wave and ray directions for light ray passing through biaxial indicatrix.

isotropic. A light ray traveling normal to one of the principal sections is broken into two polarized rays with orthogonal **E**'s and indices of refraction proportional to the semimajor and semiminor axes of the sections, α, β, or γ. Because the tangents to the indicatrix at the points of intersection of the principal axes X, Y, and Z are parallel to the principal sections, these rays behave as though they were ordinary rays with the ray direction at right angles to their **E**'s.

A ray of light for which the indicatrix is randomly oriented is broken into two rays with orthogonal **E**'s, but which are both extraordinary. However, the two rays have a common wave front, which is an elliptical section through the indicatrix and has semimajor axis γ' and semiminor axis α'. The **E**'s of the two rays are, clearly, parallel to the major and minor axes. The two ray directions may be found by constructing two more elliptical sections, one through the major and one through the minor axis of the wave-front section, both normal to the wave-front section and intersecting along the wave normal OW (Fig. 5-21c). The two ray directions, thus, lie in these two sections. The ray with the lower index of refraction has a direction defined by the radius OR_1, which is conjugate to the radius OX', defined as the intersection between the ray section and the wave-front section. The ray with the higher index has a direction conjugate to the radius OZ'.

For slices cut parallel to one of the principal sections, normally incident light is broken into two ordinary rays, one having the index and **E** of the semimajor axis and the other the index and **E** of the semiminor axis. Both rays continue to travel in the same direction, but at different velocities.

Facets, such as the two that truncate the circular section and the principal sections, XY and YZ, which are parallel to one of the principal axes but not to either of the other two, split normally incident light into two rays, one of which is ordinary and the other extraordinary. The ray that has its **E** parallel to a principal axis is an *o* ray, and the one with its **E** not parallel to a principal axis is an *e* ray.

Light, which is normally incident upon a slice cut at random with respect to the indicatrix, is split into two rays with normal **E**'s. Neither of these rays follows the ordinary laws of refraction. One of the extraordinary rays has refractive index γ' between γ and β; the other has index α' between α and β.

In summary, the behavior of light as it is transmitted by transparent materials is explained in terms of electromagnetic wave theory. Light has associated with it an electric vector (**E**) at right angles to the direction of wave propagation. When light waves pass through a transparent material, the **E** interacts with the orbital electrons of the atoms or ions that make up the material. The light wave causes the electrons to oscillate about the atomic nucleus. The electronic oscillation, in turn, generates a new light wave, which, because of the inertia of the electrons, is out of phase with the original wave and is behind it in space. These waves interfere, and the resultant wave, therefore, travels at a slower velocity than the original light wave.

The exact nature of the secondary wave generated by the electrons and the degree to which it lags behind the original wave depend upon the kinds of ions that are in the material, and upon the kinds and distribution of chemical bonds between the ions. The velocity of the resultant wave tends to decrease with increasing density of the material. Heavier ions have more electrons; therefore, there is more interference with the light wave. Very polarizable ions slow light waves more than ions with low polarizability, because the electrons are more mobile than they are in ions of low polarizability. Denser polymorphs contain more ions per unit volume than lighter polymorphs; therefore, there are more electrons to slow the light wave.

Crystals constructed so that the array of ions and chemical bonds is the same in all three crystallographic directions permit light to have a constant velocity, regardless of the direction of the **E** in the crystal. Anisotropic crystals, however, have ions and bonds arranged differently along the different crystallographic directions. Therefore, a light ray having its **E** in one crystallographic direction travels at one velocity, which is dependent upon the array of ions and bonds in that particular direction. The wave-front surface and the Fletcher indicatrix are convenient constructions, which illustrate the behavior of light in crystalline matter.

5-8. X-Ray Diffraction

The history and importance of X rays in mineralogic studies is discussed briefly in Section 1-6. This section is devoted to a theoretical treatment of the nature of X rays, their sources, and the way in which they interact with crystalline matter.

X rays are electromagnetic radiation with wavelengths between 500 and 0.005 Å (Fig. 5-1). Although commonly placed between ultraviolet light and gamma rays, X rays actually overlap both. This overlap derives from a practical, not theoretical, distinction among the different types of radiation. Ultraviolet light is generated by outer electrons, X rays by inner orbital electrons or by decelerating free electrons, and gamma rays by the atomic nucleus. If the source of radiant energy with a wavelength of 100 Å is contained in a black box, there is no way of knowing whether it is X radiation or ultraviolet light. Similarly, radiant energy with a wavelength of 0.01 Å could either be X rays or gamma rays, depending on whether the source were inner orbital electrons or the atomic nucleus.

Electrons that have been accelerated across a potential gradient have considerable kinetic energy. When these fast-moving electrons strike the anode, they are decelerated and their kinetic energy is converted to heat (more than 99 percent) and to X rays (less than 1 percent). The maximum amount of energy that an X-ray quantum can acquire occurs when the kinetic energy of a decelerated electron is converted completely to X radiation. This determines the shortest wavelength, the *short-wavelength limit* (SWL), of radiation, which is a direct function of the voltage across the potential gradient (Fig. 5-22). Partial conversion of kinetic energy to X rays yields radiation of lower energy and longer wavelength, and the radiation that comes from an X-ray tube is a mixture of these wavelengths, called *white* radiation.

As the potential gradient that accelerates the electrons is increased, the

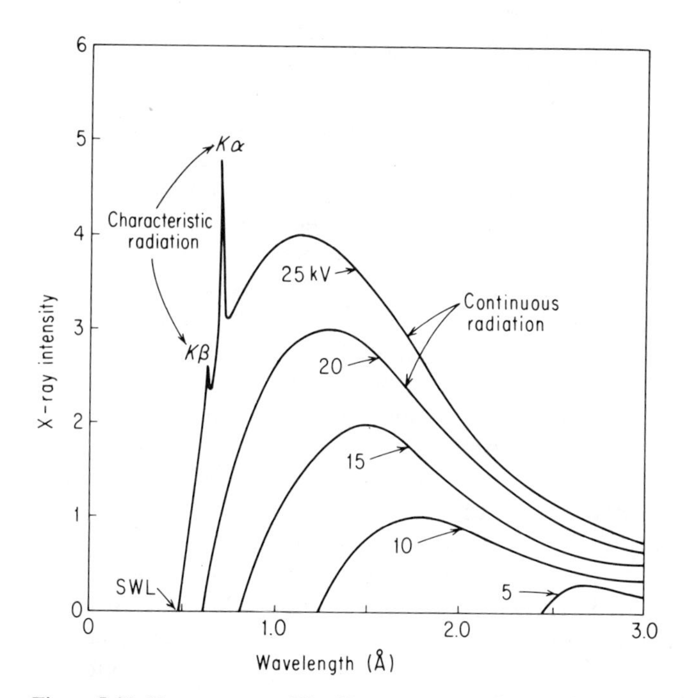

Figure 5-22 X-ray spectra. The X-ray spectrum for Mo is shown for several different voltages. When the applied voltage is greater than 25,000 V, K electrons are removed from the Mo target and the characteristic K radiation appears superimposed on the continuous radiation. (After Cullity, 1956.)

SWL is decreased. If the potential gradient is large enough, the kinetic electrons have sufficient energy to expel K electrons from the target metal. When this occurs, the metal ions of the target begin to emit their own *characteristic* radiation (Section 5-4).

The characteristic radiation of the target metal is used in the analysis of the structure of crystalline material. A beam of monochromatic (characteristic) radiation is directed to impinge on a piece of crystalline material. The X-ray quanta are scattered by the ions in the crystal in much the same way that photons of visible light are scattered by motes of dust, but there are two differences between these two modes of scattering. The wavelengths associated with X-ray quanta are much shorter than those associated with photons; and the ions in a crystal have a periodic and ordered array, which dust motes do not. From here on, the argument must be couched in terms of wave theory. Each atom or ion in the crystal lattice may scatter the incident quanta. Although the scattering may be random in direction, radiation occurs only in those directions where there is constructive interference of the waves associated with the X rays (Fig. 5-11). Destructive interference destroys rays in all other directions.

Conditions for constructive interference of scattered radiation, that is, for diffraction, are stated by Bragg's law, $n\lambda = 2d \sin \theta$ (Fig. 5-23). Consider two X rays, *1* and *2*, impinging on a crystal lattice at an angle θ (θ is measured from the surface, not from the surface normal as is the angle of incidence). Line AB represents the wave front at the time when ray *1* is scattered by atom A. At a later time ray *2* is scattered by atom C, and ray *2* interferes with ray *1*. If the added distance that ray *2* travels from the wave front AB to the wave front AD is exactly one wavelength of the X ray, the interference is constructive in the new direction that is normal to AD. If the added distance is not exactly one wavelength, or any integral multiple of the wavelength, the interference is destructive, and the ray does not exist. The diffracted beam results from reinforcement at the Bragg angle of scattered radiation from all the atoms in a crystal lattice.

Diffraction may be thought of as coming from lattice planes, which have a high atomic density, that is, planes having Miller indices. These are commonly referred to as *reflecting planes*, as though X rays were reflected by them. This term *reflecting* is unfortunate in that it erroneously suggests a process somewhat akin to light reflection. There are three main differences between reflecting surfaces for light and reflecting planes for X-ray diffraction: First, the diffracted beam from a crystal is built up of rays scattered by all the atoms in the crystal, whereas reflection of light is from the surface only. Second, diffraction of monochromatic X rays occurs at Bragg angles only, whereas light is reflected at all angles. Third, visible-light reflection by a good mirror surface is close to 100 percent efficient, whereas only a very small percentage of an incident X-ray beam is diffracted. However, the usage of reflecting plane for diffraction is well established, although the use of one word to cover two ideas does not reduce the two ideas to one.

Bragg's law for X-ray diffraction may be derived with reference to Fig. 5-23. The distance BC plus CD represents the path difference between the distance traveled by rays *1* and *2*, and this distance must be an integral multiple of the wavelength for reinforcement and diffraction to occur. The distance between the scattering ions A and C is the distance between lattice planes in the crystal structure, and is called the d spacing. Under conditions of diffraction, the angle of incidence θ is geometrically related to the d spacing as

$$\frac{BC}{d} = \sin \theta \quad \text{or} \quad BC = d \sin \theta$$

and

$$\frac{CD}{d} = \sin \theta \quad \text{or} \quad CD = d \sin \theta$$

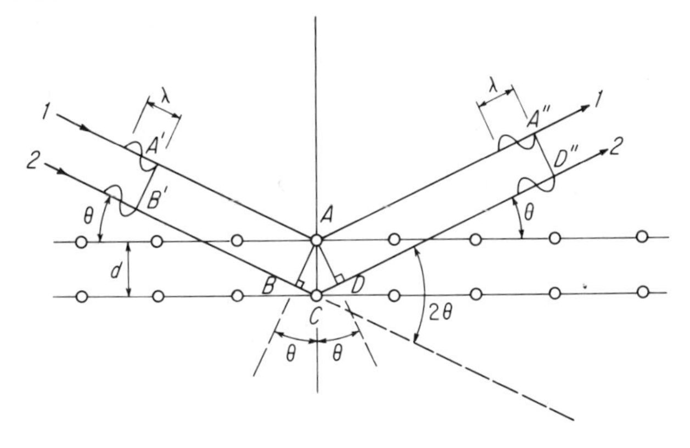

Figure 5-23 Conditions for X-ray diffraction. Diffraction occurs only when the path-length difference BCD is an integral multiple of the wavelength of the incident radiation. The wavelength λ, the interplanar spacing d, and the angle at which diffraction occurs 2θ are related by the Bragg equation $n\lambda = 2d \sin \theta$.

For diffraction to occur, the distance BC plus CD must be an integral multiple of the wavelength λ. By setting the integral multiple of the wavelength $n\lambda$ equal to the path distance, expressed in terms of the d spacing and the angle of incidence θ, the equation known as Bragg's law results:

$$n\lambda = 2d \sin \theta$$

The angle θ, for which diffraction occurs, is known as the *Bragg angle*.

The utility of this relation should be immediately obvious. The Bragg angle can be measured experimentally with any device that detects X rays. The wavelengths of characteristic radiation can be determined from a crystal structure with known d spacing. The d spacing of an unknown crystal structure can be measured from characteristic radiation of known wavelength. And ionic size may be estimated from the measured d spacings of a mineral (Section 1-7).

The number n in the Bragg equation refers to the order of the diffracted beam. For $n = 1$ the path difference is equal to one wavelength of the radiation used, for $n = 2$ the difference is two wavelengths, and so forth. Geometrically, a second-order diffraction is the same as having an additional layer of ions, which is identical to those causing diffraction, but which is located midway between them. Although this extra layer does not exist in the crystal lattice, it is convenient to visualize it as being there. When this sort of conceptual tool is used, the order of diffraction n is incorporated into the value for the d spacing. Thus, the second order of diffraction for the (100) plane comes from a hypothetical plane called the (200), and the third order of diffraction for the (111) plane becomes the (333) plane.

There are geometric limitations inherent in the Bragg equation. The values for the sine of an angle are always less than 1. Therefore, by rearranging the equation, it becomes obvious that $n\lambda/2d$ is also less than 1. For first-order defraction, for which $n = 1$, it follows that λ must be less than $2d$. The d spacings of crystals are mostly less than 3 Å, and, therefore, radiation must be less than 6 Å in wavelength to be useful.

FURTHER READINGS

BURNS, R. G., *Mineralogic Applications of Crystal Field Theory*. New York: Cambridge University Press, 1970.

CULLITY, B. D., *Elements of X-ray Diffraction*. Reading, Mass.: Addison-Wesley Publishing Company, Inc., 1956.

HARTSHORNE, N. H., and A. STUART, *Practical Optical Crystallography*. London: Edward Arnold (Publishers) Ltd., 1964.

HERZBERG, G., *Atomic Spectra and Atomic Structure*, 2nd ed. New York: Dover Publications, Inc., 1944.

OLDENBERG, O., *Introduction to Atomic Physics*, 2nd ed. New York: McGraw-Hill Book Company, 1954.

ZUSSMAN, J. (ed.), *Physical Methods in Determinative Mineralogy*. New York: Academic Press, Inc., 1967.

6

The Phase Rule and Phase Diagrams

A popular, although undoubtedly futile, pastime is to debate about who has made the single greatest achievement in a particular field of endeavor. Neither science in general, nor mineralogy in particular, is free from such interesting comparisons. Be that as it may, the contributions of Josiah Willard Gibbs to scientific thought have probably had as much impact on modern science and technology as those of any other individual. He, in one elegant publication, shaped the twentieth century course of experimental mineralogy, and that of metallurgy, ceramic science, physical chemistry, and materials science in general, as well.

The influential paper, "On the Equilibrium of Heterogeneous Substances," appeared in two parts, in 1876 and 1878, in the *Transactions of the Connecticut Academy*. There it languished for many years, not only because of the relative provinciality of the journal, but more because its mathematical rigor was beyond the training of most of the chemists and mineralogists of the time. It was eventually translated into German by Ostwald (1891) and into French by LeChâtelier (1899). Both translators gave as their reason for undertaking that task the importance of the theoretical questions posed, but at that time not yet worked out experimentally. Many scientists have elaborated on the work, including Roozeboom and Schreinemakers, but Gibbs' original work remains the ultimate arbiter in questions of the science of materials.

This chapter is devoted to two themes. First, the phase rule is derived to show the rigor and the complete generality of its statements about matter that is in equilibrium. Second, graphical representations of states of matter are introduced in the form of phase diagrams by means of which associations and origins of minerals may be represented and analyzed. Hypothetical systems of components A, B, and C are used to illustrate the geometrical properties of phase diagrams, but the particular systems have been selected to correspond to real systems of mineralogic importance (Chapter 7).

6-1. Phases, Components, and Intensive Variables

Before proceeding with a theoretical analysis of the conditions which limit the number of minerals that can be found in association with one another, rigorous definitions of the concepts of phase, component, and intensive variable are necessary.

A *phase* is a homogeneous, physically distinct, and mechanically separable part of a system. A phase may be solid, liquid, or gaseous. A mineral is an example of a solid phase, and each mineral species in an assemblage, or system, is a separate phase. Grain boundaries between minerals of the same species are not taken into account, because the theoretical treatment given to a mineral species is the same whether the mineral is a single crystal or a mosaic of many crystals. Minerals are homogeneous, by definition, and they are physically distinct. They are mechanically separable on the basis of their differing physical properties.

A liquid or a gas is also a phase. There is only one gas in any system, because part of the definition of a *gas* is that it is completely miscible with all other gases. There may be more than one liquid phase in a system, since some liquids mix and others do not. Alcohol and water can be mixed in any proportion, and, therefore, form a single phase. Oil and water do not mix; they form a two-phase system. (The statement that alcohol and gasoline do not mix is a legal and moral one, but not a physical one.)

The concept of a *phase* is an easy one to grasp, for it reflects a distinction that is based on everyday experience. This is not true of the concept of a *component*. The concept of *component* is not identical with that of chemical element, chemical compound, or constituent. An element or a compound may be a component, but not necessarily so. Components are only those constituents the concentration of which can undergo independent variation in the different phases. The components of a system are chosen from the constituents as the smallest number necessary to represent all phases in a system.

The concepts of phase and component may be illustrated with a few examples. Consider a flask of boiling H_2O. There is nothing in the flask but liquid water, one phase, and water vapor, a second phase. If the flask is sealed off and the heat removed, there remain the same two phases. As the flask cools the proportions of the phases will change and the pressure will change, but the number of phases remains at two. Within the flask there is a one-component system, since H_2O represents the composition of each phase.

If the flask of water is chilled sufficiently, crystals of ice will form. The number of components remains at one, while the number of phases has gone from two to three. Further removal of heat from the flask results in an increase in the proportion of ice at the expense of the other two phases. The temperature and pressure remain the same as long as there are the three phases of the single component H_2O present in the flask. This is the *triple point* (t.p.) for water.

Consider the mineral olivine. It is a crystalline solution with a composition that is intermediate between fayalite and forsterite. The single phase, olivine, can be represented by the proportions of the two end members of the crystalline solution present. These end members, Mg_2SiO_4 and Fe_2SiO_4, are the two components of the system. If a crystal of olivine is heated until it begins to melt, two phases, one crystalline and the other liquid, coexist. They have different compositions, but each is represented in terms of the two components.

Calcite, when heated, dissociates into lime and carbon dioxide gas according to the chemical equation

$$CaCO_3 \rightleftharpoons CaO + CO_2$$

Three phases coexist when this dissociation is partly complete. These phases do not have the same chemical composition, but the compositions of the phases are not independently variable. Thus, the equation given above represents a system that contains three chemical compounds and three constituents, but which is represented in terms of two components. If the two components CaO and CO_2 are chosen, the third constituent may be defined by the following chemical equation:

$$CaCO_3 = CaO + CO_2$$

Similarly, if $CaCO_3$ and CaO are chosen, the third constituent may be defined by

$$CO_2 = CaCO_3 - CaO$$

(Note that negative quantities of the components may be used to define the constituents of the phases in a system.)

Other variables necessary to defining a system include mass, volume, heat content, temperature, pressure, and chemical potential. These variables are differentiated according to whether or not they are divisible among the parts of the system. Clearly, if a system is divided, the mass, volume, and heat content are apportioned out to the different parts of the system. These are called *extensive* variables. *Intensive* properties are not divisible. Dividing a system does not alter the temperature of the different parts, or the pressure and chemical potentials. The analysis of mineral associations, which is set forth in the next section, is concerned with the number of different phases that can coexist, and, therefore, deals with intensive variables only.

Pressure is defined as force per unit area, and is measured, in metric units, in bars (1.02 kg/cm^2) or kbars (1000 bars). Actual measurement of pressure is usually indirect. The measurements include use of the elastic constant, which relates force to displacement of a particular metal, and the use of piezoelectric materials, which relate pressure to electric resistance. Pressure-registering instruments utilizing these materials are calibrated to dead-weight testing devices and to defined phase transformations (Table 6-1).

Table 6-1

PRACTICAL PRESSURE SCALE

Transition	*Pressure (kbars)*	*Temperature (°C)*
Mercury, m.p.	7.569	0
Bismuth, I–II	25.50	25
Thallium, I–II	36.7	25
Barium, I–II	55	25
Bismuth, III–V	77	25

Although *pressure* is simple in concept and difficult in actual measurement, *temperature* is a rather complex subject on both counts. *Temperature* is usually defined as dependent upon the amount of heat in a particular body. However, *heat* is generally defined in terms of relative temperature, so that the definitions are circular. Most temperature measurements are in terms of other properties of a system. In the thermodynamics of ideal gases, the temperature, pressure, and volume of a system are related according to the equation $PV = nRT$. Thus, the pressure or the volume is arbitrarily held constant, and temperature can be measured, with a gas thermometer, in terms of the other variable. The electrical conductivity of many substances is dependent upon temperature, so that electrical resistance of a heated wire can be used to calculate its temperature. A thermocouple operates because two different metals have electric potentials which vary differently with change in temperature such that there is a measurable electrical potential between them when they are joined. The optical pyrometer relates flame temperature to flame color.

In precise modern work, certain phase transformations occur at precisely measured temperatures, and these transformations are used to define a temperature scale (Table 6-2). Temperature values between these reference points are established by interpolation on the basis of one of the principles given in

Table 6-2

INTERNATIONAL PRACTICAL CELSIUS TEMPERATURE SCALE[a]

Transition	*Degrees C*
Oxygen, b.p.	−182.97
Water, m.p.	0.00
Water, t.p.	0.01
Water, b.p.	100.00
Zinc, m.p.	419.505
Sulfur, b.p.	444.6
Silver, m.p.	960.8
Gold, m.p.	1063.0
Palladium, m.p.	1552
Platinum, m.p.	1769

[a]Pressure for all, except the triple point of water, is assumed to be 1.013 bars (1 standard atmosphere).

the preceding paragraph. (Above the melting point of Au there is some controversy about precise definitions, and the reader should look for specific statements defining points in this range.)

The concepts behind the notion of *chemical potential* are beyond the scope of this book, and the reader is referred to more advanced texts in geochemistry or physical chemistry. To a first approximation, chemical potential may be thought of as the tendency of a component or a constituent of a system to go from one phase to another. This is analogous to electrical potential, which is a measure of the tendency for electrons to flow from one point to another. A useful fiction would be to think of a chemical "pressure" that must be equalized.

Even this very simple statement about chemical potential allows stipulation of one of the fundamental limitations of the theoretical study of mineral assemblages. The systems to be considered throughout the remainder of this chapter are assumed to be at equilibrium. *Equilibrium conditions* exist when pressure, temperature, and chemical potential are uniform throughout the system. The idea of uniform pressure and temperature is easy to comprehend.

Consider again the three systems discussed under phases and components, for the purpose of illustrating what is meant by chemical potential and equilibrium. In the example of boiling water in a closed and sealed container, the chemical potential of water in the vapor phase is equal to that in the liquid phase. At equilibrium, exactly the same amount of water evaporates as condenses, and the system is in balance. At the triple point, the chemical potential of H_2O is the same in the solid, liquid, and vapor phases. The amount of heat in the system can alter the volume or the relative proportions of the phases, but these are extensive variables.

Olivine in equilibrium with its melt has a composition different from the melt, but the fayalite component of the liquid solution has the same chemical potential as the fayalite component of the crystalline solution, and the forsterite component also has the same chemical potential in both phases. The partitioning of the two components between the two phases may be different, but the potentials are the same when the system is in equilibrium.

At equilibrium in the system $CaO–CO_2$, the CO_2 has the same chemical potential both in the gas and in the calcite. The CaO has the same chemical potential both in the lime and in the calcite.

In summary, one condition of equilibrium is that there be no gradients of temperature, pressure, or chemical potential. As a consequence, the system has no tendency to change as long as the intensive conditions of equilibrium do not change.

There are two distinct kinds of equilibrium, *homogeneous* equilibrium and *heterogeneous* equilibrium. The first, which is the one students become familiar with in chemistry laboratory courses, refers to equilibrium within a single phase. In introductory chemistry laboratories this is most generally water or some other solvent, but the definition does not restrict the concept to fluid phases. *Heterogeneous* equilibrium refers to a multiphase system in which the chemical potential of each constituent is the same in every phase. Mineral associations are treated in terms of heterogeneous equilibria.

Four statements may be made about heterogeneous equilibria, on the basis of the above definition and discussion. First, equilibrium is sensitive to change in external intensive conditions. If, for example, the temperature or the pressure of the system is changed from outside, there will be a compensatory change in the equilibria of the system. Second, the concentrations of the various constituents in the several phases are independent of time. That is, if left alone, the compositions of the phases will not change. Third, equilibrium is independent of the masses or the volumes of the phases. An example of this is a saturated salt brine, wherein the composition of the brine is the same whether there be 1 mg or 1 kg of solid halite in equilibrium with it. Finally, heterogeneous equilibrium is approachable from both directions by reversible change. This concept may be illustrated with the one-component system H_2O. Water vapor at 101° and at 1 standard atmosphere has no liquid in equilibrium with it. If cooled by infinitely small increments, a cooling that could be halted and reversed at any step along the way, liquid water appears at exactly 100° and the temperature remains constant so long as there is any water vapor present (remember that pressure, not volume, is held constant). Conversely, liquid water at 99° and 1 standard atmosphere has no vapor in equilibrium with it. If heated isobarically at infinitely small increments, that is, reversibly, the vapor appears at exactly 100°. As more heat is added, the temperature remains constant and only the proportion of liquid to vapor changes.

The word *system* has been used throughout this section without explicit definition. As defined by the American Ceramic Society, a system is any portion of the material universe which can be isolated completely and arbitrarily from the rest of the universe for consideration of the changes that may occur in it under varying conditions. In specific terms, a system is identified with reference to its components. The systems H_2O, CaO–CO_2, and Mg_2SiO_4–Fe_2SiO_4 have been referred to for illustrative purposes in this section. In general terms, systems may be referred to by the number of components they contain. Thus, H_2O is a one-component system, whereas CaO–CO_2 and Mg_2SiO_4–Fe_2SiO_4 are two-component systems. Systems of 3, 4, . . . , n components exist.

6-2. The Phase Rule

The *phase rule* is an absolutely rigorous and completely general statement about the numbers of different phases that may coexist under conditions of heterogeneous equilibrium. The question of whether or not real mineral assemblages form under equilibrium conditions is a practical one and is discussed in Chapter 7. The theoretical treatment of equilibria among phases, known as the *phase rule*, is the topic of this section. The phase rule was worked out in a purely formal and mathematical way by Gibbs.

A system, in the specific sense set forth in Section 6-1, is completely defined if all its chemical and physical properties are known. Since chemical and physical properties are related by known laws, knowledge of one allows the calculation of others in accordance with these laws. Returning to the example of the system H_2O, the intensive properties of pure water vapor, such as its density or its thermal conductivity, cannot be known until the intensive variables of temperature and pressure are stated. Put in another way, the temperature and the pressure of water vapor can be changed arbitrarily over a wide range, and there will still be only water vapor in the system. Similarly, a system containing both liquid and vapor is indeterminant as to its intensive physical and chemical properties. However, in this case, stipulating either pressure or temperature fixes the other as long as both phases are present. The same may be observed for liquid water and ice or for water vapor and ice. However, at the triple point of water—where ice, water, and vapor coexist—simple stipulation of the fact that all three phases are present at equilibrium fixes for all time the temperature, the pressure, and the chemical potential of the components of the system. All other intensive properties of each phase are also fixed.

In the system H_2O, when one phase is present, two variables must be arbitrarily fixed to define the system. Such a system is said to have *two degrees of freedom*, or to be *divariant*. When two phases are present, only one variable is arbitrary, and the system has *one degree of freedom* and is *univariant*. There are *no degrees of freedom*, and the system H_2O is *invariant* at the triple point. In general, the *degree of freedom*, the *variance*, of a system is the number of intensive variables that must be specified arbitrarily before the system is defined. The phase rule is a completely general statement that relates the variance of a system to the number of components and the number of phases.

Consider a system of C different components existing in P different phases. The question to be answered is: How many degrees of freedom F does the system have? The composition of each phase P in a system of C components is defined by stipulating $C - 1$ of its components (the last one being arrived at by difference). Since each phase is defined by $C - 1$ variables and there are P phases, there is a total of $P(C - 1)$ composition variables. Two other variables, temperature and pressure, must be added, which brings the total to $PC - P + 2$ independent variables.

There have to be as many equations of state relating the variables as there are variables in order to define uniquely the state of a system. (This is a simple algebraic requirement that n equations are needed to solve for n unknowns.) If there are fewer equations than variables, the system has degrees of freedom, and the values for some intensive properties must be stipulated in order to define the system.

The next problem is to determine the number of equations of state that exist for any system. Under conditions of equilibrium, the chemical potential of each component is the same in all phases. Therefore, by selecting one phase that contains all the components as a standard phase, the chemical potentials of all the phases can be related to those of the standard phase by an equation of state. For a system of P phases there exist $P - 1$ restricting conditions for each component. Since there are C components, there are $C(P - 1)$ equations of state in all.

Since the number of degrees of freedom F, or variance, of a system is the difference between the number of independent variables and the number of equations of state,

$$F = (PC - P + 2) - (CP - C) = C - P + 2$$

which is the phase rule.

The system H_2O now can be analyzed in terms of the phase rule. In the case of a single phase—be it ice, water, or vapor—P and C are both 1 and the equation becomes $F = 1 - 1 + 2$ or $F = 2$. As has been shown, both pressure and temperature must be stated in order to define the system. If two phases are present, such as water and vapor, the equation becomes $F = 1 - 2 + 2$ or $F = 1$. The system is then univariant, and stipulation of either pressure or temperature defines the state of the system. When all three—ice, water, and vapor—coexist, the equation becomes $F = 1 - 3 + 2$ or $F = 0$, and the system is invariant.

It should be noted, by way of a summary, that the derivation of the phase rule is completely general and without reference to any chemical or mineralogic reality. And this is what makes it such a powerful tool in studying mineral associations.

In dealing with real mineral systems that have originated by natural processes, a purely pragmatic restriction may be made. Most mineral systems are formed over a range of temperatures and pressures, so that P and T are not fixed for any real mineral assemblage. Thus, most mineral assemblages are formed under conditions where $F = 2$. If $F = 2$, then $P = C$. This latter relation has been called the *mineralogic phase rule*, and was first stated by V. M. Goldschmidt.

Simple chemical or mineralogic systems, that is, systems with four components or fewer, are most easily represented graphically on what is known as a *phase diagram*. The phase diagram for the system H_2O is shown in Fig. 6-1. It is common practice to plot temperature as the horizontal axis increas-

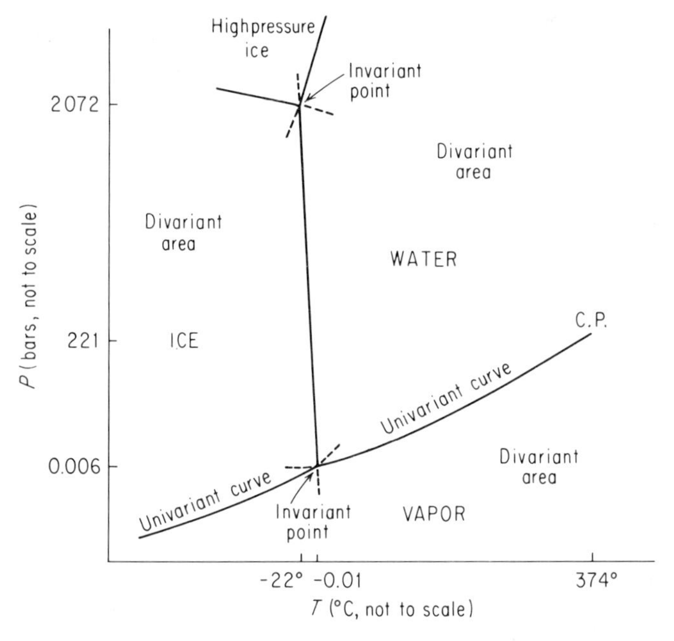

Figure 6-1 Partial phase diagram for water. Four phases are shown. Where only one phase appears, both pressure and temperature may be varied independently without changing the number of phases present. Where two phases coexist along a univariant curve, either pressure or temperature, but not both, may be varied without changing the number of phases. At the triple points where three phases coexist, neither the pressure nor the temperature may be changed without causing one of the phases to disappear. Metastable extensions of univariant lines are dashed.

ing to the right and pressure as the vertical axis increasing to the top of the page. Many geologists, in an attempt to represent phase relations more *realistically*, plot pressure as increasing to the bottom of the page in simulation of pressure increasing with depth in the earth. Confusing as it may seem at the outset, the reader must learn to deal with both conventions, simply because both conventions are used.

Referring again to Fig. 6-1, note that conditions of divariant equilibria are represented by areas in the phase diagram, univariant equilibria by lines, and invariance by a point, in this case the triple point in the one-component system H_2O. The line that indicates solid–vapor equilibria extends to absolute-zero temperature and zero pressure. The line which indicates solid–liquid equilibria extends to another triple point between liquid, normal ice, and a high-pressure polymorph of ice. The line indicating liquid–vapor equilibria extends out to a *critical end point*, where the distinction between liquid and vapor ceases to exist. Water is referred to simply as a *fluid* beyond its critical end point.

Each univariant line has a metastable extension beyond an invariant point. Metastability exists because some additional energy is required to nucleate, or to begin, a new phase. The rules of heterogeneous equilibria also apply to metastability in the absence of the nucleation energy to start the formation of the stable phase. When a univariant line terminates at an invariant point, its metastable extension *must* extend into the area of the phase which is *not* present on that univariant line.

6-3. Binary Systems

A *binary system* is one in which there are two components as well as the variables, pressure and temperature. By applying the phase rule, it can be seen that invariance ($F = 0$) occurs when four phases ($P = 4$) coexist, univariance ($F = 1$) when three phases coexist ($P = 3$), divariance ($F = 2$) when two phases ($P = 2$) coexist, and trivariance ($F = 3$) when one phase ($P = 1$) exists in the system. Trivariance means that there are three variables which may be varied arbitrarily and independently of one another. The variables are pressure, temperature, and the relative amount of one of the components (the relative amount of the other component is arrived at by difference and is not independently variable). Graphic representation of three independent variables requires three orthogonal coordinates, which can only be represented in three-dimensional space.

One variable, thus, must be held constant in order to portray a two-component system on a two-dimensional sheet of paper. Any of the variables—pressure, temperature, or composition—may be fixed, but the most common procedure is to hold the pressure constant (Fig. 6-2a). The result is an isobaric, polythermal, two-component diagram.

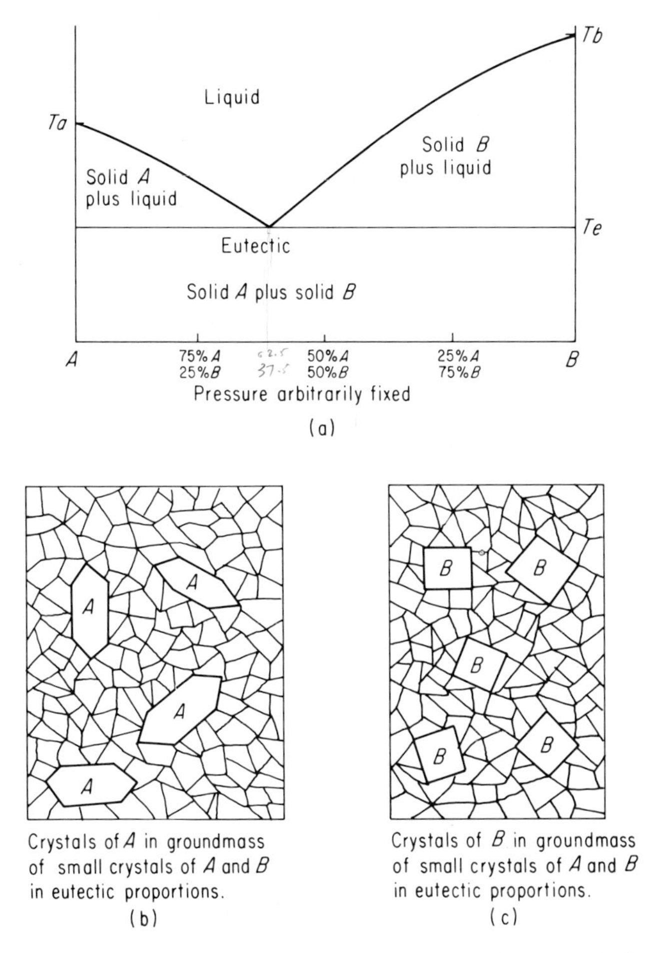

Figure 6-2 Two-component system with a eutectic. (a) Isobaric temperature–composition phase diagram for two-component system *AB* with complete liquid miscibility and no crystalline miscibility. (b) Texture of crystals having a bulk composition on the *A*-rich side of the eutectic. Crystals of *A* in a ground mass of small crystals of *A* and *B* in eutectic proportions. (c) Texture of crystals having a bulk composition on the *B*-rich side of the eutectic. Crystals of *B* in a groundmass of small crystals of *A* and *B* in eutectic proportions.

Two components may be immiscible in the solid state (that is, no crystalline solution), they may be partially miscible, or complete crystalline solution can exist between them. A phase diagram for two totally immiscible crystalline materials is shown schematically in Fig. 6-2a. Pure component *A* is shown on the left and pure component *B* on the right. Different admixtures of the two solids are indicated along the base line between them. The melting point of pure component *A* is shown at *Ta* and the melting point of *B* at *Tb*. Mixtures of *A* and *B* melt at temperatures lower than the melting point of either end member, however, owing to a mutual solution effect.

A liquid with the composition 75*A*, 25*B* (that is, 75 percent *A* and 25 percent *B*) may be cooled until it reaches the liquidus curve between the area labeled *liquid* and the area labeled *solid A plus liquid*, at a temperature between *Ta* and *Te* (the eutectic temperature). At this point, crystals of pure *A* appear in the liquid, thereby shifting the composition of the liquid away from *A* and toward *B*. Continued temperature decrease causes further crystallization of pure *A*, while the composition of the liquid changes along the liquidus curve toward the eutectic. When the temperature has dropped to *Te*, the composition of the liquid, which is in equilibrium with crystals of *A*, has shifted to the eutectic composition—in this case 60 percent *A* and 40 percent *B*. Further removal of heat from the system causes simultaneous crystallization of *A* and of *B* without further change in temperature.

The texture of the resulting mixture of crystals of *A* and *B* might look like the diagram in Fig. 6-2b. The large crystals of *A* are those which crystallized from the melt above the eutectic temperature. They are surrounded by a mat of smaller crystals of *A* and *B*, which crystallized at the eutectic temperature. The composition of the groundmass would be the eutectic composition, 60 percent *A* and 40 percent *B*. The composition of the groundmass plus the larger crystals would be that of the starting melt, 75*A*, 25*B*.

A liquid with composition 75*B* and 25*A* may be cooled until it reaches the liquidus curve at some temperature between *Tb* and *Te*. When this curve is reached on cooling, crystals of pure *B* precipitate from the melt. Further cooling increases the amount of *B* crystals, and drives the composition of the remaining melt toward *A*. Crystallization continues until the temperature drops to *Te*, where the melt has the eutectic composition of 60 percent *A* and 40 percent *B*. At this point, *A* and *B* crystallize simultaneously and isothermally at *Te* until the entire system is crystallized.

The resulting crystalline texture might look like the one shown in Fig. 6-2c, large crystals of *B* encased in a groundmass of crystals of *A* and of *B* in the eutectic proportions. The compositions of the groundmass in Figs. 6-2b and c are identical, 60*A*, 40*B*. The bulk composition in Fig. 6-2c, however, is 25*A*, 75*B*.

The equilibrium melting relations shown in Fig. 6-2a are the geometric reverse of the crystallization sequence. Any mixture of crystals of *A* and of *B* may be heated to *Te*. At that temperature, melting occurs isothermally until either all the *A* crystals or all the *B* crystals have melted to a liquid having the eutectic composition. Depending upon the original composition of the starting crystals, that is, upon which side of the eutectic composition the starting composition lies, there will be either crystals of *A* or crystals of *B* in equilibrium with the eutectic melt. Further increase in temperature results in the decrease in the amount of remaining crystals, with the melt composition following the liquidus curve either toward *Ta* or *Tb*, according to whether crystals of *A* or *B* remain. When the composition of the melt reaches the composition of the original crystal mix, melting is complete.

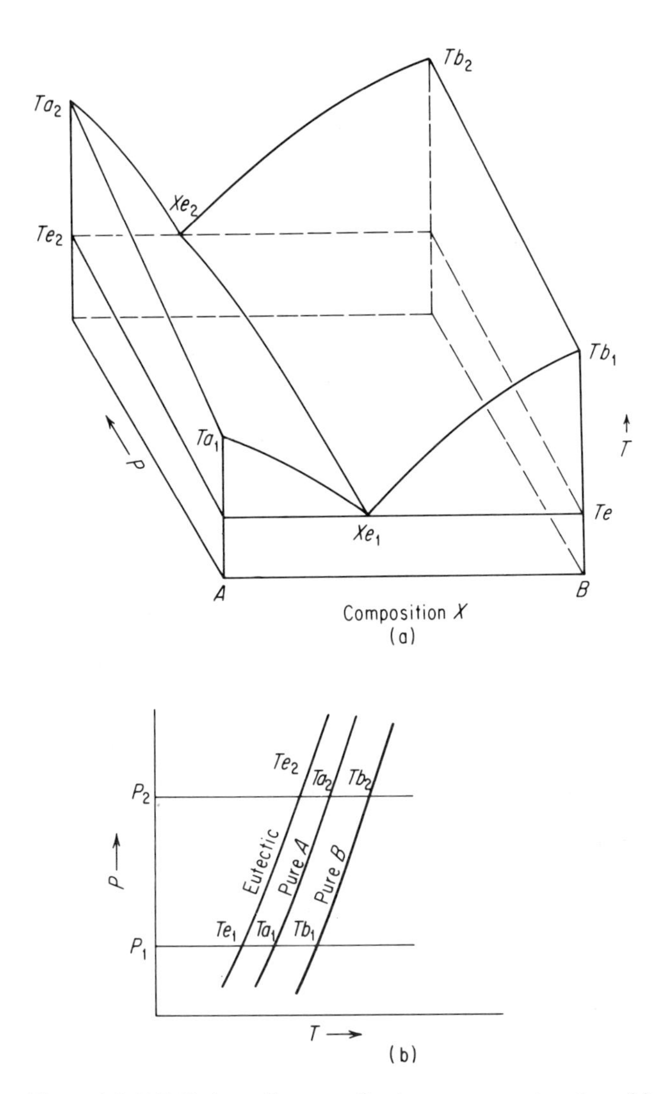

Figure 6-3 *P-T-X* phase diagram of a two-component system. (a) Pressure–temperature–composition prism for a two-component system as in Fig. 6-2. Note that the eutectic composition, as well as the temperature, changes with changing pressure. (b) Projections of the melting curves onto the *P-T* side of the *P-T-X* prism.

So far, the system *A-B* has been considered isobarically, that is, at constant pressure. Change in pressure will change the melting point of the pure components (*Ta* and *Tb*), the eutectic melting point (*Te*), and the eutectic composition. A part of the condensed pressure–temperature–composition prism is shown in Fig. 6-3a. The melting points of *A*, of *B*, and of the eutectic mixture increase with increasing pressure, but not necessarily at the same rate. Also, the eutectic composition shifts toward *A* with increasing pressure. These melt curves may be projected as lines in *P-T* space (Fig. 6-3b).

In two-component systems, as shown on pressure–temperature–composition (*P-T-X*) plots, trivariance is indicated by volumes, divariance by surfaces, monovariance by lines, and invariance by points (not shown for the system *A-B*). The binary systems with complete liquid miscibility and complete crystalline immiscibility, as exemplified in Figs. 6-2 and 6-3, that are of geologic interest include albite–quartz, anorthite–quartz, albite–diopside, anorthite–diopside, anorthite–forsterite, and portions of orthoclase–quartz.

A second type of binary system of geologic importance is one in which there is complete miscibility both in the liquid and in the crystalline phases. The phase relations in such a system are illustrated in Fig. 6-4. As was the case in Fig. 6-2, *Ta* and *Tb* are the melting points of pure *A* and pure *B*, respectively. However, because of the mutual solubility of crystalline *A* and crystalline *B*, the addition of small amounts of *B* into *A* does not bring about melting point depression.

The phase relations during the crystallization of a liquid intermediate in composition between *A* and *B* can be illustrated with a liquid mixture of composition 60*A*, 40*B*. This liquid can be cooled to point *1* on the liquidus curve, which is between *Ta* and *Tb*. At point *1*, crystals appear in equilibrium with the melt, but they are neither pure *B* nor the composition of the melt. Rather, they have the composition shown by *1′*, 25*A*, 75*B*. As the temperature of the system drops, the *B*-rich crystals drive the composition of the liquid along the liquidus curve toward *2*, while the crystals react with the melt and change composition along the solidus curve toward point *2′*. Since the temperature of the liquid and of the crystals is the same, by definition, the crystal composition is related to the liquid composition by a horizontal tie line through the two-phase region. The liquid is used up as the composition approaches *2* on the liquidus, and, finally, crystals with the composition shown at *2′*, which is identical in composition to point *1*, the original composition, are the only phase in the system.

Throughout the above discussion, it has been assumed that the system is always at equilibrium, that crystals with compositions indicated by the solidus curve are continuously reacting with liquids with compositions indicated by points on the liquidus curve. However, since the crystals change composition during the equilibrium crystallization process, there is the alternative possibility that they are somehow separated from the system so that they do not react with the liquid.

The process of fractional crystallization is illustrated in Fig. 6-4b. It is assumed that the crystals do not react with the liquid, but become effectively isolated from the system. This may be accomplished, in reality, by having the denser crystals settle out of the lighter liquid as soon as they form, or by having the crystallization occur at a more rapid rate than that

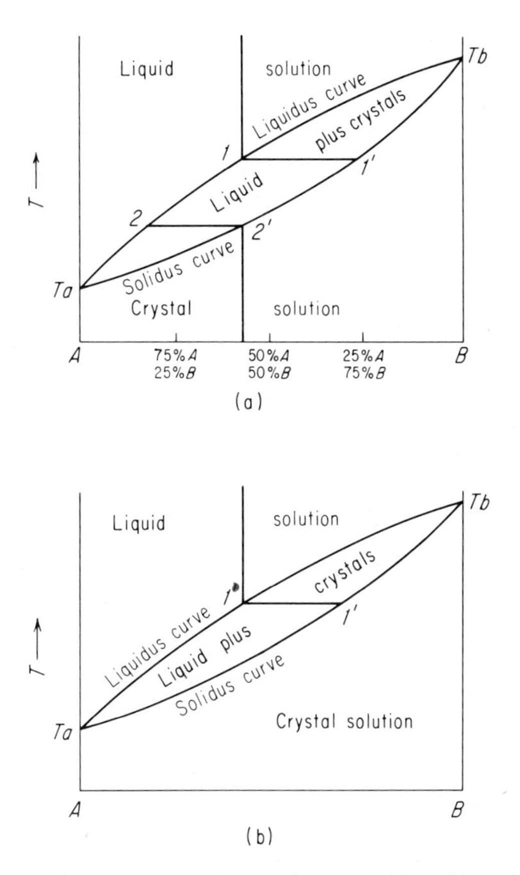

Figure 6-4 Binary system with complete miscibility of both liquid and crystal phases. (a) A liquid of composition 60*A*, 40*B* is cooled to the liquidus curve at point *1*, where it is in equilibrium with crystals of composition at point *1′*. On cooling the crystals react with the melt such that the liquid composition follows the liquidus curve and the crystal composition follows the solidus curve. When liquid composition reaches point *2*, the crystal composition reaches *2′*, the original composition of the melt, and the last melt is used up. (b) Fractional crystallization. Crystals of composition *1′* form from liquid of composition *1*. If the crystals are continuously separated from the system, the liquid composition follows the liquidus curve all the way to *Ta*.

at which the already formed crystals can react with the melt. Either way, isolation of the crystals from the melt effectively changes the composition of the system.

Beginning again with a liquid of composition 60*A*, 40*B*, cooling to point *1* (Fig. 6-4b) causes the precipitation of crystals having a composition represented by point *1′* (25*A*, 75*B*). Subsequent cooling and further crystallization of *B*-rich crystals drive the composition of the melt along the liquidus curve. However, once formed, the *B*-rich crystals do not react with the melt, and the melt is depleted in *B* at a higher rate than if the crystals reacted with it. In effect, the composition of the system shifts progressively toward pure *A*. Crystallization continues with falling temperature until the last liquid has the composition of pure *A* at a temperature of *Ta*, and crystallization is complete. The resulting crystals are not in mutual chemical equilibrium, and range in composition from 25*A*, 75*B* to 100*A*, 0*B*. The average composition of all the crystals, however, is that of the original melt, 60*A*, 40*B*.

Both crystallization under complete equilibrium conditions and crystallization with complete fractionation are relatively rare in natural systems. Most crystallization of real mineral systems takes place along some course that is intermediate between equilibrium crystallization and total fractionation (Section 7-2).

Under conditions of heterogeneous equilibrium, melting follows a path that is the geometric reversal of the crystallization sequence, as was the case with immiscible crystals. Crystals with composition 60*A*, 40*B* begin to melt when the temperature is raised to point *2′* (Fig. 6-4a). The first liquid has a composition represented by point *2*. With further rise in the temperature of the system, melting progresses, with the composition of the melt following the liquidus curve from *2* to *1*, and the composition of the crystals changing along the solidus curve from *2′* to *1′*. At the temperature represented by *1-1′*, the last bit of crystalline matter with a composition at point *1′* dissolves into liquid having the composition *1*, the original composition. This is equilibrium melting.

Fractional melting, which could occur if the liquid were squeezed out of the crystalline phase as soon as it formed, would follow a different path. Because there would be no *A*-rich liquid for the crystals to react with, increase in temperature would cause the crystals to change composition along the solidus curve until the last crystal left would have the composition of pure *B*. Therefore, liquids produced by fractional melting range in composition from *2* (Fig. 6-4a) to pure *B*.

The preceding discussion has assumed phase reactions taking place under isobaric conditions. Increase in pressure would raise the temperature of melting of both pure *A* and pure *B*, as well as the temperatures all along the liquidus–solidus loop.

Binary systems with complete miscibility both of the liquid and of the crystalline components that are of geologic interest include albite–anorthite and forsterite–fayalite.

Between the two extremes of total crystalline immiscibility and complete crystalline solution there is the intermediate possibility of partial crystalline miscibility. This is represented by a *solvus dome* in the phase diagram (Fig.

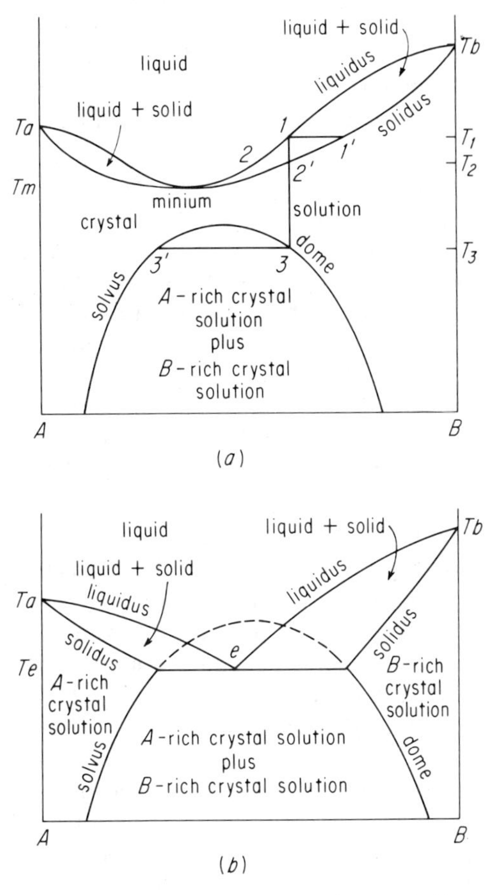

Figure 6-5 Binary systems with a solvus dome. (a) Binary system with a subsolidus solvus dome and a minimum in the liquidus–solidus loop. Equilibrium crystallization of a liquid at point *1* gives crystals with a composition at point *1′*. Subsequent crystallization proceeds toward point *2* for the liquid and point *2′* for the solid, where the last of the liquid is used up. Further cooling brings the crystal solution to the solvus at point *3*, where crystals with a composition at point *3′* begin to separate from the parent phase. (b) Binary system with a solvus intersecting the solidus–liquidus loop. The minimum is converted into a eutectic. Further raising of the solvus evolves a phase diagram such as that in Fig. 6-2.

6-5). There is a two-phase region under the dome and a crystalline solution region outside the dome. At any given temperature, the composition of the two phases, one an *A*-rich crystalline solution and one a *B*-rich crystalline solution, is given by the intersection of the isothermal line and the two sides of the solvus dome. The top of the solvus is thermally below the solidus in Fig. 6-5a, but intersects the solidus and the liquidus in Fig. 6-5b.

Liquidus and solidus curves with a minimum (instead of a eutectic) are associated with a subsolidus solvus (Fig. 6-5a). Paths of crystallization on either side of the minimum are similar to the crystallization paths discussed under complete crystalline miscibility (Fig. 6-4) and may be either fractional or equilibrium.

Consider the crystallization of a melt of composition 40*A*, 60*B* (Fig. 6-5a). When the melt has been cooled to the temperature of the liquidus for that composition (T_1), crystals (30*A*, 70*B*, point *1′*) appear in equilibrium with the melt. Further equilibrium crystallization changes the composition of the liquid along the liquidus from point *1* to point *2* (50*A*, 50*B*), and the composition of the crystalline phase along the solidus from point *1′* to point *2′* (60*A*, 40*B*), at which point all the liquid is used up.

On cooling the crystals from T_2 to T_3, no further reaction takes place. At T_3 the composition of the crystals intersects the solvus dome and a second crystalline phase (70*A*, 30*B*) exsolves from the originally homogeneous crystalline phase. On further cooling, the compositions of the two crystalline phases change along the solvus. Both crystalline phases are crystalline solutions, but the solvus dome represents a miscibility gap between an *A*-rich crystalline solution and a *B*-rich crystalline solution.

The binary system of greatest geologic interest with phase relations similar to those displayed in Fig. 6-5a is the high-pressure part of the system orthoclase–albite. However, this system is not completely analogous, since water is a component of the melt (Section 7-2).

Intersection of the solvus with the liquidus and solidus curves changes the minimum into a eutectic (Fig. 6-5b). Further raising of the solvus or lowering of the liquidus and solidus curves would swing both the solvus lines and the solidus lines out toward the pure *A* and pure *B* boundaries until the phase diagram approaches the example shown in Fig. 6-2a.

So far the discussion has excluded binary compounds. The existence of a binary compound which melts congruently (that is, to a liquid of the same composition) is illustrated in Fig. 6-6. The result of having an intermediate compound A_2B is to divide the system *A-B* into two systems, $A\text{-}A_2B$ and $A_2B\text{-}B$. Each of these subsystems may then be treated separately and may show any of the binary phase relations that have been illustrated in the foregoing figures. Systems of geologic interest that have binary compounds include periclase–(forsterite)–(enstatite)–quartz, and corundum–(kyanite, andalusite, sillimanite)–(mullite)–quartz.

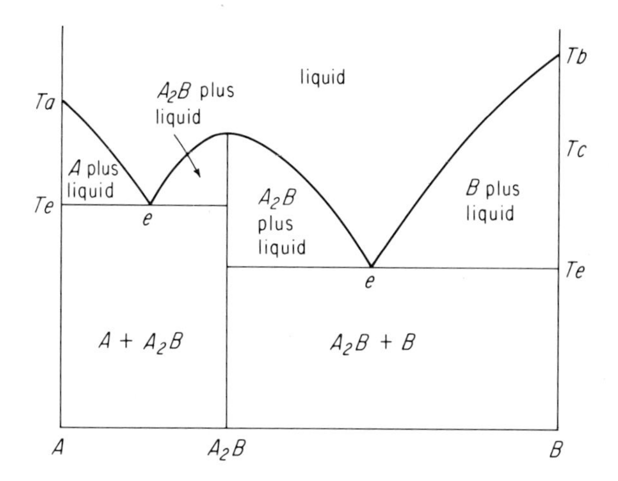

Figure 6-6 Binary system with congruently melting binary compound. A congruently melting binary compound permits the system *A-B* to be treated as two separate systems, A-A_2B and A_2B-B.

There is also the possibility that the binary compound does not melt congruently; that is, it dissociates and melts incongruently (Fig. 6-7). The binary compound A_2B, when heated to the temperature *Tp*, dissociates to pure *A* plus a liquid of composition 50*A*, 50*B*. Further heating causes the progressive melting of *A*, while the melt shifts composition along the liquidus curve toward higher temperature and the original composition.

Consider a liquid with the composition 80*A*, 20*B*. Upon cooling to the liquidus curve, crystals of pure *A* separate from the melt, driving the composition of the melt toward *B*. This continues until the temperature drops to *Tp*, the temperature of the peritectic, which represents the upper thermal stability limit of the compound A_2B. At this point the peritectic liquid reacts with the crystals of *A* to form crystals of A_2B. The liquid is used up before the crystals of *A* are, the result being a mixture of crystals of *A* with crystals of A_2B.

If the original liquid has the composition 60*A*, 40*B*, the crystals of *A* will be used up in the peritectic reaction before the liquid is. After that happens, the temperature can drop along the liquidus with further crystallization of A_2B until the composition of the liquid reaches the eutectic, where A_2B and *B* crystallize simultaneously.

The systems showing incongruent melting and peritectic relations of chief geologic interest are leucite–(orthoclase)–quartz and periclase–(magnesite)–carbon dioxide.

Many other types of binary systems with different phase relationships are known, and many of them are of significant geologic interest. However, those

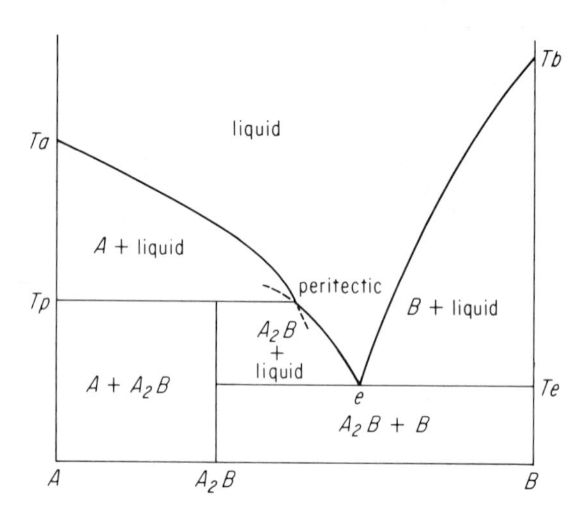

Figure 6-7 Binary system with incongruently melting binary compound. The binary compound A_2B dissociates at the peritectic temperature to crystals pure A and liquid of the peritectic composition. Liquids richer in A than A_2B have crystals of pure A in equilibrium with melt above the peritectic temperature. At the peritectic temperature the liquid reacts with the crystals of A to form a mixture of A and A_2B. Liquids with composition between A_2B and the peritectic form crystals of pure A above the peritectic temperature. At the peritectic temperature the liquid reacts with the crystals of A to form A_2B plus liquid. Further crystallization takes the liquid composition to the eutectic.

covered in the preceding pages represent most of the types involving common rock-forming minerals, and they suffice for the discussion on the genesis of these minerals in Chapter 7.

6-4. Liquidus Surfaces in Ternary Systems

Although it is possible to represent ternary compositions on two-dimensional paper, the variables of pressure and temperature require either four dimensions for full representation of phase relations or simplified sections and projections. The latter course usually is followed.

Ternary compositions are best represented on triangular graphs (Fig. 6-8). The binary composition sidelines are read in the same manner as binary compositions in binary systems. Thus, point *1* has a composition 60*B*, 40*C* with no *A*, and point *2* has a composition 60*A*, 40*C* with no *B*. Each sideline is 0 percent of the component represented at the opposite corner. Ternary compositions are found by reading the percentage of each component by the distance of the composition from the opposite sideline. Thus, point *3* represents 40*A*, 60*B*, 10*C*, point *4* represents 70*A*, 20*B*, 10*C*, and point *5* represents 30*A*, 20*B*, 50*C*. Note that the sum of the three percentages must equal 100 percent.

One method of representing melting and crystallization relations in ternary systems is to superimpose isothermal contour lines, which represent the liquidus surface, on the ternary graph of the compositions (Fig. 6-8b). The boundary binary systems are similar to those illustrated in Fig. 6-2a, where total immiscibility of the crystalline phases is assumed. That is, the boundary binaries *A-B*, *B-C*, and *A-C* in Fig. 6-8b are similar to the binary *A-B* in Fig. 6-2a, as viewed from above. The liquidus surface for ternary compositions is depicted by means of isothermal contour lines. Each binary eutectic is the head end of a thermal valley, which terminates in the ternary eutectic.

Some of the relationships included in a liquidus surface diagram may be illustrated with the crystallization sequence of selected compositions. The composition represented at point *1* (Fig. 6-8b) is 10*A*, 50*B*, 40*C*. On cooling to 900°, the liquid composition intersects the liquidus surface, and crystals of pure *B* precipitate in equilibrium with the melt. Further cooling increases the amount of *B* crystals, which drives the composition of the liquid directly away from the *B* corner of the ternary diagram. The liquid composition continues to change in this direction until the composition intersects the thermal valley at 15*A*, 35*B*, 50*C* at 860°, between the binary eutectic on the *B-C* sideline and the ternary eutectic. At this point, crystals of *C* coprecipitate with crystals of *B* in a ratio such that the composition of the liquid is driven away from the *B* and the *C* corners and down the thermal valley. The trend continues until the liquid composition reaches the eutectic composition (40*A*, 30*B*, 30*C*) at 760°, at which point *A*, *B*, and *C* crystallize simultaneously and isothermally until all the liquid is gone.

Crystallization of the composition represented at point *2* (Fig. 6-8b) begins at 920°. Crystals of pure *A* drive the liquid composition directly away from the *A* corner. The liquid composition reaches the thermal valley at 780°, between the *A-C* binary eutectic and the ternary eutectic. The simultaneous crystallization of *A* and *C* drives the liquid composition down the thermal valley toward the ternary eutectic, where *A*, *B*, and *C* crystallize isothermally until all the liquid is gone.

Equilibrium melting is the geometric reverse of equilibrium crystallization. However, fractional melting in a ternary system with one ternary eutectic produces quite different results. Consider the fractional melting of composition *1* (Fig. 6-8b). Melting begins at the temperature of the ternary eutectic. If liquid is continuously removed from the system, melting proceeds

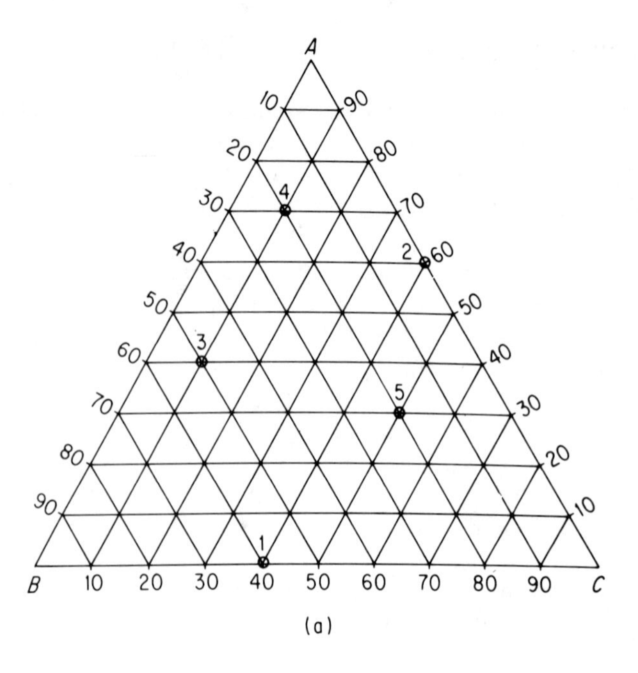

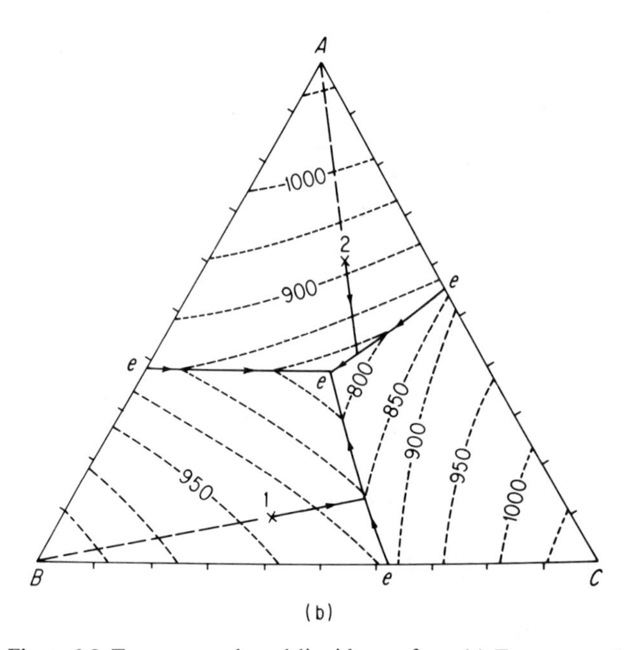

Figure 6-8 Ternary graph and liquidus surface. (a) Ternary graph used for representing three-component systems. Compositions: (1) 0*A*, 60*B*, 40*C*; (2) 60*A*, 0*B*, 40*C*; (3) 40*A*, 50*B*, 10*C*; (4) 70*A*, 20*B*, 10*C*; (5) 30*A*, 20*B*, 50*C*. (b) Ternary liquidus surface with ternary eutectic. The liquidus surface is shown by isothermal contour lines. Thermal valleys connect each of the binary eutectics with the ternary eutectic. See text for crystallization sequences.

isothermally, while the crystal mixture changes composition along a straight line away from the ternary eutectic and toward the *B-C* join. When the crystal composition reaches the *B-C* join, melting ceases completely while the temperature rises to that of the binary (*B-C*) eutectic. At that point, isothermal fractional melting resumes until all the crystals of *C* are consumed. The temperature of the system then rises without further melting until the temperature of fusion of pure *B* is attained, and the melting of the remaining *B* crystals occurs isothermally. Thus, fractional melting produces three distinct liquids—one having the composition of the ternary eutectic, a second having the composition of the binary eutectic, and a third of pure *B*.

Most ternary systems that are of geologic interest are more complicated than the hypothetical system *A-B-C*, as outlined in the preceding paragraphs. Perhaps the simplest example is that of silica [SiO_2]–anorthite [$CaAl_2Si_2O_8$]–pseudowollastonite [$CaSiO_3$].

If there is complete crystalline miscibility in one of the binary systems (*B-C*), and total crystalline immiscibility in the other two (*A-B* and *A-C*), there is but one thermal valley in the liquidus surface of the ternary system (Fig. 6-9a). The boundary binary systems *A-C* and *A-B* are of the same type as that shown in Fig. 6-2a, whereas the binary *B-C* is similar to that shown in Fig. 6-4. Crystallization of the composition represented by point *1* begins at 1020° with the precipitation of crystals of pure *A*, which drives the liquid composition along the liquidus surface directly away from the *A* corner until it intersects the thermal valley at approximately 960°. At this point, the crystals of *A* are joined by crystals of a *BC* crystalline solution, which is richer in *C* than is the liquid (Fig. 6-4). With falling temperature, the composition of the liquid follows the thermal valley while the *BC* crystalline solution crystals react with the liquid, becoming increasingly *B* rich in the process, until all the liquid is used up.

A liquid composition represented at point *2* in Fig. 6-9a would begin to crystallize at 980° with the precipitation of *BC* solution crystals having a composition represented by point *3*. Further cooling drives the composition of the liquid away from the *B-C* sideline toward the thermal valley. Because reaction between the crystals and the liquid causes the crystals in equilibrium with the liquid to become more *B* rich, *BC* crystals change composition from point *3* to point *4*, while the liquid changes composition along the curve labeled *equilibrium*. When the liquid composition reaches the thermal valley, both crystals of *BC* and crystals of pure *A* precipitate simultaneously, and the liquid composition follows the thermal valley. Crystallization becomes complete when the composition of the *BC* crystals arrives at point *5*, which is colinear with the original composition (point *2*) and the *A* corner. The composition of the final liquid will lie somewhere in the thermal valley below 900°.

As was the case in the example sketched in Fig. 6-4, fractional crystallization is possible where there is crystalline miscibility. If the *BC* solution crys-

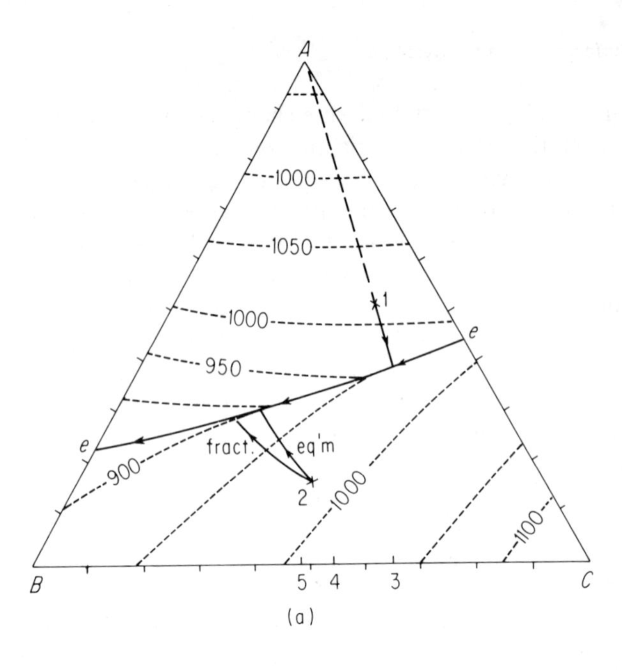

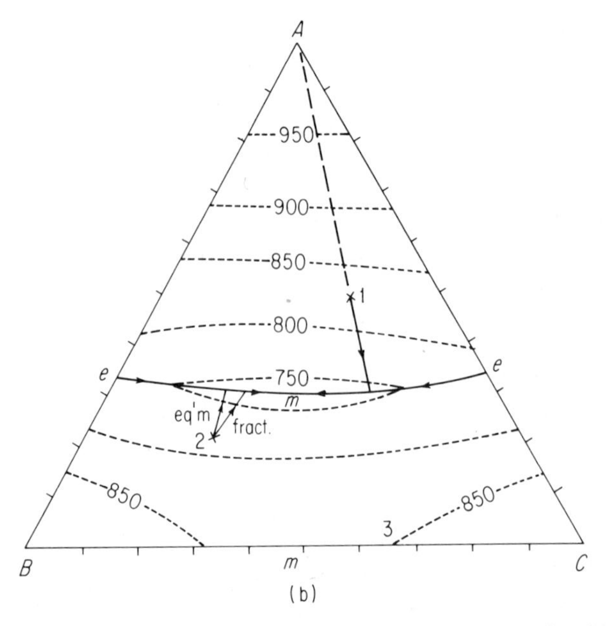

Figure 6-9 Ternary liquidus surfaces without ternary eutectics. (a) Ternary liquidus surface with a thermal valley. The thermal valley connects the two binary eutectics. Arrows indicate falling temperature. (b) Ternary liquidus surface with a thermal valley with a minimum. Arrows in the thermal valley indicate falling temperatures. See text for crystallization sequences.

tals do not react with the melt, then the composition of the liquid follows the path across the liquidus surface that is labeled *fractional*, while the precipitating crystals of *BC* change composition to a point between *3* and *4*. Again, when the liquid composition reaches the thermal valley, *A* coprecipitates with *BC* crystalline solution. However, the composition of the liquid is driven all the way to the *A-B* binary eutectic, because fractionation concentrates *C* in the early *BC* solution crystals.

It should be noted that either fractional or equilibrium crystallization can occur with composition *1* after the thermal valley has been reached. Systems of geologic interest that have a liquidus surface with this type of thermal valley include diopside–albite–anorthite, the synthetic simulated basalt system.

The system illustrated in Fig. 6-9b has a thermal valley with a minimum on its liquidus surface. The *A-B* and *A-C* sidelines represent binary systems in which there is no crystalline miscibility, while the *B-C* sideline represents a binary system having a liquidus with a minimum above a solvus dome (Fig. 6-5a). Liquids with composition at point *1* begin to crystallize pure *A* at 830°. The composition of the liquid is driven toward the thermal valley by further cooling. At this point, *C*-rich *BC* crystalline solution coprecipitates with the *A*, driving the liquid composition toward the minimum. Reaction of the liquid with the *BC* crystals causes them to change composition until all the liquid is used up, which is when the crystal solution composition reaches point *3* (colinear with point *1* and *A*). Fractional crystallization would permit the final liquid to have the composition of the ternary minimum.

Liquids with the composition represented at point *2* begin to crystallize a *BC* solution crystal having a composition 90*B*, 10*C*. Cooling and equilibrium reaction between liquid and crystals cause the crystals to change composition to 75*B*, 25*C*, while the liquid changes composition along the curve labeled *equilibrium* to the thermal valley. Coprecipitation, then, of *A* and of *BC* solution crystals continues until the liquid is used up and the *BC* crystals have a composition 70*B*, 30*C*. The final liquid composition does not quite reach the ternary thermal minimum.

In the case of complete fractional crystallization, the liquid composition follows the curve labeled *fractional* across the liquidus surface until it reaches the thermal valley where the liquid is in equilibrium with *BC* crystals having composition 65*B*, 35*C*. Continued fractional crystallization drives the liquid composition clear to the thermal minimum, where crystallization is complete. The system albite–orthoclase–quartz is the one having a liquidus surface of similar configuration that is of chief geologic interest.

Thus far, compounds have been ignored in the discussion of ternary systems. In the case of binary systems, compounds having congruent melting permit the system to be treated as two independent binary systems (Fig. 6-6). This being the case, if the compound A_2B melts congruently in the system

A-B-C, the system may be divided into the subsystems A_2B-*B-C* and *A*-A_2B-*C*, each of which may be treated independently.

A congruently melting ternary compound *ABC* (Fig. 6-10a) leads to the division of the system *A-B-C* into three ternary subsystems, *A-ABC-B*, *A-ABC-C*, and *B-ABC-C*. The liquidus surface relationships of each of these subsystems may be replotted as a separate system, as is done with the system *B-ABC-C* in Fig. 6-10b.

Graphical representation and interpretation become more involved in cases in which one or more compounds melt incongruently. The case of a binary compound *AB* melting incongruently is represented in Fig. 6-7. The system illustrated in Fig. 6-11 represents a possible effect of an incongruently melting binary compound on the shape of the liquidus surface of a ternary system. As in the system shown in Fig. 6-8b, each binary eutectic heads a valley that descends thermally to a ternary eutectic. Between the two ternary eutectics is another thermal valley, but one which has a maximum where it crosses the *AB-C* join. The peritectic point of the binary system heads a thermal inflection, which with decreasing temperature evolves into a thermal valley leading to one of the ternary eutectics.

Liquid compositions and crystallization sequences for compositions lying in the *A* field follow the same principles as those outlined for the system diagrammed in Fig. 6-8b. Upon cooling to the temperature of the liquidus surface, crystals of *A* form and drive the liquid composition directly away from the *A* composition corner toward one of the thermal valleys. When the liquid composition reaches the thermal valley, either crystals of *C* or crystals of *AB* come into equilibrium with it, driving the liquid composition to the ternary eutectic, where *A*, *AB*, and *C* crystallize simultaneously and isothermally until all the liquid is consumed.

The primary fields of *AB*, of *B*, and of *C* are each divided by the *AB-C* join. To the *A* side of the join, no crystals of *B* appear in the final assemblage, and to the *B* side, no crystals of *A*. Thus, crystallization sequences on the join and on each side of the join must be considered separately.

Liquids having compositions on the *AB-C* join within the primary field of *C* follow a unique crystallization path. Crystals of *C* drive the liquid composition toward the maximum in the thermal valley between the two ternary eutectics. There *AB* and *C* crystallize isothermally until the liquid is consumed; the liquid composition never reaches a ternary eutectic. In the primary field of *C*, liquids having compositions on the *A* side of the *AB-C* join first yield crystals of *C*, which drive the liquid composition directly away from that corner and toward the thermal valley between the fields of *C* and of *AB*. Then simultaneous crystallization of *AB* and *C* pushes the liquid composition toward the eutectic where *A*, *AB*, and *C* crystallize isothermally.

On the *B* side of the *AB-C* join, equilibrium crystallization of *C* drives the liquid composition to the thermal valley between *C* and *B* or between *C* and

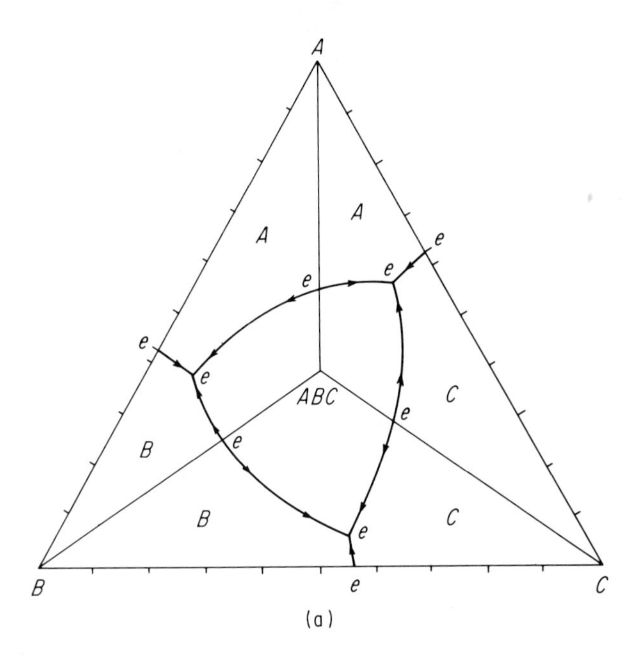

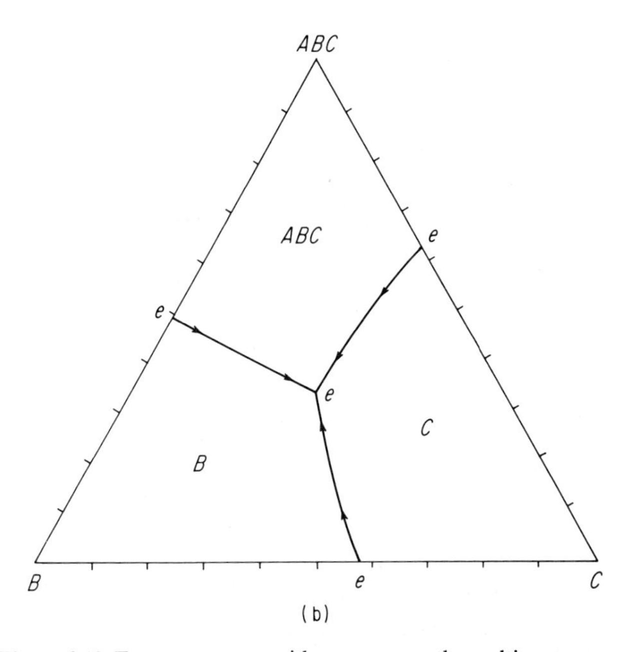

Figure 6-10 Ternary system with a congruently melting ternary compound. A congruently melting ternary compound divides the system into three ternary subsystems. (a) System with three ternary eutectics. (b) One of the three subsystems redrawn to the standard ternary graph.

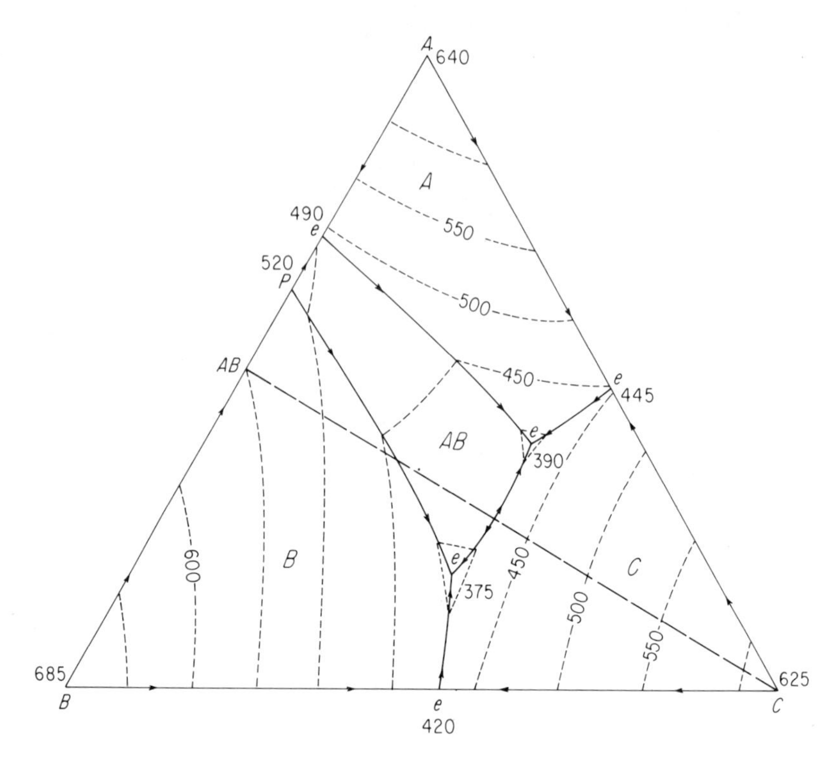

Figure 6-11 Ternary system with an incongruently melting binary compound. Arrows indicate directions of falling temperatures. Liquidus surface indicated by dotted contour lines. Compound *AB* melts incongruently to *B* plus liquid *p*. (After E. F. Osborn, *Personal Communication*, 1962.)

AB, where further crystallization moves the liquid composition to the eutectic of *AB*, *B*, and *C*.

Turning now to the primary field of *B*, on the *B* side of the *AB* join, crystals of *B* drive the liquid composition to a field boundary, where they are joined by crystals of *C* or of *AB*, thence to the eutectic involving *AB*, *B*, and *C*, where the last liquid crystallizes isothermally.

Liquid compositions in the *B* field of primary crystallization, but directly on the *AB*-*C* join, follow another unique crystallization path across the liquidus surface. Crystallization of *B* drives the liquid composition away from the *B* composition corner, toward the thermal inflection between the fields of

B and of *AB* on the liquidus surface. At that point crystals of *AB* become stable. Upon further cooling, the crystals of *B* react with the melt, being resorbed as the amount of *AB* increases. The liquid composition, thus, follows the thermal inflection to the point where it crosses the *AB-C* join and where all crystals of *B* are resorbed by reaction with the liquid. From this point, further crystallization of *AB* causes the liquid to migrate along the *AB-C* join, across the *AB* field, to the maximum in the thermal valley between the two ternary eutectics. There, *AB* and *C* crystallize isothermally until the liquid is consumed.

The field of primary crystallization of *AB* is similarly divided by the *AB-C* join. Liquids starting to crystallize on the *B*-rich side of the join first yield crystals of *AB*, which are subsequently joined by crystals of *C*, until, with falling temperature, the liquid composition migrates to the eutectic, where *AB*, *B*, and *C* crystallize isothermally. Liquids starting to crystallize on the *A* side of the join first yield *AB*, then *C*, and the liquid composition descends to the eutectic involving *A*, *AB*, and *C*, where crystallization becomes complete.

A few generalizations concerning crystallization sequences and results may be observed from the foregoing discussion. First, the primary phase field, which is bounded by thermal valleys or inflections, determines the phase to crystallize first on cooling of a liquid to the liquidus surface. Second, subsequent crystallization always shifts the liquid composition directly *away* from the point representing the crystal composition. Third, the temperature of the liquid never rises during the crystallization sequence. Fourth, there are no abrupt changes in liquid composition—it changes incrementally during crystallization. Finally, regardless of the phase to crystallize initially in equilibrium with the melt, the final assemblage will be made up of the phases defined by the joins between the phases. Stated differently, the final crystal assemblage must have the same composition as the initial liquid, if crystallization proceeded under conditions of heterogeneous equilibrium.

The system illustrated in Fig. 6-12 shows an effect on the shape of the liquidus surface produced by an incongruently melting ternary compound. Very clearly, the location of the primary field of crystallization for the ternary compound, with respect to the joins between the ternary compound and the apices of the triangle, profoundly affects the shape of the liquidus surface. Although only one possibility is illustrated, the principles of crystallization sequences are invariably the same in all cases.

All three sides of the system (Fig. 6-12) are binaries with complete immiscibility of the crystalline phases (Fig. 6-2a). Each binary eutectic heads a thermal valley leading to one of two ternary eutectics in the system. Because the primary field of crystallization of the ternary compound *ABC* does not include the composition *ABC*, there is also a ternary peritectic point. The

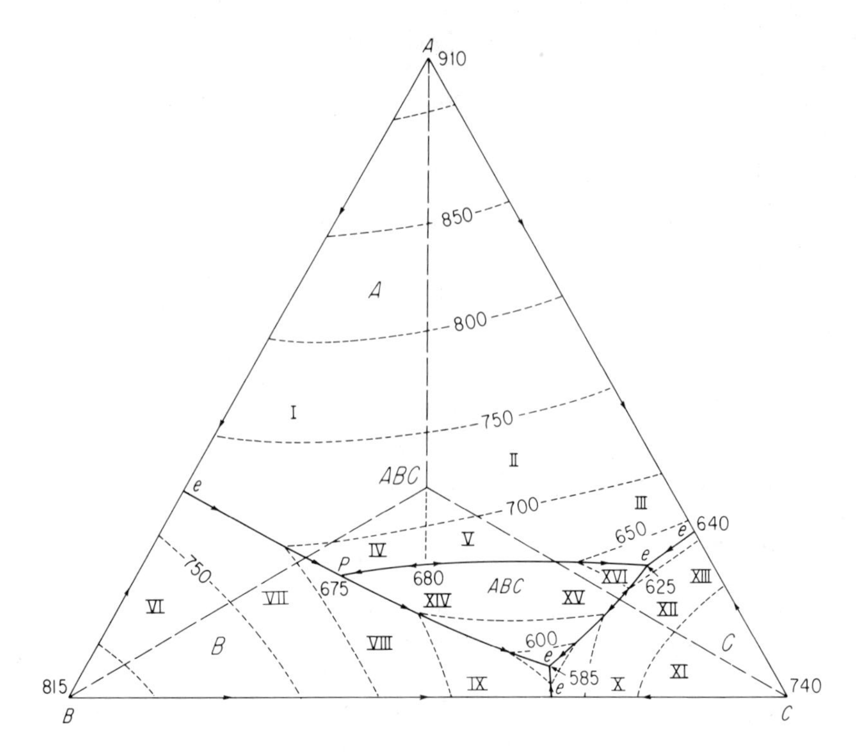

Figure 6-12 Ternary system with an incongruently melting ternary compound. Arrows indicate directions of falling temperatures. Liquidus surface indicated by dotted isothermal contour lines. Compound *ABC* melts incongruently to *A* plus liquid. Crystallization sequences for numbered compositions are given in Table 6-3. (After E. F. Osborn, *Personal Communication*, 1962.)

boundary between the primary field of *A* and that of *ABC*, unlike the thermal valleys that constitute the other field boundaries, is an inflection in its mid-region, but becomes a thermal valley at either end. It has a maximum where it crosses the extension of the *A-ABC* join. Crystallization sequences for selected compositions in this system are given in Table 6-3.

Examples of mineralogically interesting ternary systems with incongruently melting binary phases include carnegieite–kalsilite–silica, with potassium feldspar melting incongruently, and anorthite–forsterite–silica, with pyroxene melting incongruently. The system periclase–corundum–silica has two incongruently melting binary compounds, mullite and clinoenstatite, and one incongruently melting ternary compound, cordierite.

Table 6-3

CRYSTALLIZATION SEQUENCES FOR COMPOSITIONS IN FIG. 6-12

Comp	*Temp* (°C)	*Phases*	*Comp*	*Temp* (°C)	*Phases*	*Comp*	*Temp* (°C)	*Phases*
I	>760	L	II	>725	L	III	>680	L
	760–720	$A + L$		725–670	$A + L$		680–635	$A + L$
	720–675	$A + B + L$		670–625	$A + ABC + L$		635–625	$A + C + L$
	675	$A + B + ABC + L$		625	$A + C + ABC + L$		625	$A + C + ABC + L$
	<675	$A + B + ABC$		<625	$A + C + ABC$		<625	$A + C + ABC$
IV[a]	>690	L	V[a]	>690	L	VI	>780	L
	690–677	$A + L$		690–678	$A + L$		780–720	$B + L$
	677–675	$A + ABC + L$		678	$A + ABC + L$		720–675	$A + B + L$
	675	$A + B + ABC + L$		678–600	$ABC + L$		675	$A + B + ABC + L$
	675–585	$B + ABC + L$		600–585	$B + ABC + L$		<675	$A + B + ABC$
	585	$B + C + ABC + L$		585	$B + C + ABC + L$			
	<585	$B + C + ABC$		<585	$B + C + ABC$			
VII[a]	>740	L	VIII	>690	L	IX	>645	L
	740–690	$B + L$		690–655	$B + L$		645–600	$B + L$
	690–675	$A + B + L$		655–585	$B + ABC + L$		600–585	$B + C + L$
	675	$A + B + ABC + L$		585	$B + C + ABC + L$		585	$B + C + ABC + L$
	675–585	$B + ABC + L$		<585	$B + C + ABC$		<585	$B + C + ABC$
	585	$B + C + ABC + L$						
	<585	$B + C + ABC$						
X	>620	L	XI	>710	L	XII	>680	L
	620–590	$C + L$		710–630	$C + L$		680–650	$C + L$
	590–585	$B + C + L$		630–585	$C + ABC + L$		650–625	$C + ABC + L$
	585	$B + C + ABC + L$		585	$B + C + ABC + L$		625	$A + C + ABC + L$
	<585	$B + C + ABC$		<585	$B + C + ABC$		<625	$A + C + ABC$

Table 6-3 (Cont.)

Comp	*Temp* (°C)	*Phases*	*Comp*	*Temp* (°C)	*Phases*	*Comp*	*Temp* (°C)	*Phases*
XIII	>680	L	XIV	>670	L	XV	>660	L
	680–635	$C + L$		670–645	$ABC + L$		660–600	$ABC + L$
	635–625	$A + C + L$		645–585	$B + ABC + L$		600–585	$C + ABC + L$
	625	$A + C + ABC + L$		585	$B + C + ABC + L$		585	$B + C + ABC + L$
	<625	$A + C + ABC$		<585	$B + C + ABC$		<585	$B + C + ABC$
XVI	>655	L	*ABC*[a]	>720	L	*A-ABC*	(720–910)[b]	L
	655–645	$ABC + L$		720–680	$A + L$		(720–910)–680	$A + L$
	645–625	$C + ABC + L$		680	$A + ABC + L$		680	$A + ABC + L$
	625	$A + C + ABC + L$		<680	ABC		<680	$A + ABC$
	<625	$A + C + ABC$						

[a]Sequences in which the first-formed crystal phase is subsequently resorbed by reaction with the liquid.
[b]Exact temperature of beginning of crystallization depends upon *A* to *ABC* ratio.

6-5. Ternary Isothermal and Isobaric Sections

Although isothermal contouring of the ternary isobaric liquidus surface portrays melting and crystallization relations rather clearly, it does not show solvus domes or any other subsolidus relationships. One method of representing subsolidus relationships in ternary systems is by means of isobaric, isothermal sections through the pressure–temperature–composition hyperprism. All phase relations in ternary systems may be displayed graphically with a series of different isobaric, isothermal sections for different pressures and temperatures.

Although invariance in ternary systems requires the coexistence of five phases, such relations exist only at a specific temperature and pressure, and do not generally appear when temperature and pressure are arbitrarily specified. Univariance requires the coexistence of four phases (for example, the ternary eutectic liquid and three crystalline phases), but it, too, is restrictive and does not appear generally. Thus, three-, two-, and single-phase regions are what appear on a typical isothermal, isobaric diagram.

An arbitrary isobaric, isothermal section through the system *A-B-C*, represented by the liquidus surface diagram of Fig. 6-8b, is shown in Fig. 6-13a. This particular section intersects the liquidus surface above the ternary eutectic, but below the binary eutectics. Thus, the central region of Fig. 6-13a is all liquid, the boundaries of the liquid single-phase region (*a-b-c*) being isotherms on the liquidus surface (Fig. 6-8b). The area outside the liquid region *a-b-c* is all subsolidus.

There are three three-phase regions in the subsolidus (Fig. 6-13a), $A + B +$ liquid c, $A + C +$ liquid b, and $B + C +$ liquid a. The three phases represented at the corners of the three-phase triangles all coexist within these regions, their relative proportions being given by the position of the composition with respect to the three corners. That is, each of the three-phase regions could be replotted on its own ternary graph (Fig. 6-8a).

Each three-phase region is bounded by a two-phase region, $B +$ liquid $a{-}c$, $A +$ liquid $b{-}c$, $C +$ liquid $a{-}b$, $A + B$, $B + C$, and $A + C$. The last three are boundary binaries of the ternary system. The first three two-phase regions are those in which a single crystalline phase coexists with a liquid of variable composition. Single-phase regions exist only for the three components and for the liquid region.

Note that on a ternary isothermal section of the sort diagrammed in Fig. 6-13 adjacent areas differ in number of phases present by only one. Thus, each one-phase area, such as the liquid area, is bounded by two-phase areas, such as $A +$ liquid. Similarly, the two-phase area $A +$ liquid is bounded by the one-phase area liquid and by the three-phase areas $A + B +$ liquid c and $A + C +$ liquid b. The statement of this pattern is called *Schreinemakers' rule*.

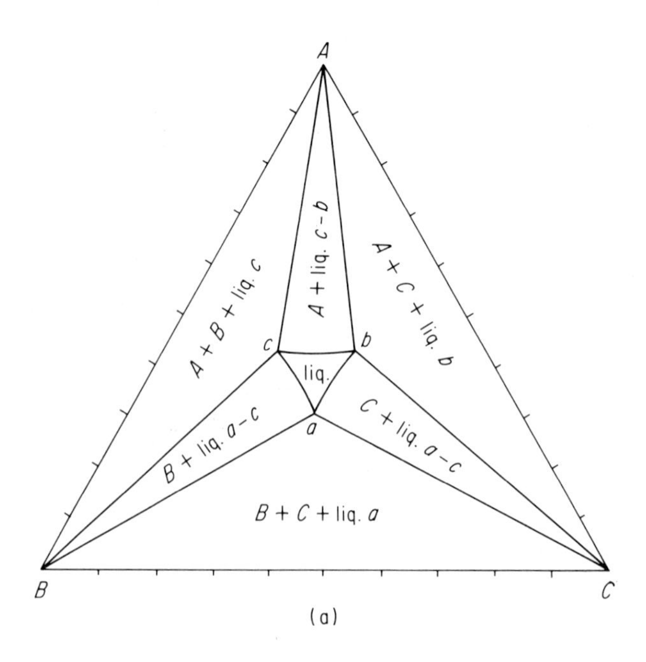

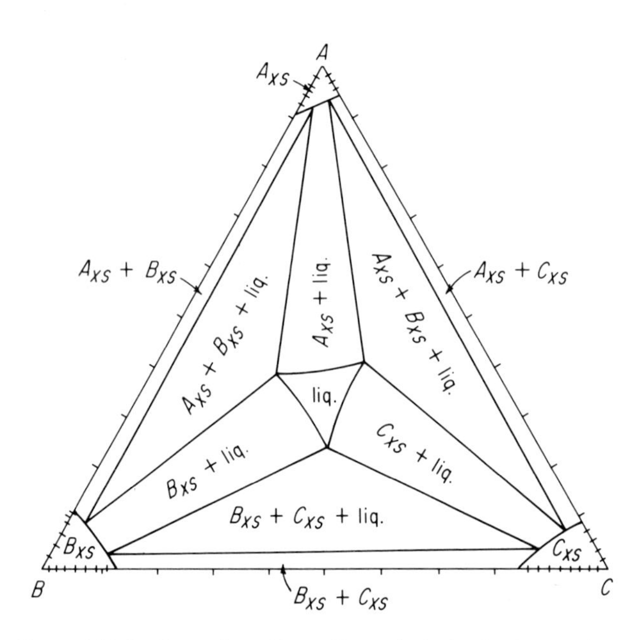

Figure 6-13 Isothermal sections through isobaric temperature–composition prism. (a) System with no ternary crystalline solution. The isothermal section intersects the liquidus surface along the closed curve *abc*, inside the curve being above the liquidus surface and outside the curve being below. (b) System with partial ternary crystalline solution. See text for details.

The liquid region shrinks toward a point and the two-phase regions shrink toward lines as lower temperature sections are taken successively through the isobaric prism. At the eutectic temperature, there is an isobarically invariant point involving the coexistence of four phases, *A*, *B*, *C*, and the eutectic liquid. At all lower temperatures, the ternary two-phase regions disappear, and three crystalline phases coexist throughout the system.

Many components that are of geologic interest have variable compositions, and partial miscibility of the crystalline phases is the norm, rather than the exception. Binary solvus domes are discussed with relation to Fig. 6-5. Ternary crystalline solutions are shown in isothermal, isobaric section in Fig. 6-13b, where they are indicated with hatch marks along the binary. The major difference between the system shown in Fig. 6-13a and that in Fig. 6-13b is that in the latter the binary lines expand into areas and the two-phase areas have both crystal and liquid compositions which are variable. However, where two crystal solutions are in equilibrium with a liquid, the compositions of the two crystal phases and of the liquid phase are unique for the temperature and pressure.

The liquid region reduces to a point in the plane at the eutectic temperature. Three crystalline solution phases of fixed composition may coexist with liquid of the eutectic composition. Below the eutectic temperature, the center of the ternary diagram represents a single three-phase region, and the border areas represent two-phase regions. Although each of the components shows crystalline solution toward each of the other two components, there is a unique component composition for any equilibrium assemblage in the system. Within the two-phase regions, each component composition is uniquely related to the other component composition by a tie line. Only in the one-phase regions of crystalline solution is phase composition independently variable.

6-6. Quaternary System with a Vapor Phase

Quaternary systems in which one of the components is a vapor phase, under the pressure and temperature conditions of interest, are usually treated as pseudoternary systems by projecting quaternary compositions onto the vapor-free base of the composition tetrahedron. Compositions of vapor-free phases are, of course, plotted directly. Components that have the vapor constituent as an essential part of their composition are projected geometrically by extending a line from the vapor corner of the tetrahedron through the component to the point where it intersects the opposite face. The projected phase may be a vapor-saturated liquid phase or a crystalline phase. An example of the latter would be a carbonate mineral in a system where CO_2 is the vapor component.

Although such projections are common practice for the representation of quaternary systems, the reader should keep in mind that the presence of vapor is stipulated. Phase relations in the vapor-free system may be quite different from the projected phase relations of the system with a vapor phase, however similar the two diagrams may appear. For instance, compare the phase diagram for SiO_2 (Fig. 1-18) with a projection of the system SiO_2–H_2O (Fig. 4-1). Furthermore, in considering applications of the phase relations in real mineralogic situations, the composition of the vapor phase must be taken into account. The vapor composition will not be that of the pure vapor end member, and the proportions of anhydrous constituents dissolved in it generally will not be the same as the proportions of those consitituents in other phases with which it is in equilibrium.

Some of the preceding discussion may be illustrated with reference to Fig. 6-14, a schematic polybaric, polythermal phase diagram of the system albite–orthoclase–quartz–water. The ternary faces of the composition tet-

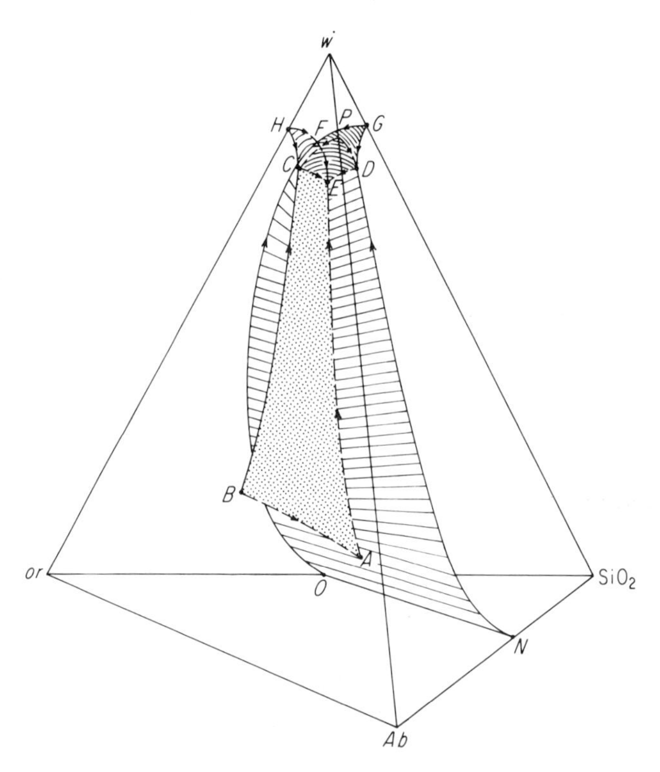

Figure 6-14 Schematic equilibrium phase diagram for the quaternary system albite (*ab*)–orthoclase (*or*)–silica (SiO_2)–water (*w*). (After O. F. Tuttle and N. L. Bowen, *Origin of Granite in the Light of Experimental Studies*, Memoir 74. New York: The Geological Society of America, 1958, Fig. 29.)

rahedron may be considered separately in preparation for study of the quaternary relationships. The phase relations of the anhydrous base are similar to those illustrated in Fig. 6-9b. The curve *ON* in Fig. 6-14 is the same as the field boundary represented by the thermal valley in Fig. 6-9b.

The ternary systems albite–quartz–water and orthoclase–quartz–water are similar to the system outlined in Fig. 6-8b, with a ternary eutectic involving the two silicate phases and ice. Partial miscibility of the silicate phases makes the albite–orthoclase–water face somewhat more difficult to visualize. Point *B* represents the composition of the ternary liquid that is in equilibrium with the binary feldspar at the maximum on the solvus (Fig. 6-5a). Point *C* is the ternary eutectic involving orthoclase crystalline solution, albite crystalline solution, and ice.

The quaternary relationships may be considered with regard to the ternary faces. As the water content of the liquids in equilibrium with feldspar crystalline solution and quartz (*ON*) is increased, a boundary plane is generated in the tetrahedron representing the compositions of silicate melts with dissolved water in equilibrium with the anhydrous phases.

Point *A* represents the composition of the quaternary liquid that is in equilibrium both with the binary feldspar at the maximum on the solvus and with quartz. The plane *ABCE* represents liquid compositions that are in equilibrium with orthoclase crystalline solution and with albite crystalline solution. Point *E* is the quaternary invariant point (eutectic) involving six phases: albite, orthoclase, quartz, ice, water, and vapor. Arrows indicate directions of falling temperature.

The phase relationships in the upper part of the tetrahedron are generally of less mineralogic interest than those in the lower part. However, it should be clear that information is lost when these phase relationships are projected onto the ternary composition diagram of the anhydrous base. The principles outlined in the foregoing discussion apply equally well when the vapor phase is CO_2 or any other volatile constituent.

FURTHER READINGS

EHLERS, E. G., *The Interpretation of Geological Phase Diagrams.* San Francisco: W. H. Freeman and Company, 1972.

FINDLAY, A., A. N. CAMPBELL, and N. O. SMITH, *The Phase Rule and Its Applications*, 9th ed. New York: Dover Publications, Inc., 1951.

GIBBS, J. W., *Scientific Papers*, vol. 1. New York: Dover Publications, Inc., 1961.

LEVIN, E. M., C. R. ROBBINS, and H. F. MCMURDIE, *Phase Diagrams for Ceramists.* Columbus, Ohio: The American Ceramic Society, 1964.

MASING, G., *Ternary Systems.* New York: Dover Publications, Inc., 1960.

RICCI, J. E., *The Phase Rule and Heterogeneous Equilibrium.* New York: Van Nostrand Reinhold Company, 1951.

7

Mineral Genesis

Any individual who has contemplated an apparently endless array of minerals in a museum or in a laboratory may have the erroneous idea that the mixtures of minerals which one may expect to encounter in nature, or to grow in the laboratory, may be even more endless. Study of the wide variety of particles to be found in beach sands or in such synthetic products as concrete and terraza could fortify this notion. In actual fact, the variety in any given mineral assemblage—be it in common rock or in a mining district—is quite restricted in comparison with all the possible combinations of elements.

The narrow restriction in the numbers of different minerals found in most rocks of the crust of the earth derives from two sources: first, the distribution of elements in the crust of the earth, and, second, the reactions that occur between minerals under conditions of heterogeneous equilibrium (Chapter 6). A mineral does not merely exist. Its very existence implies an origin, which can be inferred from the mineral itself, from its composition and minute variations in that composition, from its environs, and from the minerals with which it is associated.

7-1. Element Abundance

A glance at any modern periodic chart of the elements (Fig. 1-4) reveals that there are in excess of 100 different chemical elements. Of these, only 90 appear in the crust of the earth. The transuranium elements have half-lives so short that any which might have been present at the early stages of the formation of the crust of the earth have completely decayed over geologic time. The half-lives of U and Th are sufficiently long that a measurable portion of the primordial U and Th is still present to form minerals. The other elements between Bi and U occur only as daughter products of U and Th, and are found in association with them. Two lighter elements, Tc (number 43) and Pm (number 61), are absent from any naturally occurring assemblage owing to the short half-lives of their isotopes and to their lack of long-lived parent isotopes.

Crustal abundances of the elements are shown schematically in Fig. 7-1. Some interesting observations about the crustal abundances of the chemical elements may be made. Only eight elements (Table 7-1) account for 98.5 percent by weight of the crust of the earth. In terms of atomic percentage, nine out of ten ions are either O^{2-}, Si^{4+}, or Al^{3+}. And O^{2-} alone makes up nearly 94 percent of the volume of the crust of the earth. These figures should lead to the conclusion that most of the bulk of the minerals of the crust are oxides and silicates.

Table 7-1

MOST COMMON ELEMENTS OF THE CONTINENTAL CRUST[a]

Element	*Weight percent*	*Atom percent*	*Volume percent*
O	46.60	62.55	93.77
Si	27.72	21.22	0.86
Al	8.13	6.47	0.47
Fe	5.00	1.92	0.43
Mg	2.09	1.84	0.29
Ca	3.63	1.94	1.03
Na	2.83	2.64	1.32
K	2.59	1.42	1.83
	98.59	100.00[b]	100.00[b]

[a]After B. Mason, *Principles of Geochemistry*, 3rd ed. (New York: John Wiley & Sons, Inc., 1966).

[b]Result caused by rounding off numbers.

A second datum to be noticed about the relative abundances of the chemical elements is that many exotic elements are far more abundant than many technologically common elements. Rb is more common than Ag or Zn, Ce is more common than Ne, and Li is more common than Hg. Au, which was perhaps the first element worked by man, ranks in abundance behind such *space-age* elements as Ga, Ge, Cd, Hf, and all the stable rare earths.

The explanation for this lack of congruence between abundance and familiarity lies in the fact that some ions are similar enough to other more common ions in size and charge that they occur mainly in the minerals of the

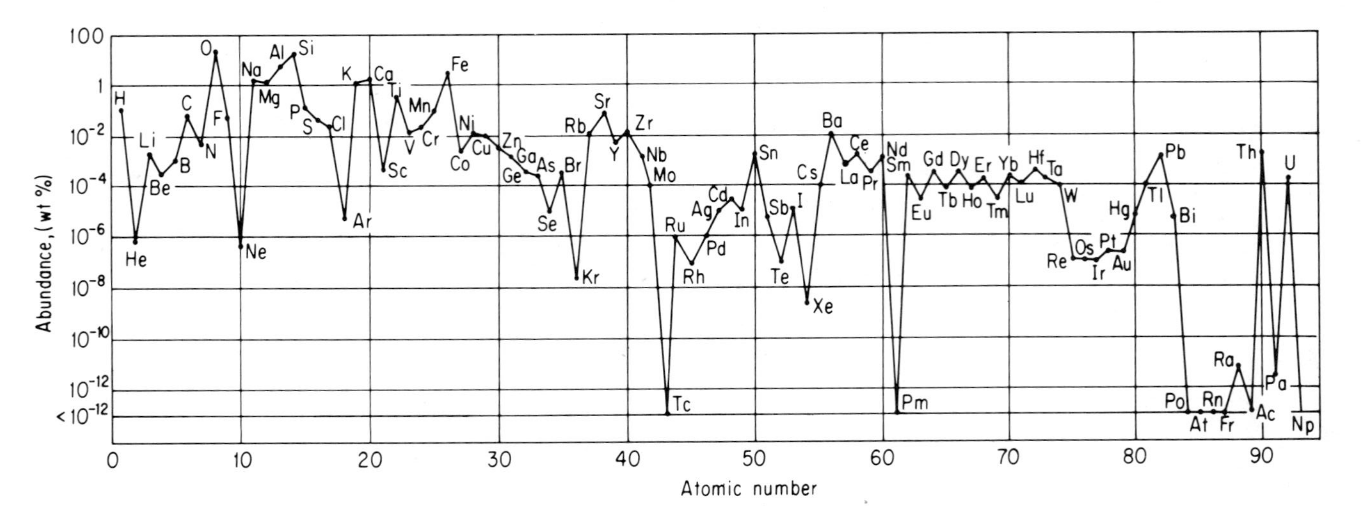

Figure 7-1 Continental crustal abundances of the chemical elements. Only eight elements appear in the continental crust in abundances greater than 1 percent. Note that many space-age elements occur in greater abundances than well-known elements, such as Ag and Au. (After B. Mason, *Principles of Geochemistry*. New York: John Wiley & Sons, Inc., 1966, p. 47.)

common ions, and are rarely concentrated in their own phases. Such elements are *dispersed.* Prime examples of dispersed elements are Ga, which follows Al; Ge, which follows Si; and Hf, which follows Zr. The more familiar elements are those which form their own, easily recognizable minerals, such as Cu, Au, and Pb.

Two of the limiting factors which contribute to the simplicity of mineral assemblages in the crust of the earth are, first, that only a few elements are truly abundant in the crust and, second, that many of the less abundant elements are rarely or never concentrated into their own discrete mineral phases.

7-2. Magmas and Crystallization

A *rock* is a mass of mineral matter, which constitutes an appreciable part of the crust of a planetary body. The term is commonly extended to cover rock samples. An *igneous rock* is one which has crystallized from a *magma*, that is, a melt of composition similar to that of the final rock. (See Chapter 1 for criticism of generic definitions. The generic classification of rocks has led to endless controversy, much of it semantic.)

Interpretation of geophysical data indicates that nearly all the crust of the earth is solid, presumably crystalline, and that the mantle from a depth of approximately 60 to several hundred kilometers contains a small percentage of magma. Local concentrations of heat, sufficient to keep rocks molten, must exist to provide for volcanic activity. Interpretation of various rock textures and structures, plus a large amount of laboratory experimental data, leads to the inference that many rocks have formed within or upon the crust of the earth by crystallization from a magma. These interpretations of field observations and of experimental results also lead to the conclusion that the rules of heterogeneous phase equilibria can apply to such rocks.

As a starting point in the study of rocks, it is convenient to ignore, temporarily, the origin of magmas, and to proceed with a description of their behavior once they have formed and have begun to cool.

Before discussing crystallization trends of magmas, it is necessary to establish to terminology of igneous rocks. Two criteria are used to classify rocks, texture and mineralogy (Fig. 7-2). Texturally, a rock is classified as *phaneritic* if its individual crystals are visible to the unaided eye, and *aphanitic* if they are not. Phaneritic and aphanitic rocks with the same mineral composition are given different rock names. Since grain size is not significant in discussions of heterogeneous equilibria, that distinction is ignored throughout most of the following discussion. In the rocks classified in Fig. 7-2, silicate minerals predominate, and most igneous rocks are predominantly silicate. Magmas that are predominantly carbonate or magnetite crystallize to

Feldspar composition \ Silica content	Quartz >10%	Feldspathoids or quartz <10%	Feldspathoids >10%
or ≳ *ab* >>> *an*	GRANITE rhyolite	SYENITE trachyte	NEPHELINE SYENITE phonolite
ab > *or* > *an*	ADAMELLITE rhyodacite	MONZONITE latite	
ab > *an* > *or*	GRANODIORITE dacite		
ab > *an* >> *or*	QUARTZ DIORITE quartz andesite	DIORITE andesite	
an ≳ *ab* >>> *or*	QUARTZ GABBRO quartz basalt	GABBRO basalt	

Figure 7-2 Classification of silicate igneous rocks. Felsic minerals constitute more than 65 percent of the rocks. Phaneritic (capitals) and aphanitic (lowercase) names are given for various mineral associations. Boundaries between rock types are arbitrary.

carbonatites or to magmatic iron deposits. The most common igneous minerals are listed in Table 7-2.

Silica and alumina are the two most abundant oxides in the crust of the earth. One parameter that is used in the chemical characterization of rocks and their constitutent minerals is *silica saturation*, a rock being oversaturated with silica if quartz occurs as a constituent. The minerals that occur with free quartz in a rock are said to be *saturated* with silica. This may be illustrated with reference to the system $NaAlSiO_4$–$KAlSiO_4$–SiO_2 (Fig. 7-3). Only K-feldspar and albite coexist stably with quartz in this system, and the join between them is the *silica saturation line*. All minerals below that line—nepheline and leucite being petrologically the most important—are silica undersaturated minerals. The rocks in which these minerals occur are also silica undersaturated.

Alumina saturation is also based on feldspar compositions. All rocks in which the mole ratio of (K + Na)/Al is greater than 1 are alumina undersaturated. If the ratio is less than 1, they are oversaturated. Muscovite is an indicator of alumina oversaturation, although biotite is not. The presence of riebeckete or of aegirine indicates that a rock is undersaturated with respect to alumina.

Magmas exist at 1 bar pressure in the form of lava flows, and are inferred from geophysical data to constitute between 0.001 and 1.0 percent of subcrustal rocks, to depths of several hundred kilometers. The magmatic temperature range extends from approximately 650°, a temperature at which

Table 7-2

IGNEOUS MINERALS

Silica	
Quartz,[a] tridymite, cristobalite	SiO_2
Feldspar	
Sanidine, *orthoclase*, *microcline*	$KAlSi_3O_8$
Plagioclase crystalline solution series	
Albite	$NaAlSi_3O_8$
Anorthite	$CaAl_2Si_2O_8$
Feldspathoid	
Nepheline	$KNa_3(AlSiO_4)_4$
Leucite	$KAlSi_2O_6$
Sodalite	$Na_8(AlSiO_4)_6Cl_2$
Cancrinite	$Na_8(AlSiO_4)_6(HCO_3)_2$
Olivine crystalline solution series	
Fayalite	Fe_2SiO_4
Forsterite	Mg_2SiO_4
Tephroite	Mn_2SiO_4
Pyroxene	
Enstatite	$MgSiO_3$
Hypersthene	$(Mg, Fe)SiO_3$
Augite	$Ca(Fe, Mg)(SiO_3)_2$
Aegirine	$NaFe(SiO_3)_2$
Spodumene	$LiAl(SiO_3)_2$
Amphibole	
Hornblende	$NaCa_2(Mg, Fe, Al)_5(Al, Si)_8O_{22}(OH)_2$
Riebeckite	$Na_2Fe_3^{2+}Fe_2^{3+}Si_8O_{22}(OH)_2$
Mica	
Muscovite	$KAl_2(AlSi_3O_{10})(OH)_2$
Biotite	$K(Mg, Fe)_3(AlSi_3O_{10})(OH)_2$
Lepidolite	$KLi_2Al(Si_4O_{10})(OH)_2$
Accessory	
Apatite	$Ca_5(PO_4)_3(OH, F, Cl)$
Corundum	Al_2O_3
Sphene	$CaTiSiO_5$
Fluorite	CaF_2
Zircon	$ZrSiO_4$
Magnetite	$FeFe_2O_4$
Ilmenite	$FeTiO_3$
Pyrite	FeS_2

[a]The names of the more common minerals of each group are set in italics.

granite can be melted in the laboratory, to something greater than 1200°, as actually measured in lava pools in active volcanos. The possible range of pressure and temperature in the crust of the earth is given in Fig. 7-4. The observation that 10 km of rock burial equals approximately 3 kbars may be used to assist in the interpretation of the diagram.

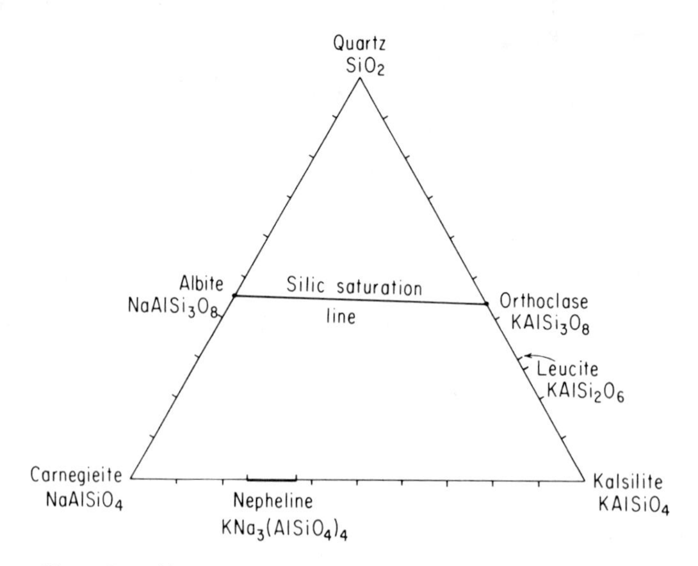

Figure 7-3 Silica saturation in the system carnegieite–kalsilite–quartz. Rocks containing free quartz are defined as oversaturated with silica; those containing carnegieite, nepheline, kalsilite, or leucite are undersaturated.

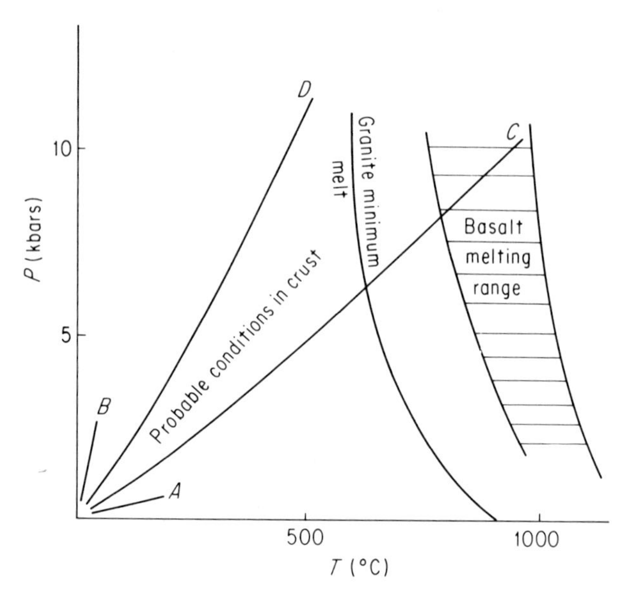

Figure 7-4 Pressures and temperatures in the earth's crust. The temperature of the rocks in the earth's crust increases with depth. Curves *A* and *B* represent extreme values for measured thermal gradients, *A* being for areas of recent vulcanism and *B* for geosynclinal areas of rapid sediment deposition. Probable conditions range between *C* and *D* in most continental areas. The granite minimum melting curve and the basalt melting range are shown for reference.

Basalt is the most abundant igneous rock in the crust of the earth, making up almost the entirety of the oceanic crust and occurring as extensive flows on the continental crust. Gabbro, its phaneritic counterpart, is found intrusive into crustal rocks as sills, dikes, and plutons in many parts of the crust. The essential minerals for the formation of basalt–gabbro are plagioclase and pyroxene; and olivine, magnetite, or quartz may occur as accessories. Consideration of the minor mineral constituents is not necessary to a qualitative analysis of how a magma of this composition crystallizes. Therefore, crystallization of a basaltic magma may be described in terms of the three components: albite [$NaAlSi_3O_8$], anorthite [$CaAl_2Si_2O_8$], and diopside [$CaMgSi_2O_6$]. Iron is the only major element in basalt that is ignored in this simplified basalt system (Fig. 7-2). However, it is present mostly as Fe^{2+}, which means that it behaves like Mg^{2+} (Section 1-13 and Table 1-8). Thus, the simplified system shown in Fig. 7-6 is a reasonable first approximation of real basalts.

Before discussing the ternary system *ab-an-di*, the three boundary binary systems, *ab-an*, *ab-di*, and *an-di*, must be understood. Figure 7-5a displays the phase relations in the binary system *ab-an* at 1 bar pressure. Although the details of the phase diagram are different at higher pressures with water dissolved in the liquid, the general relationships are the same and may be applied both to intrusive and extrusive rocks. There is complete liquid and

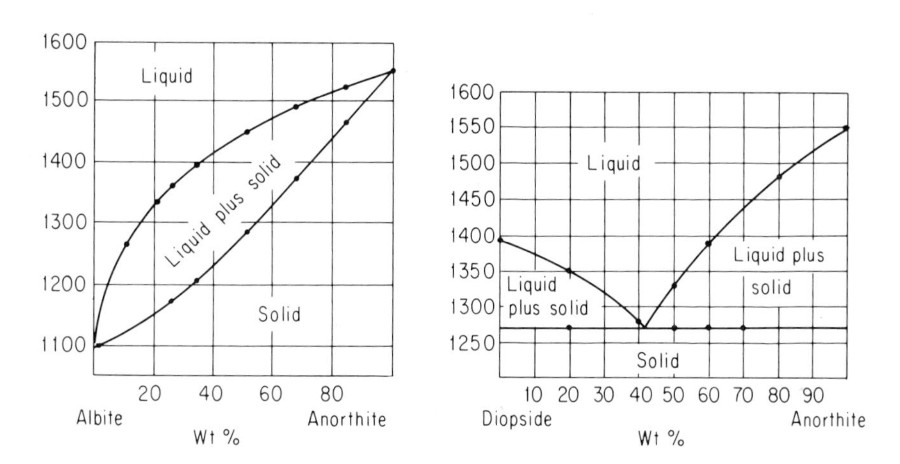

Figure 7-5 Binary systems albite–anorthite and anorthite–diopside. (a) Phase relationships for plagioclase at atmospheric pressure. The upper curve is the liquidus, representing the highest temperature at which solids can exist, and the lower curve is the solidus, representing the lowest temperature at which liquids exist. (b) Melting curves for diopside–anorthite at atmospheric pressure. The eutectic temperature *E* represents the lowest temperature at which liquids exist in this system. Crystallization drives liquid compositions toward this point. (After Bowen, 1956.)

crystal miscibility between *ab* and *an* in this system at the temperatures and pressures under discussion (compare with Fig. 6-4 for crystallization sequences).

The phase relationships in the system *an-di* (Fig. 7-5b) are shown to be a simple eutectic between two completely immiscible crystalline end members (compare with Fig. 6-2 for crystallization paths). The phase relations between *ab* and *di* are similar to those between *an* and *di*, except that the eutectic is at almost pure *ab* at a temperature of 1100° at 1 bar pressure.

Having considered the boundary binary systems, it is now possible to consider the ternary crystallization relations in the system *ab-an-di*. Figure 7-6 shows the liquidus surface by means of isothermal contour lines (compare

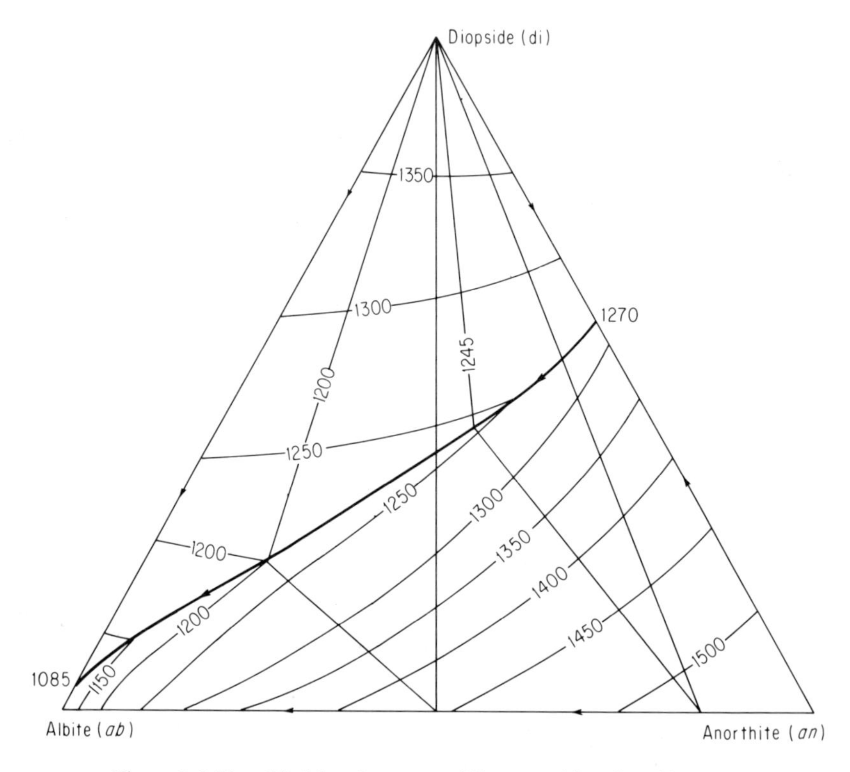

Figure 7-6 Simplified basalt system albite–anorthite–diopside. The liquidus surface for the simplified basalt system is shown by isothermal contour lines. A thermal valley, which represents the boundary between the field of primary crystallization of plagioclase and the field of primary crystallization of diopside, extends from the *di-an* eutectic to the *di-ab* eutectic. Two isothermal three-phase triangles are shown. (After Bowen, 1956.)

with Fig. 6-9a). A thermal valley divides the diagram into two fields of primary crystallization. On the *ab-an* side of the thermal valley, a plagioclase crystallizes first and drives the liquid composition toward the thermal valley. Similarly, melt compositions on the *di* side are driven toward the thermal valley by crystallization of *di*. A melt composition in the thermal valley is in equilibrium with two crystals, pure *di* and a plagioclase (see Fig. 6-9a for crystallization paths).

The main point to be derived from the *simplified basalt system ab-an-di* is that liquids with a composition of an almost pure alkali feldspar can be derived in small amounts from the fractional crystallization of a basaltic magma. The addition of other components, such as are found in real basaltic magmas, changes the details, but does not alter the basic pattern of the crystallization process. The addition of FeO would alter the composition of the pyroxene. The addition of K_2O would alter the composition of the residual melt, in that it would not take part in the phases formed in the system, although it could substitute in the final alkali feldspar. Silica, in excess of the amount needed to form plagioclase and diopside, would also be concentrated in the melt phase.

Combining observations in the field with interpretation of the liquidus surface in the system *ab-an-di*, N. L. Bowen proposed a *reaction principle* and the reaction series that bears his name (Fig. 7-7). During crystallization of a magma of basaltic composition, the first crystals to appear in the melt upon

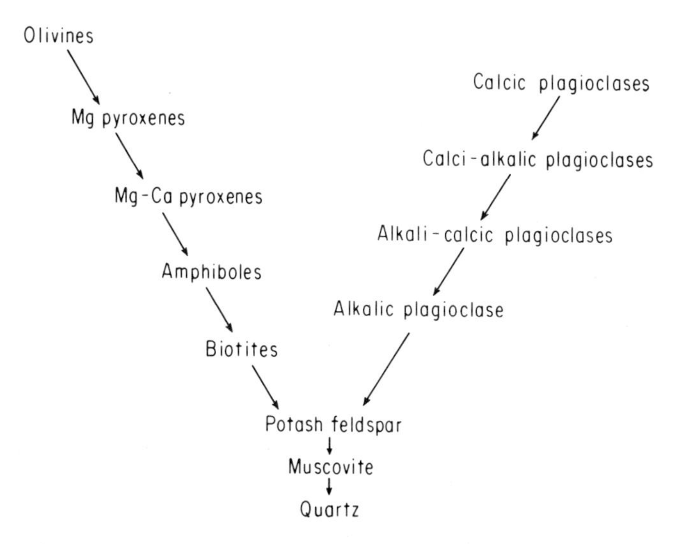

Figure 7-7 Bowen's reaction series. Fractional crystallization of a magma with falling temperature produces a series of minerals. The first minerals to crystallize with falling temperature are at the top (After Bowen, 1956.)

cooling to liquidus temperatures are those at the top of the two series—olivine and calcic plagioclase. These two minerals head the discontinuous and the continuous reaction series, respectively. Upon further cooling and continued crystallization, the minerals at the top of the two series react with the residual melt, to form the next mineral in each of the two series. Reaction between early crystals and the residual melt continues until the melt is consumed, as would be the case under equilibrium conditions, or until the bottom of the series is reached, if crystallization is fractional.

The evolution of mineral assemblages and the rocks that they comprise are illustrated in Fig. 7-8. As minerals high in the reaction series are separated

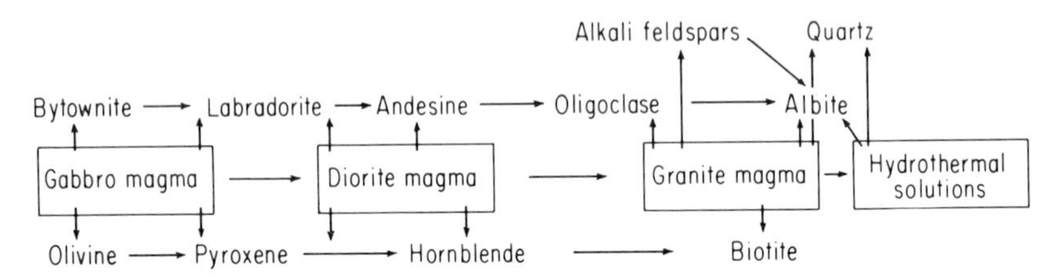

Figure 7-8 Magmatic evolution. Minerals at the top of Bowen's reaction series crystallize from a gabbroic magma, causing it to change composition until it evolves into a dioritic magma. Crystallization of minerals from the middle of the reaction series causes further magmatic composition change until a granitic magma is evolved. Additional crystallization of anhydrous phases finally produces hydrothermal solutions.

from the initially basaltic magma, it first evolves into one of dioritic composition, and then into one of granitic composition.

Fractional crystallization in the system *ab-an-di*, which is a simplified basalt, leads to a liquid rich in alkali feldspars. Any K_2O, excess SiO_2, or H_2O in the original melt would also be concentrated in the residual melt. This concentration of silica and alkali feldspar constituents in low-temperature silicate melts, which are derived from magmas predominantly made up of pyroxene and calcic plagioclase, has led to the concept of *Petrogeny's Residua System*, first set forth by Bowen (Fig. 7-9). In principle, fractional crystallization of a magma of basaltic composition, or of any composition intermediate between that of basalt and that of granite, produces a residual liquid of granitic composition. However, beginning with a basaltic magma, the amount of granite that could be so produced is quite small, representing only a few percent of the total material.

The most abundant plutonic rock in the continental crust is granite, which, taken together with the closely related rock, granodiorite, constitutes the major rock type of batholiths, which are the exposed roots of eroded mountain chains. The essential minerals for these rocks are potassium feldspar, sodic plagioclase, and quartz, with muscovite, biotite, hornblende, or

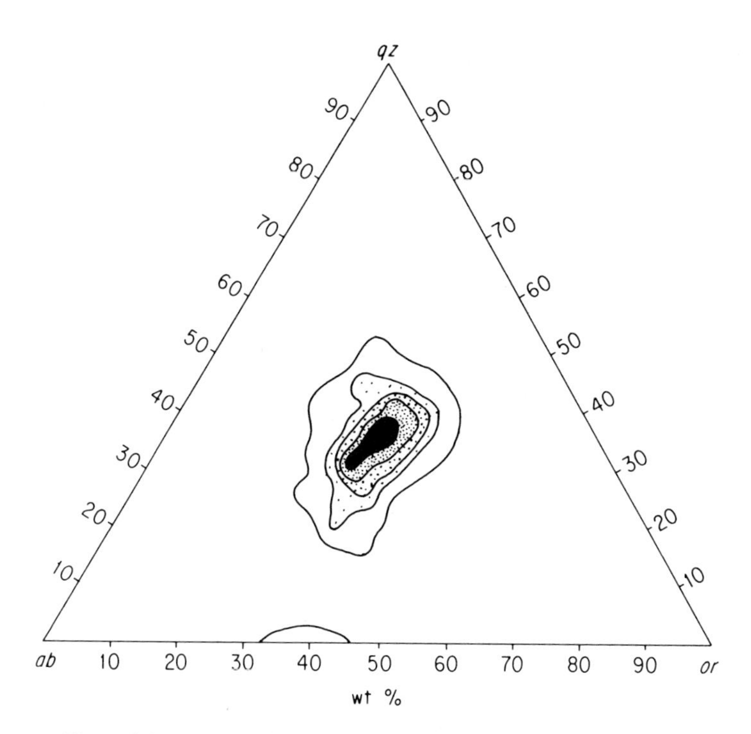

Figure 7-9 Contour diagram for the composition of siliceous plutonic rocks. Rocks selected on the basis of having more than 80 percent *ab* + *or* + *qz* as calculated from chemical analysis. Compositions have been recalculated to 100 percent *ab* + *or* + *qz*. Most of the rock compositions are concentrated near the center of the diagram, which is best explained if the granites are formed by fractional crystallization or fractional melting of the parent rock (see Fig. 7-11).

magnetite occurring as common accessories. A contour plot of chemically analyzed plutonic rocks reduced to calculated amounts of albite, orthoclase, and quartz is given in Fig. 7-9. Only two other criteria were used to select analyses for this diagram: (1) the rocks all contain more than 80 percent *ab* + *or* + *qz*, and (2) the rocks contain no nepheline. Analyses of extrusive rocks selected on the same criteria show a similar plot.

Granite is commonly distinguished from granodiorite on the basis of the *or*: *ab* ratio, granite having more *or* and granodiorite having more *ab*. This distinction rather arbitrarily bisects the concentration of analyzed rocks in the center of Petrogeny's Residua System. Since the clustering of compositions supports the idea that these rocks have a common origin in terms of heterogeneous equilibrium, the term granite will be used henceforth to name all plutonic rocks having *ab*, *or*, and *qz* in roughly equal amounts.

An explanation for the concentration of compositions in the system

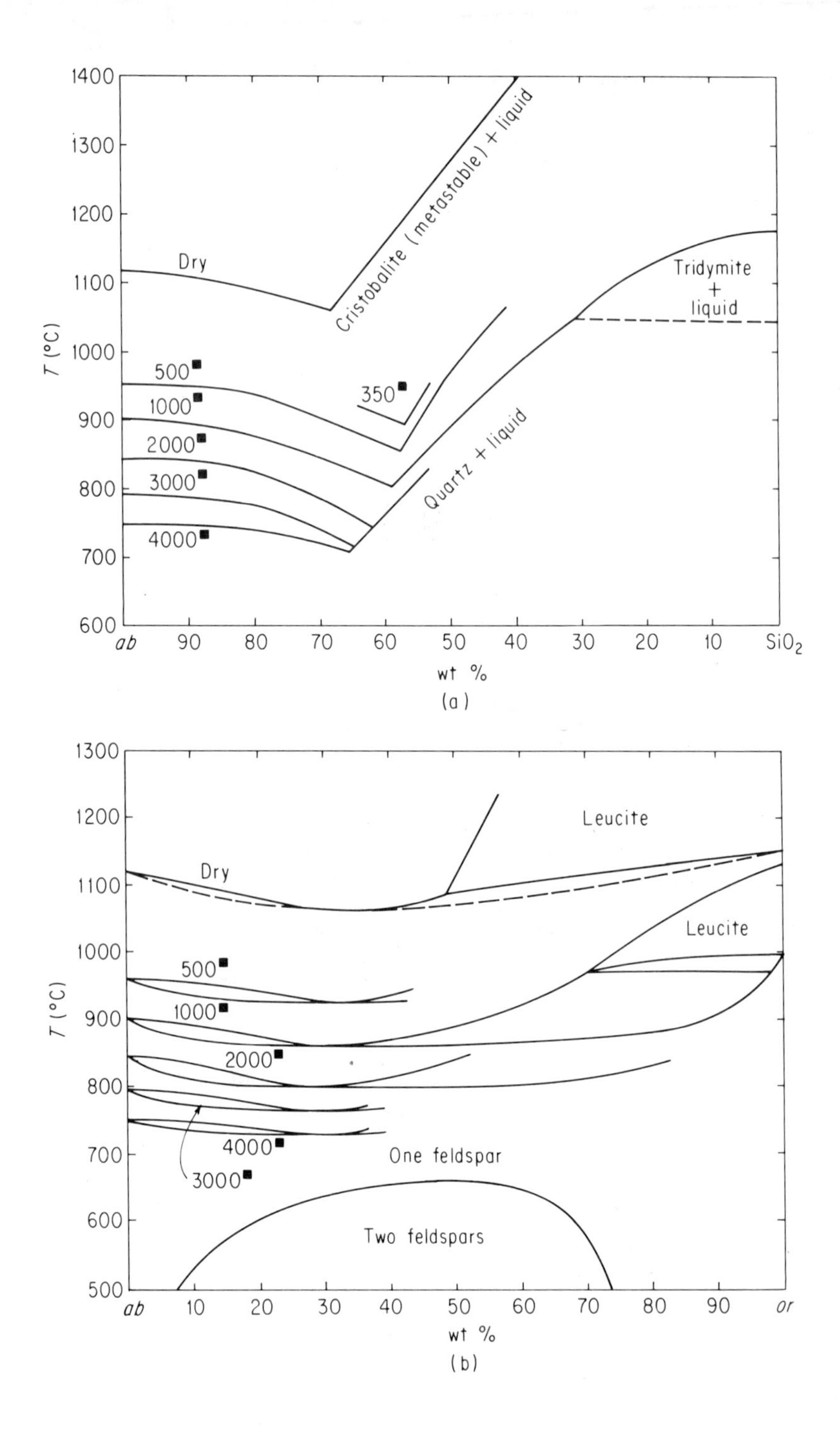
1400
1300
1200
1100
1000
900
800
700
600
T (°C)
Dry
Cristobalite (metastable) + liquid
Tridymite
+
liquid
500
1000
2000
3000
4000
350
Quartz + liquid
ab
90
80
70
60
50
40
30
20
10
SiO2
wt %
(a)
1300
1200
1100
1000
900
800
700
600
500
T (°C)
Leucite
Dry
Leucite
500
1000
2000
4000
3000
One feldspar
Two feldspars
ab
10
20
30
40
50
60
70
80
90
or
wt %
(b)

ab-or-qz may be interpreted from inspection of Fig. 7-11, which shows the liquidus surface, at 1 kbar pressure, in the system *ab-or-qz-w* as projected onto the anhydrous base of the composition tetrahedron (compare with Figs. 6-9b and 6-14). There is a thermal minimum in the liquidus surface, which is approximately coincident with the highest density of analyzed granitic rocks.

Throughout the discussion in Chapter 6 concerning the reading of phase diagrams, pressure was used to mean hydrostatic pressure, without any reference to the pressure-transmitting medium. However, in many cases it is convenient, experimentally, to use water as a pressure-transmitting medium. Since water dissolves in the melt and may react with one or more of the anhydrous phases to form a hydrous mineral in the assemblage, addition of water adds another component to the system, thus adding one more dimension to the graphical representation of the system (Fig. 6-14).

One method employed to circumvent the additional complexities of an additional component is to plot the anhydrous side of the phase diagram with the compositions of the melt and the hydrous phases projected onto it. When this is done, the presence of water as a component is represented by incorporating it into the statement of pressure—P_{H_2O}. Most commonly P_{H_2O} is the same as total pressure, although this is not necessarily the case when the phase relations with undersaturated liquids are being explored. Thus, Fig. 7-11 is the projection of the phase relations represented in Fig. 6-14 onto the anhydrous base of the composition tetrahedron.

The convention of using a volatile component as an intensive variable is not at all unrealistic, since all magmas have some dissolved water, and many granitic magmas are saturated or nearly saturated with water. However, phase diagrams utilizing the convention of P_{H_2O} (or the pressure of any other volatile component) are not precisely the same as those discussed in Chapter 6, and the phase relationships of a hydrous system may be quite different from those of an anhydrous one, since the presence of excess water in the melt lowers the melting points of the solid phases as the pressure (and the water content of the melt) is increased.

Interpretation of Fig. 7-11 can be made with respect to the ternary systems *ab-or-w* and *ab-qz-w* (Fig. 7-10). Figure 7-10a gives the liquidus curves

---

Figure 7-10 Systems boundary to Petrogeny's Residua System. (a) The system *ab-qz-w* projected onto the anhydrous plane. This system shows a eutectic that shifts to lower temperatures and toward the albite corner with increased pressure in the presence of excess water. (b) The system *ab-or-w* projected onto the anhydrous plane. This system shows a minimum that shifts to lower temperatures and toward the albite corner with increased pressure in the presence of excess water. Note the presence of an exsolution dome in the subsolidus region. Pressure on the water-saturated liquid in bars (■). (After O. F. Tuttle and N. L. Bowen, *Origin of Granite in the Light of Experimental Studies*, Memoir 74. New York: The Geological Society of America, 1958, Figs. 17, 20.)

projected onto the anhydrous side in the system *ab-qz-w* for several pressures with maximum dissolved water in the melt ($P_{H_2O} = P_{total}$). The pseudobinary system has a simple eutectic (compare with Fig. 6-2). Slight increase in pressure, and, thus, in the amount of dissoved water in the melt, drastically reduces liquidus temperatures. In addition, at higher pressure and at lower temperature the eutectic shifts to more albitic compositions.

The boundary system *or-qz-w* is basically similar to the system *ab-qz-w*, but it is complicated at lower pressures and higher temperatures by the appearance of a stability field for leucite. The eutectic in it, also, shifts away from quartz with higher pressures.

The boundary system *ab-or-w* (Fig. 7-10b) is somewhat more complicated than the other two. First, there is a stability field for leucite; second, it has a minimum melting composition instead of a eutectic; and third, there is partial miscibility of the alkali feldspars at low temperatures (compare with Fig. 6-5a).

The liquidus surface of Petrogeny's Residua System (Fig. 7-11) may now be explored with reference to the boundary systems. The eutectics in the systems *ab-qz-w* and *or-qz-w* are connected by a thermal valley. Like the boundary system *ab-or-w*, this valley in the full quaternary system has a thermal minimum in it, and the minimum shifts away from the quartz corner with increased pressure and increased water in the melt (compare with Fig. 6-9b for crystallization sequences).

Crystallization of anhydrous phases accounts for the feldspar and the quartz in the system *ab-or-qz-w*, but not for the water. The limit of solubility of water in a silicate melt of granitic composition is approximately 10 percent. Clearly, if the melt is saturated with water when it reaches the liquidus surface, formation of crystalline phases causes supersaturation and the separation of a water-rich vapor phase. Even if the original melt is not saturated with water, at some point in the crystallization history it will become saturated. This point is followed by the separation of a vapor phase (see Section 7-5).

Below the solidus line in Fig. 7-10b there is a solvus, indicating that homogeneous alkali feldspars separate into two phases, one sodic and the other potassic, with further drop in temperature. The existence of this solvus means that intermediate alkali feldspar crystalline solutions, which are stable at the liquidus surface, become unstable at some temperature below the liquidus and tend to exsolve into two phases (see Fig. 6-5 and discussion).

The foregoing discussion on crystallization from silicate melts has leaned heavily on the results of experimental studies in simplified systems. While real rocks are more complex in that they contain other components and other phases, the general trend in crystallization sequences, as indicated in the foregoing discussion, fits well with petrographic observations of various rock sequences in the crust of the earth.

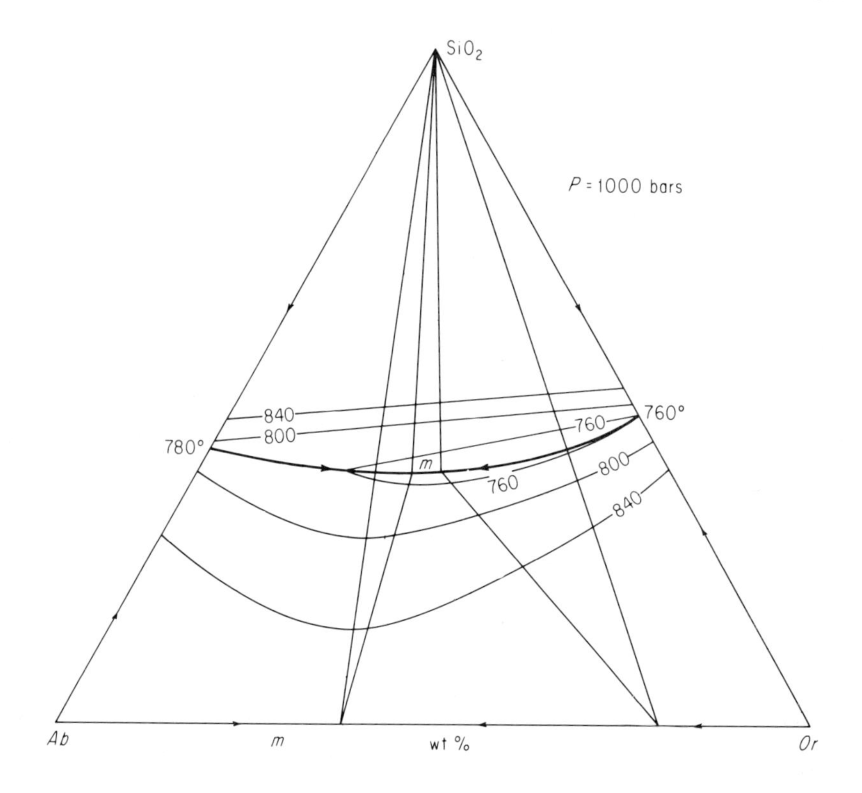

Figure 7-11 Liquidus surface for the system *ab-or-qz-w*. Liquid compositions are projected onto the anhydrous base of the composition tetrahedron. Liquidus surface in the presence of excess water at a pressure of 1 kbar. A thermal valley with a minimum connects the eutectics *ab-qz-w* and *or-qz-w*. The liquidus surface is indicated by isothermal contour lines. Two three-phase triangles for feldspar–quartz–liquid are shown, and the position of the thermal minimum in the thermal valley is between the apices of the two three-phase isothermal triangles. Increased pressure shifts the minimum away from SiO_2 and toward lower temperatures. (After O. F. Tuttle and N. L. Bowen, *Origin of Granite in the Light of Experimental Studies*, Memoir 74. New York: The Geological Society of America, 1958, Fig. 23.)

Carbonate rocks of igneous origin have been given scant discussion and little credence until recently, owing to (1) their low abundance in the crust of the earth, (2) their high susceptibility to weathering in humid climates which leaves them poorly exposed, (3) the remoteness of active volcanoes producing carbonate-rich lavas (Tanzania), (4) the difficulty of conceiving a geologically reasonable and physicochemically sound mode of genesis, and (5) a lack of experimental data relating to the pertinent phase equilibria involved in their formation. Recent experimental work in silicate–carbonate–water systems demonstrates that carbonate magmas can exist at geologically reasonable temperatures and pressures, although the ultimate origin of these liquids remains unclear.

Where carbonatite lavas have been observed and where carbonatite plutons have been studied in detail they are found in association with silicate rocks that are very low in silica and rich in alkalis, and with host rocks that have been extensively altered by hydrous fluorocarbonate fluids (*fenitized*). However, the alkali magmatic rocks and the fenitized country rocks are distinct, which leads to the possible existence of three different fluid phases.

There are three immiscible fluid phases in the system $NaAlSi_3O_8$–Na_2CO_3–H_2O (Fig. 7-12): silicate melt saturated with vapor and carbonate, carbonate melt saturated with silicate and vapor, and vapor saturated with silicate and carbonate. The two liquids could crystallize as an alkali-rich igneous rock and a carbonate-rich igneous rock, while the vapor fenitizes the country rock with Na_2O, H_2O, and CO_2 (plus F, which is generally associated with such processes, but ignored in the experimental treatment).

Although a magmatic origin for the Kiruna (Sweden) magnetite deposits has been hypothesized by geologists working in the field, until recently there has been some doubt about the possibility of the existence of magnetite-rich liquids at geologically reasonable temperatures and pressures. This doubt has continued to be voiced even after the observation of a magnetite flow, which, like the Tanzanian carbonatite lavas, is conveniently remote (northern Chile). Although the debate about magmatic origin of magnetite rock bodies will probably continue, there is experimental evidence that magnetite-rich magmas can exist in the crust of the earth.

The phase relations on the liquidus surface of the system Fe–C–O at 500 bars of pressure, shown in Fig. 7-13, indicate the existence of magnetite-rich melts at temperatures as low as 815°. Crystallization of melts having this composition yields an igneous rock composed predominantly of magnetite with a small amount of graphite. At temperatures well below the solidus, the graphite can react with the CO_2-rich vapor to form siderite, which is a common accessory mineral in magnetite deposits of alleged igneous origin.

All of which does not prove the existence of carbonatites and magmatic magnetite deposits, but it does make them plausible if field observations warrant such a conclusion.

The foregoing considerations demonstrate that the associations of minerals found in igneous rocks may be interpreted as the result of crystallization from a magma according to the laws of chemistry and physics, and

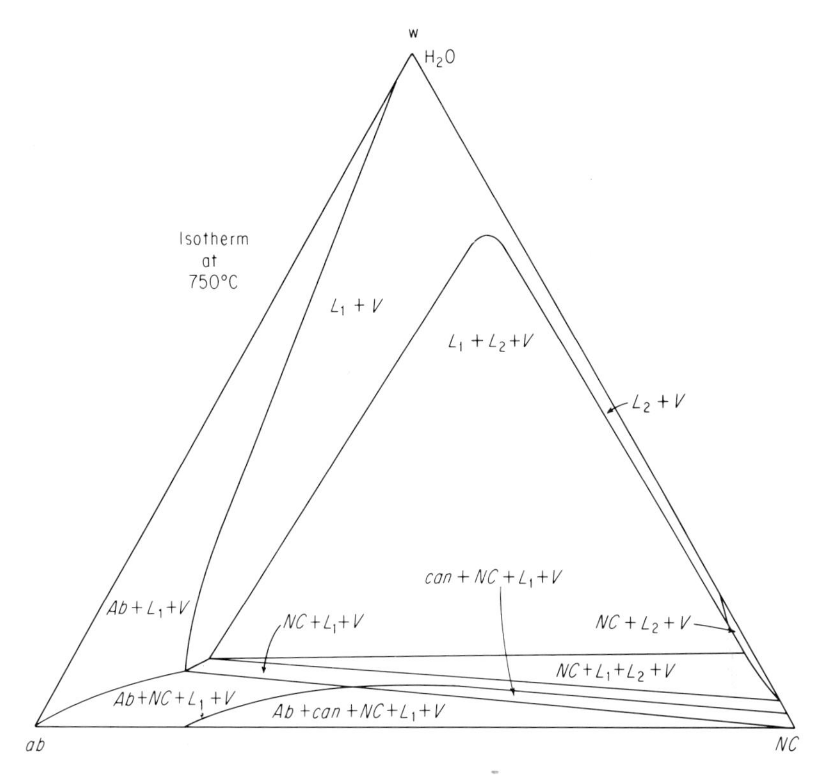

Figure 7-12 Immiscibility between silicate and carbonate magmas. Isothermal, isobaric section through the system *ab-NC-w* at 1000 bars pressure and 750° is shown. Abbreviations: *NC* = Na_2CO_3, *can* = cancrinite, *ab* = albite, L_1 = silicate magma, and L_2 = carbonate magma. The miscibility gap between the two magmas disappears below 700°. (After A. F. K. van Groos and P. J. Wyllie, "Liquid Immiscibility in the Join $NaAlSi_3O_8$–$NaCO_3$–H_2O and Its Bearing on the Genesis of Carbonatites," *Amer. Jour. Sci. 226*, p. 957.)

that such crystallization can be duplicated and illustrated by laboratory experiments. The discussion of igneous rocks is not exhaustive, as no mention has been made of many of the igneous rocks classified in Fig. 7-2 or of the many rare igneous rocks that occur in the earth's crust. However, the laws of heterogeneous equilibrium may be used to account for their origins as well. Nor are these principles confined to the earth, since much of what is known

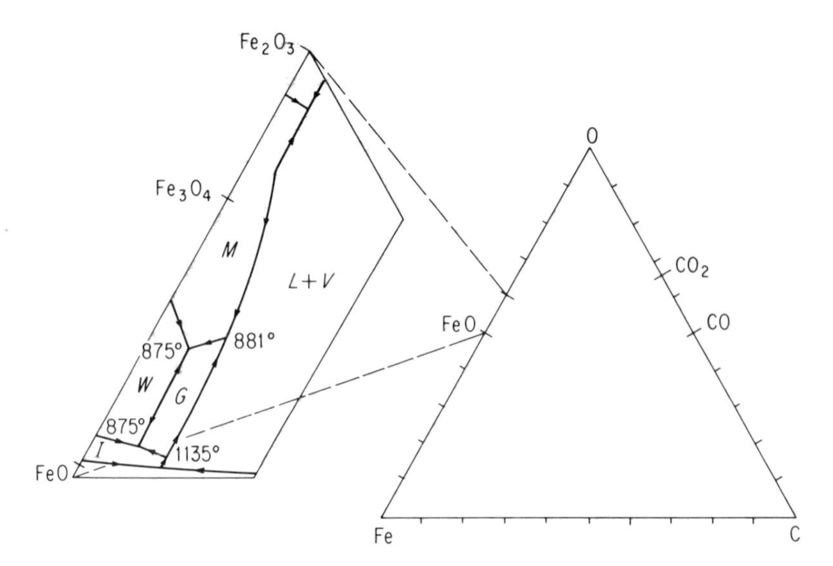

Figure 7-13 Liquids in the system Fe–C–O. Abbreviations: I = iron, G = graphite, W = wustite, M = magnetite, and H = hematite. Thermal valleys are shown with arrows indicating falling temperatures. The low temperatures of the eutectics indicate that liquids exist in this system at geologically reasonable temperatures. Pressure is 500 bars. (After J. R. Weidner, Ph.D. Thesis, Pennsylvania State University, 1968.)

about the history of the moon now derives from analyses of the rock samples collected there and the simulation of these rocks experimentally in the laboratory.

Having established physicochemical controls on the manner in which magmas crystallize upon cooling, some comment may be made concerning magmatic sources. Two sites may be considered for magma genesis—in the crust and in the mantle of the earth. Many sediments (Section 7-3) contain the components of *Petrogeny's Residua System* (Figs. 6-14 and 7-11), and the granite melting curve intersects the geothermal gradient in parts of the continental crust where such sediments have accumulated to exceptional thicknesses (Fig. 7-4). Under these conditions, partial melting of the rock yields a magma having a composition corresponding to the minimum in the system *ab-or-qz-w*. If the pressure remains constant, addition of heat increases the amount of magma isothermally until, under equilibrium conditions, one of the *phases* disappears. However, if melting is fractional and the magma removed continuously as it is produced, the amount of magma increases until one of the *components* is consumed, a process which increases the amount of magma of constant composition that can be obtained from a rock. Once this has occurred, temperatures must rise substantially before melting resumes (Section 6-4).

The magmas thus generated may then rise to higher levels in the crust and there crystallize as granites, or they may be extruded as rhyolites. Crustal temperatures are too low to produce much magma of composition more mafic than granite, but crystal–liquid mushes of compositions intermediate between granite and basalt may be intruded to higher levels.

Interpretation of geophysical data indicates that interstitial molten material exists in the rocks of the mantle. The mantle is known to vary physically, both vertically and horizontally, and chemical and mineralogic inhomogeneity is inferred. The presence of a minimum, a eutectic, or a peritectic in the system would give rise to initial fusion of a unique magma from a mantle of variable composition. Indeed, given all the factors potentially capable of producing mineralogic diversity, the uniformity of composition of basaltic magmas throughout time and space supports the concept of partial fusion at an invariant point to generate them. However, once generated, basaltic magma can then fractionate, with or without compositional change resulting from assimilation of wall rock, to give rise to more felsic intermediate rocks.

A second source of magmas within the mantle may be noted. Some continental margins are characterized by subduction zones wherein oceanic basalts, together with sediments deposited upon them, are forced beneath the continental crust. Partial fusion of these materials would give rise to magmas of intermediate composition, such as andesite and dacite, which characterize the volcanic rocks of these continental margins. Were carbonate rocks to be carried beneath continental crust in a subduction zone, carbonatite magmas could also be generated by partial fusion.

Although different geologic processes are involved in magma generation, partial fusion according to the laws of heterogeneous equilibrium may be employed to account for their composition, and fractional or equilibrium crystallization processes, as outlined in this section and in Chapter 6, may be used to explain the mineral assemblages of igneous rocks.

7-3. Sedimentary Mineral and Rock Formation

Minerals that are formed in the magmatic environment are not, in general, stable at the surface of the earth in the presence of free exygen and water. Reduced ions are oxidized, anhydrous minerals are hydrated, and soluble ions are dissolved, each causing the crystal lattice of the magmatic minerals to break down. New minerals that are stable at the surface of the earth are formed from the remnants of the unstable minerals. The stable minerals of the sedimentary environment are given in Table 7-3.

An understanding of the behavior of minerals at the earth's surface may be approached from the point of view of the chemical behavior of their con-

Table 7-3
SEDIMENTARY MINERALS

Elements	
Sulfur	S_8
Silica	
Quartz,[a] chalcedony	SiO_2
Opal	$SiO_2 \cdot nH_2O$
Feldspar	
Microcline	$KAlSi_3O_8$
Albite	$NaAlSi_3O_8$
Mica	
Muscovite, sericite	$KAl_2(AlSi_3O_{10})(OH)_2$
Montmorillonite	$Al_2Si_4O_{10}(OH)_2 \cdot nH_2O$
Vermiculite	$Mg_3Si_4O_{10}(OH)_2 \cdot nH_2O$
Glauconite	$K(Fe, Mg, Al)_2(Si_4O_{10})(OH)_2$
Clay	
Kaolinite, nacrite, dickite	$Al_4Si_4O_{10}(OH)_8$
Halloysite	$Al_4Si_4O_{10}(OH)_8 \cdot nH_2O$
Oxides	
Magnetite	$FeFe_2O_4$
Hematite	Fe_2O_3
Rutile	TiO_2
Pyrolusite	MnO_2
Hydroxides	
Gibbsite	$Al(OH)_3$
Diaspore, boehmite	$AlO(OH)$
goethite, lepidocrocite	$FeO(OH)$
Manganite	$MnO(OH)$
Sulfides	
Pyrite, marcasite	FeS_2
Sulfates	
Gypsum	$CaSO_4 \cdot 2H_2O$
Anhydrite	$CaSO_4$
Halides	
Halite	$NaCl$
Sylvite	KCl
Carnallite	$KMgCl_3 \cdot 6H_2O$
Carbonates	
Calcite, aragonite	$CaCO_3$
Dolomite	$CaMg(CO_3)_2$
Siderite	$FeCO_3$
Borates	
Kernite	$Na_2B_4O_7 \cdot 4(H_2O)$
Borax	$Na_2B_4O_7 \cdot 10(H_2O)$
Phosphates	
Apatite, collophane	$Ca_4(PO_4)_3Ca(F, Cl, OH)$
Monazite	$CePO_4$
Xenotime	YPO_4

[a]The names of the more common minerals of each group are set in italic.

stituent ions. This behavior can be summarized by their *ionic potential*, which is ionic charge divided by ionic radius. The common ions of the continental crust of the earth are plotted according to their size and charge in Fig. 7-14, along with lines of equal ionic potential. Elements having an ionic potential less than 3 tend to be soluble in the waters of the earth's crust, whereas those with potentials greater than 12 tend to form soluble hydroxide complexes. Hydrolysate elements have ionic potentials falling between 3 and 12, and tend to be insoluble.

Minerals containing water-soluble elements lose those elements to rain and groundwater, causing collapse of the remaining crystal lattice. Ions

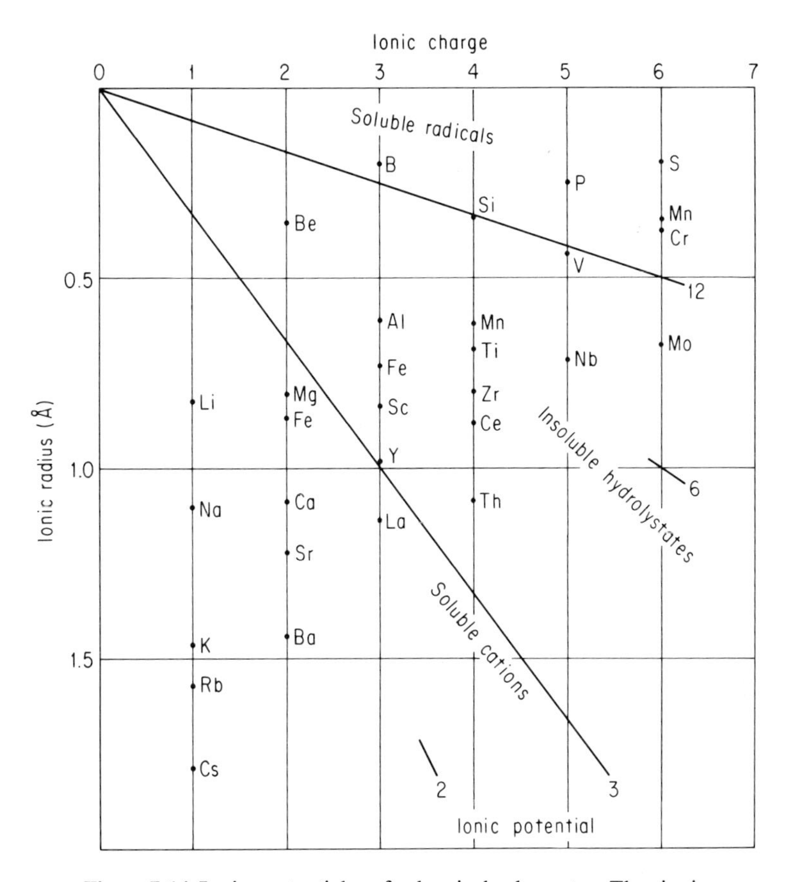

Figure 7-14 Ionic potentials of chemical elements. The ionic potential is the ionic charge divided by the ionic radius. Cations with ionic potential less than 3 are soluble as cations, and cations with ionic potential greater than 12 form water-soluble radicals. Cations with ionic potential between 3 and 12 form insoluble oxides.

having more than one valence state, of which Fe predominates, but Mn and S are also included, react with free oxygen to go from the lower to the higher oxidation state. The removal of an electron from an ion not only increases the charge on the ion, but also reduces its size. Oxidation upsets electrostatic neutrality in the lattice and changes the coordination number of the ion with a single operation, either result being sufficient to cause the breakdown of the crystal lattice.

Some compounds formed in an anhydrous environment react directly with water to form hydrated minerals. Anhydrite reacting with water to form gypsum is an example. Although such simple hydrations between magmatic minerals and water are rare, minerals formed at the surface of the earth from the breakdown products of anhydrous magmatic minerals include many hydrated silicates.

The processes that convert magmatically formed minerals to sedimentary minerals may be illustrated by the weathering of a granodiorite, which represents the average rock of the continental crust. The minerals to be considered, then, are quartz, orthoclase, plagioclase, and a ferromagnesium mineral. Quartz is relatively insoluble; Si^{4+} is a hydrolysate element. Some SiO_2 will go into solution, but in such small amounts and at such a slow rate as to be insignificant compared with the weathering of the other minerals. The Al^{3+} and the Si^{4+} in the orthoclase are hydrolysate elements, but the K^+ is a soluble cation. The removal of the K^+ from orthoclase causes the collapse of the tektosilicate structure which, under temperate conditions, rearranges itself into a clay structure. The weathering of orthoclase may be represented by the chemical reaction

$$2KAlSi_3O_8 + 2H_2O \longrightarrow Al_2Si_2O_5(OH)_4 + 2K^+ + O^{2-} + 4SiO_2$$

$$\text{orthoclase} + \text{water} \longrightarrow \text{kaolinite} + K^+ \text{ and } O^{2-}{}_{\text{(solution)}} + \text{silica}_{\text{(solution ?)}}$$

At this point in the argument, a slight digression is necessary. Hydrocarbons burn according to the equation

$$CH_4 + 2O_2 \longrightarrow CO_2 + 2H_2O$$

$$\text{methane} + \text{oxygen} \longrightarrow \text{carbon dioxide} + \text{water vapor}$$

However, as anyone who has ever watched a flame knows, real combustion is never complete. Combusion is only partial, and smoke and carbon monoxide are generated by any real flame, although varying the conditions can influence the degree to which combustion takes place. The same sort of partial reaction occurs in the real weathering of minerals. Thus, some of the common breakdown products of igneous silicates are illite, vermiculite, montmorillonite, and glauconite. Much weathered product of igneous rock is composed of such metastable layered silicates, which may react further or which may persist throughout the sedimentary process.

The weathering of plagioclase may be considered in terms of the weathering of the two end members of the crystalline solution series, albite and anorthite. The breakdown of albite parallels that of orthoclase, with the dissolution of Na^+ in place of K^+. If the reaction goes to completion, the product will be kaolinite. In reality, some of the Na^+ goes into the formation of montmorillonite and illite. However, a larger proportion of the dissolved K^+ than Na^+ is trapped in the layered silicates. The weathering of anorthite may be represented ideally with the reaction

$$CaAl_2Si_2O_8 + 2H_2O \longrightarrow Al_2Si_2O_5(OH)_4 + Ca^{2+} + O^{2-}$$

Two differences between the weathering of anorthite and the weathering of the alkali feldspars are to be noted. First, there is no leftover silica, and, second, the Ca^{2+} has a higher ionic potential than do the alkali ions. The latter means that Ca^{2+} is less soluble than the alkalis. The chemical behavior of Ca^{2+} during weathering is discussed later in terms of carbonate behavior.

The weathering of ferromagnesian minerals—olivines, pyroxenes, amphiboles, and biotites—may be considered together, owing to their chemical similarities. Although the crystal chemistry of Mg^{2+} is quite different from that of Ca^{2+}, their solution chemistries are similar. The Mg^{2+} ions are dissolved out of the crystal lattice positions and carried away in solution. Because of its higher ionic potential, Mg^{2+} is less soluble than Ca^{2+}. Furthermore, because of its size, Mg^{2+} may enter the octahedral sites in the crystal lattice of the layered silicates glauconite, montmorillonite, and vermiculite.

The Fe^{2+} in the ferromagnesian minerals is unstable in the presence of free O_2. Two separate reactions may take place. Under reducing conditions, Fe^{2+} may be taken into solution as a soluble cation, its ionic potential being between that of Mg^{2+} and Ca^{2+}. Organic acids in temperate soils keep Fe^{2+} reduced while it is carried away in solution. Under oxidizing conditions, Fe^{3+} is a hydrolysate element. Either process leads to the breakdown of the coordination polyhedra. The complete ideal oxidation of Fe^{2+} of magmatic ferromagnesian minerals may be represented by the equation

$$\begin{array}{c} 2FeO \quad + \quad \tfrac{1}{2}O_2 \quad \longrightarrow \quad Fe_2O_3 \\ \text{ferrous iron} + \text{oxygen} \longrightarrow \text{hematite} \end{array}$$

There are several partial steps possible when this reaction does not go ideally to completion. The oxyhydroxides goethite [α-FeO(OH)] and lepidocrocite [γ-FeO(OH)] may form metastably. The pseudomineral, *limonite*, is a mixture of these and other hydroxides with adsorbed water. Alternatively, the Fe^{3+} may enter into the lattice of such layered silicates as glauconite or vermiculite. Any Ca^{2+}, K^+, or Na^+ in the ferromagnesian minerals behaves in the same chemical manner as it does in the weathering of feldspars. The Al^{3+} of amphiboles and biotite combines with the Si^{4+} to form kaolinite or one of the other layered silicates.

So far in this discussion, kaolinite or a kaolin mineral has been assumed to be the stable aluminous end product in weathering reactions. This is true for chemical weathering in temperate climates and in some tropical climates. However, for the case of well-drained uplands in tropical, monsoon climates, the end product is a mixture of the hydroxide, gibbsite [$Al(OH)_3$], and the oxyhydroxides, diaspore [α-AlO(OH)] and boehmite [γ-AlO(OH)]. This mixture, known as *bauxite*, is the major ore of Al. Bauxite, limonite, and clay form the residual tropical soil known as *laterite*. The mechanism of the formation of bauxite is a matter of some controversy. Some mineralogists hold that under conditions of alternate wetting and drying in the tropics the bauxite minerals form directly from plagioclase, as represented by the reaction

$$2NaAlSi_3O_8 + CaAl_2Si_2O_8 + 6H_2O \longrightarrow 4Al(OH)_3 + 2Na^+ + Ca^{2+} + 5SiO_2$$

$$\text{plagioclase} + \text{water} \longrightarrow \text{gibbsite} + (Na^+ + Ca^{2+} + \text{silica})_{(\text{solution})}$$

Other mineralogists maintain that plagioclase first reacts to form kaolin minerals, which are then desilicified according to the reaction

$$Al_2Si_2O_5(OH)_4 + H_2O \longrightarrow 2Al(OH)_3 + 2SiO_3$$

$$\text{kaolinite} + \text{water} \longrightarrow \text{gibbsite} + \text{silica in solution}$$

There is field evidence to support both contentions, and there well may be more than one mechanism for the formation of bauxite.

The chemical reactions discussed in the foregoing paragraphs have divided the elements of igneous minerals into two groups, the soluble elements, which may be transported chemically, and the insoluble elements, which may be left behind or transported physically.

Both Ca^{2+} and Mg^{2+} are transported in solution as bicarbonate radicals. The CO_2 of the atmosphere reacts with rainwater to form carbonic acid

$$CO_2 + H_2O \rightleftharpoons H_2CO_3$$

which dissociates

$$H_2CO_3 \rightleftharpoons H^+ + HCO_3^-$$

The Ca^{2+} is leached out of plagioclase and amphibole as represented by equation

$$CaO + H_2CO_3 \longrightarrow CaHCO_3^+ + OH^-$$

Anything that tends to increase the amount of dissolved CO_2 therefore permits the solution of more Ca^{2+}. Anything that tends to decrease the amount of CO_2 causes the precipitation of calcite from a saturated solution

$$CaHCO_3^+ + HCO_3^- \rightleftharpoons CaCO_3 + CO_2 + H_2O$$

Addition of CO_2 drives the equation to the left, and removal of CO_2 drives the equation to the right. Thus it is that cooling or increasing pressure promotes the solution of Ca^{2+}, while warming or decreasing pressure causes the precipitation of calcite.

Dissolved Mg^{2+} follows the same pattern except that it is considerably less soluble. In addition, some Mg^{2+} may be taken up by layer silicates.

The sedimentary behavior of Ca^{2+} is further complicated by the existence of living creatures, which take it up either as carbonate or as phosphate for the formation of skeletal material. Indeed, many limestones seem to be composed, in large part, of the accumulated skeletal materials of such animals.

Having established the chemical reactions by which sedimentary minerals form from high-temperature and high-pressure minerals, a brief discussion of the way in which these minerals combine to form rocks is in order. The soluble cations are either tied up in intermediate minerals or are transported in solution directly to the oceans. Intermediate and end-product minerals are either concentrated as residual deposits or, as is more likely, transported by wind, water, or ice to the shallow seas of the continental margins or interiors. Some sediments, of course, are deposited subaerially on the dry part of the continent, but for most of them it is but a temporary stopping point.

The elements taken into solution remain so until the chemistry of the water changes to cause their precipitation or until they are removed by biological processes. Most of the Ca^{2+} is removed from solution by the latter process, in the form of animal or plant skeletal material. This biogenetic calcite then accumulates on the death of the animals and plants, generally in the shallow waters of the continental margins or about oceanic islands.

A sediment is not a rock. Certain changes must take place within sediment to convert it into rock. *Diagenesis*, the term covering these changes, includes compaction, cementation, and recrystallization. Clay-sized clastic particles may be deposited with more than an equal volume of water. Compaction occurs as more sediments are piled on top and the water is squeezed out. If the interstitial water is seawater, there can be a chemical reaction between the ions in the interstitial water and the sedimentary minerals. Water percolating through the sediment can leave a chemical precipitate that cements the grains together. Finally, the sedimentary minerals can recrystallize to a coarser texture or, by reacting with the interstitial water, recrystallize to new mineral species. Further burial leads to an increase in pressure and temperature, and true metamorphism begins.

7-4. Metamorphic Minerals and Rocks

Metamorphism includes a range of processes, which operate in the crust and mantle of the earth, that alter a solid rock, both mineralogically and structurally, as a result of a change from the physical and chemical conditions of its origin. Metamorphism may be *prograde*, that is, taking place in response to rising temperatures, or it may be *retrograde*, in response to falling tem-

Table 7-4

METAMORPHIC MINERALS

Mineral	Formula
Silica	
Quartz,[a] coesite, stishovite	SiO_2
Feldspar	
Orthoclase, *microcline*	$KAlSi_3O_8$
Plagioclase crystalline solution series	
Albite	$NaAlSi_3O_8$
Anorthite	$CaAl_2Si_2O_8$
Pyroxene	
Hypersthene	$(Mg, Fe)SiO_3$
Enstatite	$MgSiO_3$
Diopside	$CaMg(SiO_3)_2$
Augite	$Ca(Mg, Fe)(SiO_3)_2$
Jadeite	$NaAl(SiO_3)_2$
Amphibole	
Anthophyllite	$(Mg, Fe)_7Si_8O_{22}(OH)_2$
Cummingtonite	$(Fe, Mg)_7Si_8O_{22}(OH)_2$
Tremolite–actinolite	$Ca_2(Mg, Fe)_5Si_8O_{22}(OH)_2$
Hornblende	$NaCa_2(Mg, Fe, Al)_5(Si, Al)_8O_{22}(OH)_2$
Glaucophane	$Na_2Mg_3Al_2Si_8O_{22}(OH)_2$
Pyroxenoid	
Wollastonite	$CaSiO_3$
Mica	
Muscovite	$KAl_2(AlSi_3O_{10})(OH)_2$
Biotite	$K(Mg, Fe)_3(AlSi_3O_{10})(OH)_2$
Olivine crystalline solution series	
Fayalite	Fe_2SiO_4
Forsterite	Mg_2SiO_4
Garnets	
Almandine	$Fe_3Al_2(SiO_4)_3$
Pyrope	$Mg_3Al_2(SiO_4)_3$
Grossular	$Ca_3Al_2(SiO_4)_3$
Andradite	$Ca_3Fe_2(SiO_4)_3$
Aluminosilicates	
Kyanite, *andalusite*, *sillimanite*	Al_2SiO_5
Epidote group	
Zoisite, *clinozoisite*	$Ca_2Al_3(SiO_4)(Si_2O_7)O(OH)$
Epidote	$Ca_2(Al, Fe)_3(SiO_4)(Si_2O_7)O(OH)$
Layered silicates	
Kaolinite	$Al_4Si_4O_{10}(OH)_8$
Pyrophyllite	$Al_2Si_4O_{10}(OH)_2$
Talc	$Mg_3Si_4O_{10}(OH)_2$
Serpentine	$Mg_6Si_4O_{10}(OH)_2$
Chlorite	$Mg_5Al(AlSi_3O_{10})(OH)_8$
Carbonates	
Calcite	$CaCO_3$
Dolomite	$CaMg(CO_3)_2$
Siderite	$FeCO_3$

Table 7-4 (Cont.)

Other silicates	
Cordierite	$(Mg, Fe)_2Al_3(AlSi_5O_{18})$
Zeolites	
Analcime	$NaAlSi_2O_6 \cdot H_2O$
Chabazite	$CaAl_2Si_4O_{12} \cdot 6H_2O$
Laumontite	$CaAl_2Si_4O_{12} \cdot 4H_2O$
Heulandite	$CaAl_2Si_7O_{18} \cdot 6H_2O$
Oxides and hydroxides	
Periclase	MgO
Brucite	$Mg(OH)_2$

[a]The names of the more common minerals of each group are set in italic.

peratures. At the low-temperature end, *diagenesis*, the conversion of sediments into sedimentary rocks, is completely transitional with low-temperature metamorphism, the division between the two being arbitrary and, therefore, frequently debated. At the high-temperature end, the partial fusion of the low-temperature melting components of a metamorphic rock gives rise to material that is partly metamorphic and partly magmatic. Indeed, in rocks of mixed genesis, *migmatites*, it may be wholly impossible to determine whether a mineral crystallized from a magma or recrystallized in equilibrium with a magma without actually having been dissolved in it.

In addition to the effects of pressure and temperature in metamorphic processes, there is the effect of transport of chemical elements into or out of a particular body of rock during metamorphism. The transporting agent, *metasomatic fluid*, is predominantly H_2O with variable amounts of dissolved CO_2, HCl, HF, and H_2SO_4. Those elements most susceptible to solution during the weathering processes are the ones most commonly moved about during metasomatism.

The mineralogy of a metamorphic rock depends upon the chemistry and mineralogy of the parent sedimentary, igneous, or metamorphic rock, the total pressure under which it was formed, the temperature of formation, the amount of differential pressure or stress in the rock, the partial pressure of H_2O and CO_2, and any chemical changes wrought by metasomatism.

The control exercised by the parent rock should be clear. If no ironbearing minerals existed in the parent rock and no iron were transported in during metamorphism, then no ironbearing minerals can be expected in the metamorphic rock, regardless of the conditions under which it was formed. Typical metamorphic minerals are listed in Table 7-4. Certain minerals are favored by conditions of stress, however, and their occurrence and orientation may be influenced by differential pressures.

From the already enumerated wide variety of sedimentary and igneous rocks, the conclusion that there is an even wider variety of metamorphic rocks follows directly. In 1915, Eskola proposed the concept of *metamorphic facies* in order to classify and group metamorphic rocks in a systematic fashion. All rocks having a unique and characteristic correlation between chemical composition and mineralogical composition belong to the some facies. That is, all rocks of the same chemical composition have the same mineralogical composition if they belong to the same facies. Different chemical composition within a facies, however, leads to different mineralogical composition. The facies concept is based on the assumption that conditions of heterogeneous equilibria have existed during the metamorphic processes. The facies, then, comprises all rocks which have formed under the same set of physical conditions. The conditions for the major divisions of metamorphic facies are indicated in Fig. 7-15.

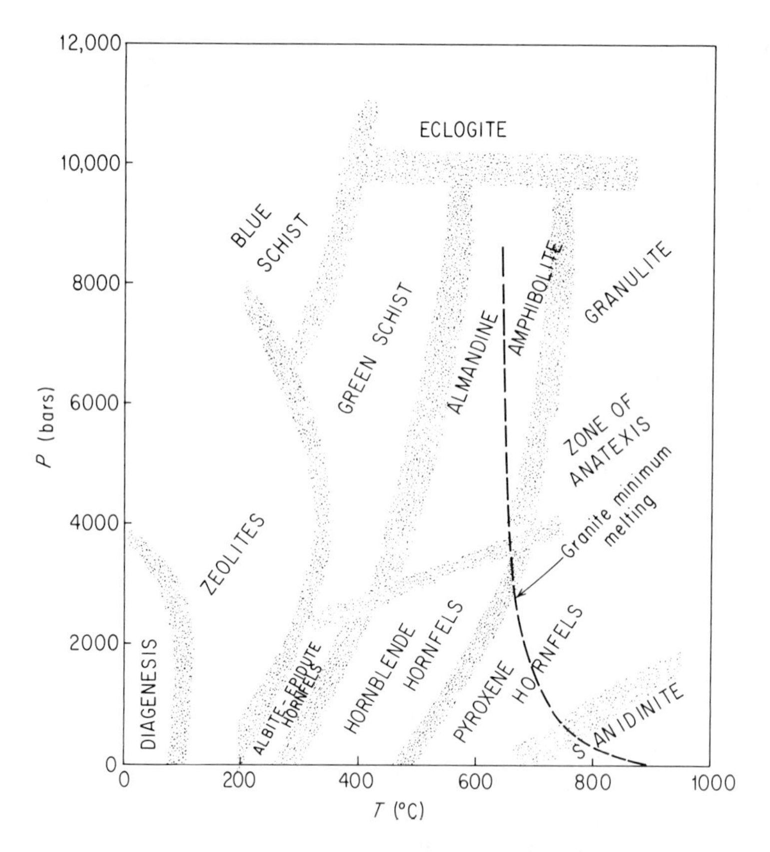

Figure 7-15 Metamorphic facies in pressure–temperature space. Facies boundaries are broad because facies are defined in terms of rock assemblages.

Metamorphic facies are named according to characteristic or distinctive mineral assemblages. The key word here is *assemblage*. The existence of a single mineral depends not only on the temperature and pressure, but also on the chemical environment. The mineral albite has a much wider stability range than does the assemblage albite plus epidote.

The concept of *facies* and the application of phase equilibria to metamorphic assemblages can be illustrated best by the progressive mineral changes during metamorphism of siliceous carbonate rocks in the presence of water. Such rocks are common, and the system $CaO–MgO–SiO_2–CO_2–H_2O$ can be represented in a relatively simple fashion. For these reasons, this system was one of the first to be studied experimentally in detail (again by Bowen) and to have the results applied to real rocks.

Under sedimentary conditions, four minerals—calcite, dolomite, magnesite, and quartz—are the only ones stable (Fig. 7-16) (compare with Fig. 6-13). Two ternary assemblages are possible, calcite–dolomite–quartz and

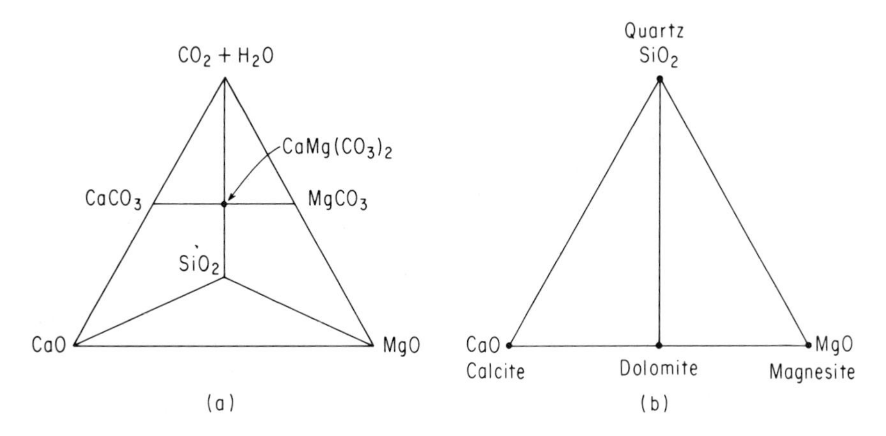

Figure 7-16 Mineral phases in the system $MgO–CaO–SiO_2–CO_2–H_2O$ at the earth's surface. (a) Composition tetrahedron showing stable phases. (b) Projection of stable phases onto the anhydrous base of the composition tetrahedron. The two possible three-phase assemblages are shown. This system approximates real siliceous carbonate rocks. (After Winkler, 1965.)

magnesite–dolomite–quartz. Calcite and magnesite cannot coexist stably because of the stable intermediate mineral dolomite. Five binary systems can exist, calcite–dolomite, calcite–quartz, dolomite–quartz, dolomite–magnesite, and magnesite–quartz. In addition, each of the four minerals can exist by itself. These relations are shown in Fig. 7-16b, where the minerals are projected onto the anhydrous, CO_2-free, triangular base.

Tremolite [$Ca_2Mg_5Si_8O_{22}(OH)_2$] and talc [$Mg_3Si_4O_{10}(OH)_2$] become additional stable minerals in the siliceous carbonate system during the first stages of metamorphism. The formation of these silicates can be expressed as

$$3MgCO_3 + 4SiO_2 + H_2O \rightleftharpoons Mg_3Si_4O_{10}(OH)_2 + 3CO_2$$

magnesite + quartz + water ⇌ talc + carbon dioxide

and

$$5CaMg(CO_3)_2 + 8SiO_2 + H_2O \rightleftharpoons Ca_2Mg_5Si_8O_{22}(OH)_2 + 3CaCO_3 + 7CO_2$$

dolomite + quartz + water ⇌ tremolite + calcite + carbon dioxide

The positions of tremolite and talc are shown in Fig. 7-17a. Each of the joins connecting two stable minerals provides a field boundary for a three-phase assemblage (four, if the fluid phase $CO_2 + H_2O$ is taken into account). The assemblages represented in Fig. 7-17a are representative of the greenschist facies (regional metamorphism) and of the albite–epidote–hornfels facies (contact metamorphism) (Fig. 7-15).

Diopside and forsterite become new stable phases in the siliceous carbonate system when the temperature and the pressure are raised to conditions of the hornblende–hornfels facies (contact metamorphism) (Fig. 7-17b). A characteristic reaction leading to this facies can be expressed as

$$Ca_2Mg_5Si_8O_{22}(OH)_2 + 3CaCO_3 + 2SiO_2 \rightleftharpoons 5CaMgSi_2O_6 + 3CO_2 + H_2O$$

tremolite + calcite + quartz ⇌ diopside + carbon dioxide + water

As can be seen from Fig. 7-17b, the three-component assemblages most likely to be encountered in real marbles are calcite–dolomite–forsterite, calcite–forsterite–diopside, and calcite–diopside–quartz.

Magnesite, dolomite, tremolite and talc become unstable as physical conditions are raised to those of the pyroxene hornfels facies; and wollastonite, periclase, and enstatite become stable along with calcite, forsterite, diopside, and quartz. The mineral assemblages characteristic of this facies may be read from Fig. 7-17c.

All the reactions considered in the metamorphism of siliceous carbonate rocks have resulted in the liberation of CO_2, and many have included hydration or dehydration as well. It must be assumed, therefore, that water is available for hydration reactions, and that the partial pressure of CO_2 in the H_2O has been sufficient to prevent the complete decarbonation of the carbonate minerals. In other words, the volatile constituents in these reactions are assumed to be available in quantities sufficient to cause the reactions as written. The marbles themselves provide the justification for these assumptions, in that the minerals discussed are precisely those found in siliceous carbonate rocks which have undergone contact metamorphism.

Temperatures and pressures have not been assigned to the reactions discussed and, in general, cannot be assigned precisely to boundaries between

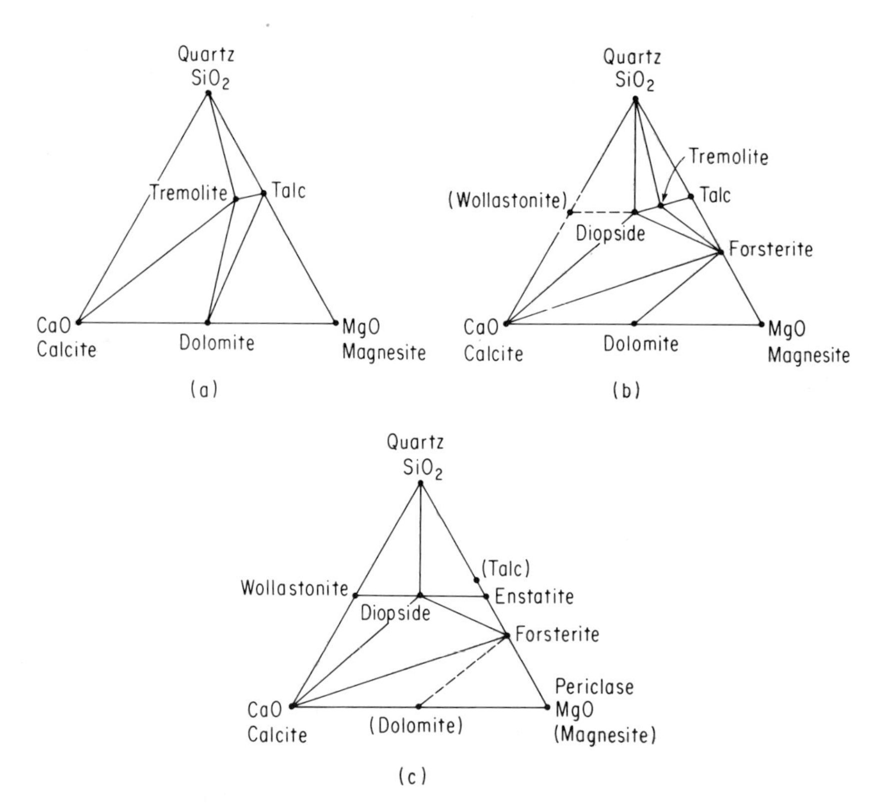

Figure 7-17 Stable phases in metamorphosed siliceous carbonate rocks. (a) Mineral assemblages of the greenschist facies. The two new mineral phases are tremolite and talc. (b) Mineral assemblages of the hornblende–hornfels facies. New mineral phases are diopside and forsterite with wollastonite appearing only at high temperatures within the facies. (c) Mineral assemblages of the pyroxene–hornfels facies. Tremolite disappears as a mineral phase; talc, dolomite, and magnesite are stable only at the low-temperature end of the facies. (After Winkler, 1965.)

facies (Fig. 7-15). One reason for this imprecision can be gleaned from Fig. 7-18. The temperatures of several of the reactions are plotted against the mole fraction of CO_2 in the fluid phase. It rarely is possible to know the composition of the fluid phase, which is long departed when the rock is exposed at the earth's surface and is sampled. The decarbonation curves shown in Fig. 7-18 are based on the assumption that the pressure on the fluid was the same as the total pressure on the rocks. This, too, is unverifiable.

The sequence of mineral assemblages for progressive contact metamorphism of siliceous carbonate rocks has been carried through in some detail to

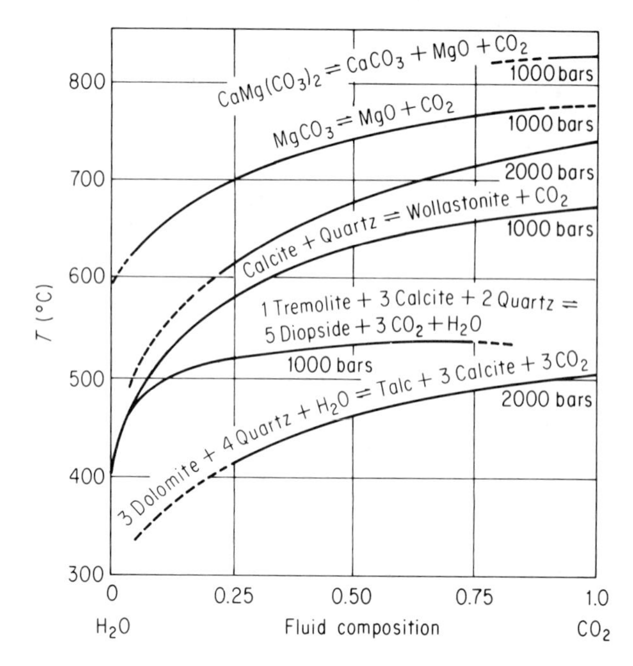

Figure 7-18 Isobaric temperatures for decarbonation reactions. Isobaric equilibrium curves for various decarbonation reactions. The fluid phase may vary from pure H_2O to pure CO_2. The curves are for reactions in the presence of excess fluid. Total pressure on the system is indicated for each curve. (After Winkler, 1965.)

illustrate the facies concept. Any sedimentary or igneous rock, however, can be subjected to these physical conditions, and the sequential mineral assemblages are different because the initial materials are different. The facies will be the same in all cases, because the pressure and temperature are the same. The mineralogical reactions leading to these different assemblages of the same facies are not treated in detail in this text because of their extreme complexity. The principles of reaction are the same regardless of the complexity involved.

Table 7-5 summarizes the minerals that may be found in the different metamorphic facies. Some minerals, such as serpentine, may, individually, be diagnostic of a particular facies. Others, such as quartz, may occur in any one of many different facies, provided that the rock has excess silica.

In cases where compositional variables are so extensive as to preclude the simple representation of the possible stable mineral assemblages as a function of composition, a kind of pseudo-phase diagram may be employed. One such is the *ACF* diagram (Fig. 7-19), first proposed by Eskola. The *ACF* diagram simplifies mineral compositions by omitting consideration of certain constituents that have little effect on the resulting representation. The *A* corner of the *ACF* diagram represents the Al_2O_3 in a mineral along with Fe_2O_3, which

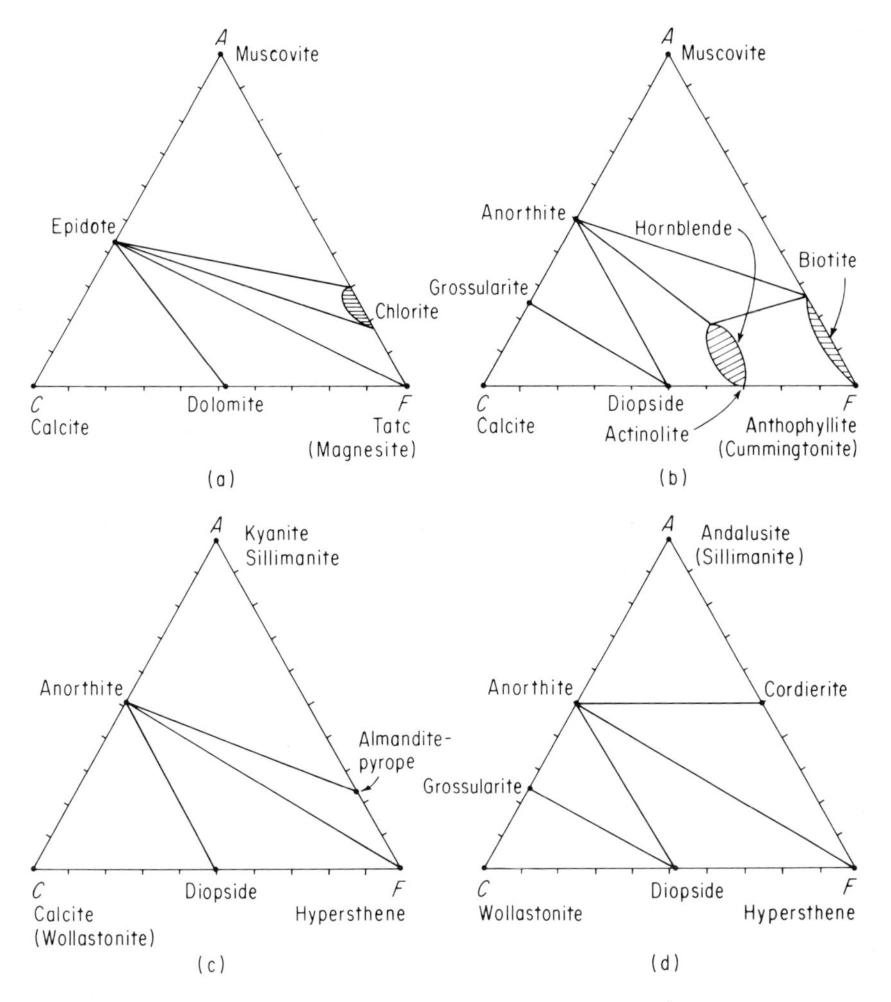

Figure 7-19 *ACF* diagrams for different facies. See text for explanation of diagram. (a) Greenschist facies assemblages for rocks with excess SiO_2. (b) Amphibolite facies assemblages for rocks with excess SiO_2 and K_2O. (c) Granulite facies assemblages for rocks with excess SiO_2. (d) Pyroxene–hornfels assemblages for rocks with excess SiO_2. (After Winkler, 1965.)

behaves very similarly mineralogically (Section 1-13). The *C* corner represents the CaO in a mineral, and the *F* corner represents the FeO and the MgO, which generally may proxy one for the other. The main constituent ignored on an *ACF* diagram is SiO_2. Since many metamorphic rocks are oversaturated with SiO_2 (Section 7-2), stipulation of this fact in the *ACF* diagram suffices. A second, and generally different, diagram may be prepared for those mineral assemblages that are undersaturated with respect to SiO_2. The *ACF* diagram

Table 7-5

MINERALOGY OF METAMORPHIC FACIES[a]

Facies	*Si*	*Al, Si*	*K, Al, Si*	*Na, Al, Si*	*Ca, Al, Si*	*Ca, Si, (CO₂)*
Greenschist	Quartz		Muscovite Microcline	Albite	Zoisite	Quartz + calcite
Epidote–amphibolite	Quartz	Kyanite	Muscovite Microcline	Albite	Zoisite	Quartz + calcite
Amphibolite	Quartz	Kyanite Sillimanite	Muscovite Microcline	Plagioclase	Zoisite Grossular	Wollastonite Quartz + calcite
Granulite	Quartz	Kyanite Sillimanite	Orthoclase	Plagioclase		Wollastonite Quartz + calcite
Pyroxene hornfels	Quartz	Andalusite Sillimanite	Orthoclase	Plagioclase	Grossular	Wollastonite
Sanidinite	Tridymite	Mullite	Sanidine	Plagioclase		Wollastonite (pseudowollastonite) Larnite Rankinite

[a]After B. Mason, *Principles of Geochemistry*, 3rd ed. (New York: John Wiley & Sons, Inc., 1966).

also ignores the alkalis, so the presence or absence of K_2O and Na_2O must also be stipulated. Minor elements are ignored in most cases.

ACF diagrams for different facies are shown in Fig. 7-19 (compare with Fig. 6-13). Interpretation of these diagrams can be made with reference to the progressive regional metamorphism of a sandy shale cemented with dolomite. The assemblage dolomite–talc–epidote–quartz will be the stable one in the greenschist facies. The first three minerals may be read from the *ACF* diagram, while the presence of quartz is inferred from the stipulation of excess SiO_2.

Further increase in grade of metamorphism to the amphibolite facies gives rise to the assemblage diopside–anorthite–hornblende–quartz, biotite–anorthite–hornblende–quartz, or biotite–hornblende–quartz. Note that the compositions of biotite and hornblende are variable, as indicated by the shaded regions in the *ACF* diagram. Not all assemblages are diagnostic of the facies.

7-5. Formation of Minerals from Aqueous Fluids

Water is ubiquitous in the crust of the earth, ranging from fractions of 1 percent in the basaltic rocks of the lower crust to over half by weight in some recently deposited sediments. Temperatures, rates of reactions, and species of

Table 7-5 (Cont.)

Si, Fe, (Mg), (CO_2)	*Si, Mg, (Fe), (CO_2)*	*Si, Mg, Ca, (CO_2)*	*Ca, Al, Mg, Si*	*Mg, (Fe), Al, Si*	*Fe, (Mg), Al, Si*	*K, Mg, Fe, Al, Si*
Siderite + quartz	Magnesite + quartz Talc	Dolomite + quartz Talc + calcite Tremolite		Chlorite	Chloritoid	Muscovite + chlorite
Cummingtonite	Talc Serpentine Anthophyllite	Tremolite	Blue-green hornblende	Chlorite	Chloritoid Almandine	Biotite
Cummingtonite	Anthophyllite Forsterite	Tremolite Diopside	Green hornblende	Cordierite	Almandine Staurolite	Biotite
Hypersthene	Enstatite Forsterite	Diopside	Augite	Pyrope–almandine		Orthoclase + pyroxene
Hypersthene	Enstatite Forsterite	Diopside	Brown hornblende Augite	Cordierite Pyrope–almandine		Biotite
Clinohypersthene	Forsterite Pigeonite Clinoenstatite	Diopside Pigeonite Melilite Merwinite	Augite	Cordierite		Orthoclase + pyroxene

minerals formed depend upon the amount and the state of the fluid in the rock. Several roles of aqueous fluids in mineral genesis have been alluded to in the sections on igneous, sedimentary, and metamorphic rocks.

Four possible sources of aqueous fluids are vapors escaping from the final crystallization of a magma, water driven off by the dehydration of hydrous minerals during late stages of progressive metamorphism, interstitial water trapped in sediments and expelled during early stages of progressive metamorphism, and meteoric water heated by passage through rocks warmed by nearby magmatic activity. Actual aqueous fluids represent various mixtures of waters from these several sources.

Identification of the source of a particular aqueous fluid is possible, since the isotopic composition of the oxygen varies with the source of the water. Average seawater contains approximately 0.2 percent O^{18}. Deviations from this standard are given as δ, where

$$\delta = \left(\frac{O^{18}/O^{16}\ \text{sample}}{O^{18}/O^{16}\ \text{standard}} - 1\right) \times 1000$$

which gives O^{18}/O^{16} enrichment or depletion, with respect to standard seawater, in per mil. Normal igneous rocks have O^{18}/O^{16} values typically 6 to 10 per mil heavier than standard ocean water, whereas groundwater of meteoric origin is 5 to 10 per mil lighter. When meteoric water reacts with igneous rock, oxygen isotopes are exchanged, depleting the rock while enriching the water in O^{18}/O^{16}. Thus, by oxygen isotope analysis, the source of aqueous fluids may be partially deduced. However, since their behavior is independent of origin, the effects on rocks and minerals produced by these fluids may be analyzed without the necessity of specifying an ultimate source.

In discussions of hydrothermal fluids and metasomatising solutions, the term water, or aqueous phase, is used without implication of a liquid, as opposed to a gaseous, state. Where these processes take place at temperatures and pressures above the critical end point for pure water (374°, 221 bars) (Fig. 6-1), the distinction between liquid and vapor has no meaning. The presence of other materials dissolved in water may alter substantially the critical end point. In the following discussion, aqueous fluid phase should be taken to mean a fluid that is predominantly H_2O, and does not necessarily show a *miniscus* (liquid–vapor interface).

The conversion of sediments into sedimentary rocks is the process of diagenesis. In this process, water with its dissolved load percolates through the sediments and reacts with them. This water was probably originally seawater, but reactions with the sediments will have changed its composition. Cementation occurs when the dissolved material precipitates in the interstices, bonding the particles together. The most common cementing materials are calcite and dolomite precipitated out by decarbonation; silica, either crystallizing as an overgrowth on quartz grains or coming out of solution as a gel, which later dehydrates to chert (Section 1-18); or hematite precipitating out of solution by oxidation to the ferric state. In addition, *authigenic minerals*, minerals that grow in the sedimentary environment, may crystallize from the interstitial waters, or form by reaction and recrystallization from metastable layered silicates such as illite, glauconite, and vermiculite. Some authigenic minerals are quartz, microcline, albite, zeolites, and muscovite.

Much dolomite in sedimentary rocks appears to be authigenic in origin, although the precise mechanism of dolomitization is not clear. Pacific Ocean atolls are formed by the biogenetic precipitation of calcite, but are progressively dolomitic with increased depth beneath the atoll. An exchange of Mg^{2+} from seawater for some of the Ca^{2+} in the calcite is postulated for the process of dolomitization of the rock. It should be noted that although surface waters in the oceans are saturated with respect to calcite, below approximately 1000 m ocean waters appear to be undersaturated.

When a fracture appears in the crust of the earth, it opens up an area of lower pressure and a channel for the transport of aqueous solutions. When precipitated material fills these fractures, they become veins. Evidence from synthesis of the vein minerals themselves, from the compositions of hot springs, and from the compositions of gases emanating from active and recently active volcanos indicates that most aqueous fluids are over 99 percent H_2O, with the remainder made up of CO_2, HCl, H_2, SO_2, H_2S, and HF. Some fluids, however, are analyzed to be one third or more CO_2, and others have several percent of various metal chlorides and appreciable amounts of sulfur oxide gases.

Perhaps as important as the composition of the aqueous solutions is their density, since, for a given fluid composition, a slight increase in density results in a substantial increase in the solubility of most ionic species. Figure 7-20 is a plot of the density of pure water as a function of pressure and temperature.

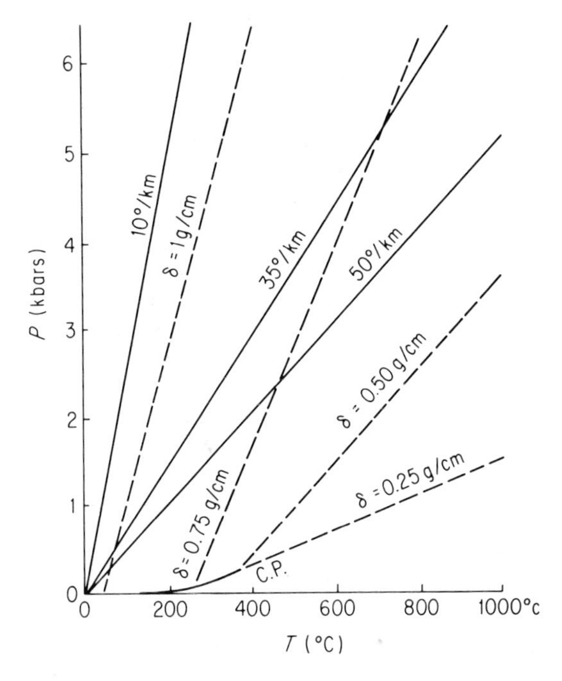

Figure 7-20 Density of H_2O as a function of pressure and temperature. Solid lines represent various geothermal gradients. Solid curve is the liquid–vapor curve for pure water. Dashed lines indicate water of equal density. Under most geologic conditions water has a density between 0.75 and 1.00 g/cm³. (After F. G. Smith, *Physical Geochemistry*. Reading, Mass.: Addison-Wesley Publishing Company, Inc., 1963.)

It is superimposed on different geothermal gradients that exist in different parts of the earth's crust. In sedimentary and in diagenetic environments, especially those in geosynclinal areas, the density of the interstitial fluids, while slightly different from that of pure water, can be expected to be approximately or slightly greater than 1 g/cm³. In high-temperature magmatic environments the density of the fluids will be less than in cooler environments, in the range from 0.50 to 0.75 g/cm^3, which is substantially greater than steam as it is known at the surface of the earth.

Aqueous fluids react in response to wall-rock chemistry and to changing physical environment, altering the mineralogy of the wall rock and precipitating vein minerals from solution. The most common vein minerals are quartz, calcite, and sulfides, of which pyrite predominates. Although most veins are barren, in the economic sense, some contain exploitable amounts of sulfides of economically valuable transition metals, such as Cu, Zn, and Pb, chlorides and tellurides of Ag, and native Au.

Magmas of granitic composition can dissolve up to approximately 10 percent by weight of H_2O, and crystallization of the anhydrous minerals of Petrogeny's Residua System concentrates H_2O and other volatiles in the magma until, at some point, it becomes saturated with vapor. Beyond that point, an aqueous fluid phase separates from the silicate magma. Experimental results indicate that all the components of the melt are to be found in the aqueous phase, but not in the same proportions as in the magma. If, however, the magma is undersaturated with respect to alumina (Section 7-2), complete miscibility is possible between silicate melt and aqueous fluid. In this case, continued crystallization would produce a progressively more hydrous silicate melt.

On the other hand, if, as is more common, the melt is saturated with alumina, a vapor phase separates and coexists with a water-saturated silicate magma. Volatiles, such as Cl^-, F^-, and CO_2, are concentrated in the vapor phase. In addition, the partitioning of the alkalis between melt and vapor results in K^+ being more concentrated in the vapor than Na^+, which means that more K-feldspar is likely to crystallize from the vapor and more albite from the silicate melt under these conditions.

Owing to its lesser density, the aqueous phase occupies a greater volume than its constitutents did while dissolved in the magma, and it is physically unstable in the crust because of its buoyancy. If the aqueous phase escapes into fractures in the rock, carrying with it some silicate melt, crystallization occurs simultaneously from the melt and from the vapor. Crystallization from a vapor phase leads to much larger crystals than does crystallization at the same rate from a melt. Some pegmatite megacrysts are thought to form by this process.

A *pegmatite* is a body of rock, presumably of igneous origin, that is composed of exceptionally large crystals, ranging from a few centimeters in diameter to many meters. Most pegmatites have compositions that are very close to the composition of the thermal minimum in Petrogeny's Residua System. In addition, many pegmatites have exotic minerals containing ions that are not tolerated in the lattices of the major silicate minerals. Included are beryl (Be^{2+}), spodumene (Li^+), and various rare-earth element minerals.

Many pegmatites are layered with the following top-to-bottom sequence:

megacrysts of microcline (triclinic K-feldspar) and cleavelandite (albite) intergrown with quartz, a massive quartz core, and a soda aplite (albite and quartz with a uniform crystalline texture made up of grains 1–2 mm in diameter). This may be interpreted as simultaneous crystallization of the megacrysts from the aqueous vapor, quartz from the vapor–melt interface, and aplite from the melt in such a ratio as to keep the composition of the melt coincident with the minimum in Petrogeny's Residua System.

Similar reasoning may be offered for the origin of aplite veins and dikes. An *aplite* is a sugary-textured crystalline rock with grains 1–2 mm in diameter that occurs in dikes generally less than 5 cm across. The composition is close to the minimum in Petrogeny's Residua System. These appear to crystallize by a rapid quench of a water-rich melt that underwent a pressure drop and consequent degassing when intruded into the country rock.

As the aqueous phase escapes farther from the magma chamber, carrying with it dissolved silicate material and escaping volatile constituents, it is capable of forming veins. Furthermore, material leached from the wall rock in one place may be redeposited under different physical and chemical conditions farther downstream in the vein (that is, at higher levels in the crust).

Evidence from the altered rocks themselves indicates that the replacement of one mineral for another does not necessarily take place on an ion-for-ion basis, but rather on a volume-for-volume basis. In many cases, relict grain structure (pseudomorphism, Section 1-17) of the original rock remains after the original minerals have been partly or completely replaced.

Although the vein-forming fluids have been treated as of magmatic origin in the foregoing discussion, the veining processes are independent of the origin of the veining fluid, and oxygen isotope studies indicate that, in some cases, the bulk of the fluid is of meteoric origin.

As noted in the discussion on metamorphism (Section 7-4), hydrated silicate minerals are formed during low-temperature metamorphic reactions. These same hydrated minerals may be formed directly from igneous minerals or from high-grade metamorphic minerals by reaction with aqueous fluids of the appropriate temperature and composition. The hydration of ferromagnesian minerals can be illustrated by the reaction

$$\underset{\text{forsterite}}{2Mg_2SiO_4} + \underset{\text{water}}{2H_2O} \rightleftharpoons \underset{\text{serpentine}}{Mg_3Si_2O_5(OH)_4} + \underset{\text{magnesia in solution}}{MgO}$$

There is also the reaction

$$\underset{\text{forsterite}}{3Mg_2SiO_4} + \underset{\text{silica}}{SiO_2} + \underset{\text{water}}{4H_2O} \rightleftharpoons \underset{\text{serpentine}}{2Mg_3Si_2O_5(OH)_4}$$

The hydration of olivine to serpentine could follow either reaction, depending upon the composition of the aqueous phase. In actuality, it probably occurs by a combination of the two.

Feldspars are also susceptible to hydrothermal alteration, as illustrated by the reaction

$$3CaAl_2Si_2O_8 + 2H_2O + K_2O \rightleftharpoons 2KAl_2(AlSi_3O_{10})(OH)_2 + 3CaO$$

$$\text{anorthite} + \text{water} + \text{potash} \rightleftharpoons \text{muscovite} + \text{lime}$$

Alkali feldspars could be altered in a similar fashion. Furthermore, the end product could be kaolinite, in which case the reaction between the hydrothermal solution and the feldspar would be the same as that outlined for the weathering of feldspars (Section 7-3). Indeed, in many cases the effects of low-temperature hydrothermal alterations are all but indistinguishable from the effects of weathering.

The preceding discussion has dealt with large amounts of aqueous fluids passing through the rock to give it a new mineralogy. There is the additional possibility that small amounts of water, expelled by the crystallization of a relatively anhydrous magma, remain in the interstices of the rock and alter its mineralogy as it cools below its final crystallization temperature. This process, known as *deuteric alteration*, could bring about reactions similar to those outlined for the hydrothermal alteration of an igneous rock, but on a much smaller scale. Many apparently unweathered igneous rocks show the effects of deuteric alteration to varying degrees.

FURTHER READINGS

Barth, T. F. W., *Theoretical Petrology*, 2nd ed. New York: John Wiley & Sons, Inc., 1962.

Bowen, N. L., *The Evolution of Igneous Rocks*. New York: Dover Publications, Inc., 1956 (originally published 1928).

Krauskopf, K. B., *Introduction to Geochemistry*. New York: McGraw-Hill Book Company, 1967.

Mason, B., *Principles of Geochemistry*, 3rd ed. New York: John Wiley & Sons, Inc., 1966.

Pettijohn, F. J., *Sedimentary Rocks*. New York: Harper & Row, Inc., 1957.

Turner, F. J., and J. Verhoogen, *Igneous and Metamorphic Petrology*, 2nd ed. New York: McGraw-Hill Book Company, 1960.

Tuttle, O. F., and N. L. Bowen, *Origin of Granite in the Light of Experimental Studies*, Memoir 74. New York: The Geological Society of America, 1958.

Winkler, H. G. F., *Petrogenesis of Metamorphic Rocks*. New York: Springer-Verlag New York, Inc., 1965.

Appendix

Appendix

Mineral Data

The following table is intended to provide a quick reference to data about a particular mineral. Free use has been made of standard publications of mineral data, including especially the A.S.T.M. X-Ray Powder Diffraction File; Deer, Howie, and Zussman; and Winchell and Winchell (see below). The 150 species are tabulated according to chemical class, and within each class, isomorphs and other structurally related minerals are grouped together. For a more exhaustive treatment of mineral data, the reader should consult one or more of the following works:

American Society for Testing Materials, *X-ray Diffraction File* and *Index to the X-ray Powder Diffraction File*. Philadelphia: A.S.T.M., 1965.

DEER, W. A., R. A. HOWIE, and J. ZUSSMAN, *Rock-Forming Minerals*, vols. 1–5. New York: John Wiley & Sons, Inc., 1962–63.

———, *An Introduction to the Rock-Forming Minerals*. New York: John Wiley & Sons, Inc., 1966, 528 pp.

PALACHE, C., H. BERMAN, and C. FRONDEL, *Dana's System of Mineralogy*, 3rd ed. New York: John Wiley & Sons, Inc., 1962.

WINCHELL, A. N., and H. WINCHELL, *Elements of Optical Mineralogy, Part II, Descriptions of Minerals*, 4th ed. New York: John Wiley & Sons, Inc., 1951.

Being a chemical classification, this tabulation of mineral data is not suitable for the purpose of determining the identity of unknown minerals, nor is it intended to be more than a quick reference. Various specialized data tables exist for the purpose of mineral identification. For identification on the basis of characteristics observed in hand specimens, see the following:

BERRY, L. G., and B. MASON, *Mineralogy*. San Francisco: W. H. Freeman and Company, 1959, 630 pp.

DIETRICH, R. V., *Mineral tables*. New York: McGraw-Hill Book Company, 1969, 237 pp.

FORD, W. E., *Dana's Textbook of Mineralogy*. 4th ed. New York: John Wiley & Sons, Inc., 1932, 851 pp.

MASON, B., and L. G. BERRY, *Elements of Mineralogy*. San Francisco: W. H. Freeman and Company, 1968, 550 pp.

For identification on the basis of optical properties, see the following:

BLOSS, F. D., *An Introduction to the Methods of Optical Crystallography*. New York: Holt, Rinehart and Winston, Inc., 1961, 294 pp.

KERR, P. F., *Optical Mineralogy*. New York: McGraw-Hill Book Company, 1959, 442 pp.

LARSEN, E. S., and H. BERMAN, *The Microscopic Determination of the Nonopaque Minerals*, 2nd ed. Washington D.C.: U.S. Government Printing Office, 1934, 266 pp.

TRÖGER, W. E., *Optische Bestimmung der gesteinsbildenden Minerale*. Stuttgart: E. Scheizerbart'sche Verlags., 1959, 147 pp.

For identification on the basis of X-ray powder data (see also the A.S.T.M. reference above), see:

Joint Committee on Powder Diffraction Standards, *Powder Diffraction File*. 3rd ed. Swarthmore, Pa., 1972.

	Mineral, composition	System, class, space group	Unit cell dimension (Å)
ELEMENTS			
1.	Copper Cu	Isometric $\frac{4}{m}\bar{3}\frac{2}{m}$ Fm3m	a = 3.615
2.	Gold Au	Isometric $\frac{4}{m}\bar{3}\frac{2}{m}$ Fm3m	a = 4.079
3.	Sulfur S	Orthorhombic $\frac{2}{m}\frac{2}{m}\frac{2}{m}$ Fddd	a = 10.45 b = 12.84 c = 24.46
4.	Graphite C	Hexagonal $\frac{6}{m}\frac{2}{m}\frac{2}{m}$ C6mmc	a = 2.461 c = 6.708
5.	Diamond C	Isometric $\bar{4}3m$ Fd3m	a = 3.5667
HALIDES			
6.	Halite NaCl	Isometric $\frac{4}{m}\bar{3}\frac{2}{m}$ Fm3m	a = 5.6404
7.	Sylvite KCl	Isometric $\frac{4}{m}\bar{3}\frac{2}{m}$ Fm3m	a = 6.2931
8.	Fluorite CaF_2	Isometric $\frac{4}{m}\bar{3}\frac{2}{m}$ Fm3m	a = 5.4630
9.	Cryolite Na_3AlF_6	Monoclinic $\frac{2}{m}$ $P2_1/n$	a = 5.40 b = 5.60 c = 7.78 β = 90°11′
SULFIDES			
10.	Chalcocite Cu_2S	Orthorhombic 2mm Ab2m	a = 11.881 b = 27.323 c = 13.491
11.	Galena PbS	Isometric $\frac{4}{m}\bar{3}\frac{2}{m}$ Fm3m	a = 5.936
12.	Sphalerite ZnS	Isometric $\bar{4}3m$ F$\bar{4}$3m	a = 5.409
13.	Pentlandite $(Ni, Fe)_9S_8$	Isometric $\frac{4}{m}\bar{3}\frac{2}{m}$ Fm3m	a = 10.196

Specific gravity (G), hardness (H)	*Cleavage*	*Luster, color, streak*	*Refractive index, optic sign, 2V*	*Paragenesis*
$G = 8.94$ $H = 2\frac{1}{2}$–3	—	Metallic, copper red, copper streak	Opaque	Reduction of hydrothermal solutions by basic rocks
$G = 19.23$ $H = 2\frac{1}{2}$–3	—	Metallic, gold	Opaque	Hydrothermal, placer
$G = 2.0$–2.1 $H = 1\frac{1}{2}$–2	{001} {110} {111} Poor	Waxy, yellow, white streak	$\alpha = 1.960$ $\beta = 2.040$ $\gamma = 2.248$ $2V = 68°58'$ (+)	Volcanic, sedimentary
$G = 2.25$ $H = 1$	{0001} Perfect	Metallic, gray, gray streak	Opaque	Metamorphosed carbonaceous material
$G = 3.52$ $H = 10$	{111} Perfect	Adamantine, colorless to pale pastel	$n = 2.417$	Volcanic necks (kimberlite), placer
$G = 2.17$ $H = 2\frac{1}{2}$	{100} Perfect	Vitreous, colorless, yellow, gray, blue, purple, white streak	$n = 1.5443$	Evaporites, sublimates
$G = 1.99$ $H = 2$	{100} Perfect	Vitreous, colorless, gray white streak	$n = 1.490$	Evaporites
$G = 3.18$ $H = 4$	{111} Perfect	Vitreous, colorless, white, yellow, green, blue, purple, white streak	$n = 1.434$	Hydrothermal, pegmatite, veins
$G = 2.97$ $H = 2\frac{1}{2}$–3	{001} {110} {$\bar{1}$01} Parting	Vitreous to greasy, colorless, white, blue, white streak	$\alpha = 1.338$ $\beta = 1.338$ $\gamma = 1.339$ $2V = 43°$ (+)	Granite, pegmatite
$G = 5.78$ $H = 2\frac{1}{2}$–3	{110} Poor	Metallic, lead gray, gray streak	Opaque	Hydrothermal supergene enrichment zones
$G = 7.5$ $H = 2\frac{1}{2}$	{100} Perfect	Metallic, gray, dark gray streak	Opaque	Hydrothermal
$G = 4.1$ $H = 3\frac{1}{2}$–4	{110}	Resinous to adamantine, black, brown, yellow, white, brown streak	$n = 2.37$	Hydrothermal
$G = 4.6$–5 $H = 3\frac{1}{2}$–4	{111} Good	Metallic, brown, black streak	Opaque	With mafic igneous rocks

	Mineral, composition	*System, class, space group*	*Unit cell dimension* (Å)
14.	Cinnabar HgS	Trigonal 32 $P3_121$ or $P3_221$	$a = 4.149$ $c = 9.459$
15.	Pyrrhotite FeS	Hexagonal $\frac{6}{m}\frac{2}{m}\frac{2}{m}$ $P6_3/mmc$	$a = 3.43$ $c = 5.68$
16.	Covellite CuS	Hexagonal $\frac{6}{m}\frac{2}{m}\frac{2}{m}$ $P6_3/mmc$	$a = 3.792$ $c = 16.344$
17.	Bornite Cu_5FeS_4	Tetragonal $\bar{4}2m$ $P\bar{4}2_1c$	$a = 10.94$ $c = 21.88$
18.	Chalcopyrite $CuFeS_2$	Tetragonal $\bar{4}2m$ $I42d$	$a = 5.299$ $c = 10.434$
19.	Pyrite FeS_2	Isometric $\frac{2}{m}\bar{3}$ $Pa3$	$a = 5.417$
20.	Marcasite FeS_2	Orthorhombic $\frac{2}{m}\frac{2}{m}\frac{2}{m}$ $Pnnm$	$a = 4.443$ $b = 5.423$ $c = 3.388$
21.	Arsenopyrite FeAsS	Monoclinic $\frac{2}{m}$ $P2_1/c$	$a = 5.744$ $b = 5.675$ $c = 5.785$ $\beta = 112.17°$
22.	Molybdenite MoS_2	Hexagonal $\frac{6}{m}\frac{2}{m}\frac{2}{m}$ $P6_3/mmc$	$a = 3.160$ $c = 12.295$
23.	Tetrahedrite Cu_3SbS_3	Isometric $\bar{4}3m$ $I\bar{4}3m$	$a = 10.327$
24.	Stibnite Sb_2S_3	Orthorhombic $\frac{2}{m}\frac{2}{m}\frac{2}{m}$ $Pbnm$	$a = 11.229$ $b = 11.310$ $c = 3.839$
25.	Realgar AsS	Monoclinic $2/m$ $P2_1/m$	$a = 9.29$ $b = 13.53$ $c = 6.57$ $\beta = 106°33'$
26.	Orpiment As_2S_3	Monoclinic $2/m$ $P2_1/n$	$a = 11.49$ $b = 9.59$ $c = 4.24$ $\beta = 90°27'$
OXIDES			
27.	Ice H_2O	Hexagonal $6mm$ $P6_3/mmc$	$a = 4.521$ $c = 7.366$

Specific gravity (G), hardness (H)	*Cleavage*	*Luster, color, streak*	*Refractive index, optic sign, 2V*	*Paragenesis*
$G = 8.1$ $H = 2$–$2\frac{1}{2}$	$\{10\bar{1}0\}$ Perfect	Adamantine, red, scarlet streak	$\omega = 2.913$ $\varepsilon = 3.272\ (+)$	Hydrothermal
$G = 4.6$ $H = 4$	{0001}	Metallic, bronze, brown, dark gray streak	Opaque	Hydrothermal
$G = 4.68$ $H = 1\frac{1}{2}$–2	{0001}	Metallic, bluish black, black streak	Opaque	Hydrothermal
$G = 4.9$–5.3 $H = 3$	{100} Imperfect	Metallic, iridescent purple tarnish, dark gray streak	Opaque	Hydrothermal
$G = 4.1$–4.3 $H = 3\frac{1}{2}$–4	{011} {111} Poor	Metallic, brass yellow with iridescent tarnish, greenish-black streak	Opaque	Hydrothermal veins
$G = 4.95$–5.03 $H = 6$–$6\frac{1}{2}$	{001} Poor	Metallic, brass yellow, black streak	Opaque	Hydrothermal veins, shales, slates, igneous accessory
$G = 4.8$ $H = 6$–$6\frac{1}{2}$	{110} Poor	Metallic, white to brass yellow, black streak	Opaque	Hydrothermal, shales
$G = 5.9$–6.2 $H = 5.5$–6	{110} Poor	Metallic, white to steel gray, black streak	Opaque	Hydrothermal
$G = 4.7$–4.8 $H = 1$–$1\frac{1}{2}$	{0001} Perfect	Metallic, bluish gray, gray streak (greenish on glazed porcelain)	Opaque	Granites, pegmatites
$G = 4.4$–5.4 $H = 3$–4	—	Metallic, steel gray, reddish-gray streak	Opaque	Hydrothermal veins
$G = 4.6$–4.7 $H = 2$	{010} Perfect {100} {110} Fair	Metallic, gray, gray streak	Opaque	Hydrothermal
$G = 3.56$ $H = 1\frac{1}{2}$–2	{010} Perfect	Vitreous, red, yellowish-orange streak	$\alpha = 2.54$ $\beta = 2.68$ $\gamma = 2.70$ $2V = 40°34'\ (-)$	Hydrothermal veins, sublimates
$G = 3.49$ $H = 1\frac{1}{2}$–2	{010} Perfect	Vitreous, lemon yellow, yellow streak	$\alpha = 2.4$ $\beta = 2.81$ $\gamma = 3.02$ $2V = 76°\ (-)$	Hydrothermal veins, sublimates
$G = 0.92$ $H = 1\frac{1}{2}$	—	Vitreous, colorless, white streak	$\omega = 1.309$ $\varepsilon = 1.311$	Sedimentary (snow), igneous (sea ice), metamorphic (glaciers)

	Mineral, composition	System, class, space group	Unit cell dimension (Å)
28.	Cuprite Cu_2O	Isometric 432 Pn3m	$a = 4.2696$
29.	Periclase MgO	Isometric $\frac{4}{m}\bar{3}\frac{2}{m}$ Fm3m	$a = 4.213$
30.	SPINEL GROUP: Spinel $MgAl_2O_4$	Isometric $\frac{4}{m}\bar{3}\frac{2}{m}$ Fd3m	$a = 8.080$
31.	SPINEL GROUP: Magnetite $FeFe_2O_4$		$a = 8.394$
32.	SPINEL GROUP: Chromite $FeCr_2O_4$		$a = 8.378$
33.	Chrysoberyl Al_2BeO_4	Orthorhombic $\frac{2}{m}\frac{2}{m}\frac{2}{m}$ Pnma	$a = 9.404$ $b = 5.476$ $c = 4.427$
34.	Corundum Al_2O_3	Trigonal $\bar{3}\frac{2}{m}$ R$\bar{3}c$	$a = 4.760$ $c = 12.98$
35.	Hematite Fe_2O_3	Trigonal $\bar{3}\frac{2}{m}$ R$\bar{3}c$	$a = 5.033$ $c = 13.749$
36.	Ilmenite $FeTiO_4$	Trigonal $\bar{3}$ R$\bar{3}$	$a = 5.093$ $c = 14.055$
37.	Perovskite $CaTiO_3$	Monoclinic $\frac{2}{m}$ P$2_1/m$	a, b, c 7.59–7.71 Pseudocubic
38.	Quartz SiO_2	Trigonal 32 $P3_121$ or $P3_221$	$a = 4.913$ $c = 5.405$
39.	Chalcedony SiO_2	Fibrous microcrystals of quartz with submicroscopic pores	
40.	Tridymite SiO_2	Orthorhombic $\frac{2}{m}\frac{2}{m}\frac{2}{m}$ Fmm, Fmmm, or F222	$a = 9.92$ $c = 81.5$ (Pseudohexagonal)
41.	Cristobalite SiO_2	Tetragonal $\frac{4}{m}\frac{2}{m}\frac{2}{m}$ $P4_12_12$ or $P4_32_12$	$a = 4.97$ $c = 6.92$
42.	Opal SiO_2	Amorphous	

Specific gravity (G), hardness (H)	*Cleavage*	*Luster, color, streak*	*Refractive index, optic sign, 2V*	*Paragenesis*
$G = 5.8–6.2$ $H = 3\frac{1}{2}–4$	{111} Good	Submetallic to earthy luster, red to black, brownish-red streak	$n = 2.849$	Oxidation of Cu ores
G = 3.58 $H = 5\frac{1}{2}–6$	{100} Perfect	Vitreous, colorless, white, yellow, green, white streak	$n = 1.732$	Contact metamorphism of dolomite
$G = 3.55$ $H = 8$	{111} Parting	Vitreous, colorless, white, red, pink, blue, white streak	$n = 1.719$	Contact metamorphic (limestone), placers
$G = 5.20$ $H = 5\frac{1}{2}$	None	Metallic, black, black streak	Opaque	Magmatic, hydrothermal, metamorphic, placers
$G = 5.09$ $H = 5\frac{1}{2}$	None	Metallic, black, brown streak	Opaque	Ultramafic igneous
$G = 3.72$ $H = 8\frac{1}{2}$	{010} Good	Vitreous, greenish, yellow-green, white streak	$\alpha = 1.745$ $\beta = 1.748$ $\gamma = 1.756$ 2V = 45° (+)	Pegmatites, aplite
$G = 3.98$ $H = 9$	{0001} {10$\bar{1}$1} Parting	Adamantine, white, gray, blue, red, yellow, green, white streak	$\omega = 1.786$ $\varepsilon = 1.760$ (−)	Silica-poor pegmatites, hydrothermal
$G = 5.25$ $H = 5–6$	{0001} {10$\bar{1}$1} Parting	Metallic-earthy, black, silver, red, blood-red streak	$\omega = 3.15–3.22$ $\varepsilon = 2.87–2.34$ (−)	Scarns, Precambrian metamorphosed jasper, sedimentary
$G = 4.70–4.78$ $H = 5–6$	{0001} {10$\bar{1}$0} Parting	Metallic, back, blackish-brown streak	Opaque	Igneous accessory
$G = 3.98–4.26$ $H = 5\frac{1}{2}$	{001} Parting	Adamantine, reddish brown to black, off-white streak	α 2.30 β 2.38 γ (+)	Igneous accessory in mafic alkaline rocks
$G = 2.65$ $H = 7$	None	Vitreous, colorless or colored, white streak	$\omega = 1.544$ $\varepsilon = 1.553$ (+)	Widely distributed except with olivine and feldspathoids
$G = 2.60$ $H = 6$	None	Resinous, colorless or colored, white streak	$\omega = 1.530–1.533$ $\varepsilon = 1.538–1.543$	Hydrothermal
$G = 2.27$ $H = 7$	None	Vitreous, colorless or white, white streak	$\alpha = 1.471–1.479$ $\beta = 1.472–1.480$ $\gamma = 1.474–1.483$ 2V = 40–90° (+)	Volcanic, meteorites
$G = 2.33$ $H = 6–7$	None	Vitreous, white or colorless, white streak	$\omega = 1.487$ $\varepsilon = 1.484$ (−)	Volcanic, meteorites
$G = 2.1–2.2$ $H = 5\frac{1}{2}–6\frac{1}{2}$	None	Vitreous, colored or colorless, white streak	$n = 1.3–1.45$	Hydrothermal and sedimentary

	Mineral, composition	System, class, space group	Unit cell dimension (Å)
43.	Rutile TiO_2	Tetragonal $\frac{4}{m}\frac{2}{m}\frac{2}{m}$ $P4_2mnm$	$a = 4.594$ $c = 2.961$
44.	Cassiterite SnO_2	Tetragonal $\frac{4}{m}\frac{2}{m}\frac{2}{m}$ $P4/mnm$	$a = 4.738$ $c = 3.188$
45.	Pyrolusite MnO_2	Tetragonal $\frac{4}{m}\frac{2}{m}\frac{2}{m}$ $P4_2/mnm$	$a = 4.338$ $c = 2.865$
46.	Psilomelane $(Ba, H_2O)_2Mn_5O_{10}$	Monoclinic $2/m$ $C2/n$	$a = 13.94$ $b = 2.846$ $c = 9.683$ $\beta = 92°32'$
HYDROXIDES			
47.	Gibbsite $Al(OH)_3$	Monoclinic $2/m$ $P2_1/n$	$a = 8.68$ $b = 5.07$ $c = 9.72$ $\beta = 94°34'$
48.	Brucite $Mg(OH)_2$	Trigonal $\bar{3}m$ $P\bar{3}m1$	$a = 3.417$ $c = 4.769$
49.	Diaspore α-AlO(OH)	Orthorhombic $\frac{2}{m}\frac{2}{m}\frac{2}{m}$ $Pbnm$	$a = 4.40$ $b = 9.42$ $c = 2.84$
50.	Goethite α-FeO(OH)	Orthorhombic $\frac{2}{m}\frac{2}{m}\frac{2}{m}$ $Pbnm$	$a = 4.596$ $b = 9.957$ $c = 3.021$
51.	Boehmite γ-AlO(OH)	Orthorhombic $\frac{2}{m}\frac{2}{m}\frac{2}{m}$ $Cmcm$	$a = 2.868$ $b = 12.227$ $c = 3.700$
52.	Lepidocrocite γ-FeO(OH)	Orthorhombic $\frac{2}{m}\frac{2}{m}\frac{2}{m}$ $Amam$	$a = 3.87$ $b = 12.53$ $c = 3.06$
53.	Manganite MnO(OH)	Monoclinic $\frac{2}{m}$ $P2_1/c$	$a = 8.94$ $b = 5.28$ $c = 5.74$ $\beta \cong 90°$
CARBONATES			
54.	CALCITE GROUP: Calcite $CaCO_3$	Trigonal $\bar{3}\frac{2}{m}$ $R\bar{3}c$	$a = 4.990$ $c = 17.064$
55.	Magnesite $MgCO_3$		$a = 4.6330$ $c = 15.016$
56.	Siderite $FeCO_3$		$a = 4.69$ $c = 15.37$
57.	Smithsonite $ZnCO_3$		$a = 4.653$ $c = 15.025$
58.	Rhodochrosite $MnCO_3$		$a = 4.777$ $c = 15.664$

Specific gravity (G), hardness (H)	*Cleavage*	*Luster, color, streak*	*Refractive index, optic sign, 2V*	*Paragenesis*
$G = 4.25$ $H = 6–6\frac{1}{2}$	{110} Good {100} Fair	Vitreous, red, brown, black, green, light brown streak	$\omega = 2.605–2.613$ $\varepsilon = 2.899–2.901$ (+)	Accessory in granites and in metamorphic rocks
$G = 6.89–7.02$ $H = 6–7$	{100} {110} Poor	Vitreous, reddish brown to black, yellowish streak	$\omega = 1.990–2.001$ $\varepsilon = 2.093–2.098$ (+)	Associated with felsic igneous rocks
$G = 5.0$ $H = 2–6$	{110} Excellent	Metallic, gray to black, black streak	Opaque	Sedimentary and weathered zone
$G = 3.3–4.7$ $H = 2–6$	None	Submetallic, dull, dark gray to black, brownish-black streak	Opaque	Sedimentary and weathered zone
$G = 2.40$ $H = 2\frac{1}{2}–3\frac{1}{2}$	{001} Perfect	Pearly, white to off-white, white streak	$\alpha = 1.566–1.568$ $\beta = 1.566–1.568$ $\gamma = 1.587–1.589$ 2V = 0–40° (+)	Tropical weathering product, in bauxites
$G = 2.39$ $H = 2\frac{1}{2}$	{0001} Perfect	Pearly to waxy, white to off-white, white streak	$\omega = 1.560–1.590$ $\varepsilon = 1.580–1.600$ (+)	Altered from periclase in marbles
$G = 3.2–3.5$ $H = 6\frac{1}{2}–7$	{010} Perfect	Vitreous, colorless, white to off-white, white streak	$\alpha = 1.682–1.706$ $\beta = 1.705–1.725$ $\gamma = 1.730–1.752$ 2V = 84–86° (+)	Tropical weathering product, in bauxites
$G = 4.3$ $H = 5–5\frac{1}{2}$	{010} Excellent {100} Fair	Adamantine-submetallic, yellow to brownish black, yellow-ochre streak	$\alpha = 2.260–2.275$ $\beta = 2.393–2.409$ $\gamma = 2.398–2.515$ 2V =0–27° (−)	Weathering product, in limonite and laterite
$G \simeq 3$ $H = 3\frac{1}{2}–4$	{010} Excellent	Earthy, white, white streak	$\alpha = 1.64–1.65$ $\beta = 1.65–1.66$ $\gamma = 1.65–1.67$ 2V ≃ 80° (+)	Tropical weathering product, in bauxites
$G = 4.09$ $H = 5$	{010} Excellent {100}, {001} Fair	Adamantine-submetallic, reddish brown, yellow-ochre streak	$\alpha = 1.94$ $\beta = 2.20$ $\gamma = 2.51$ 2V = 83° (−)	Weathing product, in limonite
$G = 4.3$ $H = 4$	{010} {110} Excellent	Submetallic, dark gray to black, streak reddish brown to black	$\alpha = 2.25$ $\beta = 2.25$ $\gamma = 2.53$ 2V = small (+)	Alteration product of veins in felsic volcanics
$G = 2.72$ $H = 3$	{10$\bar{1}$1} perfect	Vitreous, colorless or colored white streak	$\omega = 1.658$ $\varepsilon = 1.486$ (−)	Limestones, marbles, carbonatites, veins
$G = 2.98$ $H = 4–4\frac{1}{2}$	{10$\bar{1}$1} Perfect	Vitreous, white or colorless, white streak	$\omega = 1.700$ $\varepsilon = 1.509$ (−)	Metamorphic schists and marbles
$G = 3.96$ $H = 4–4\frac{1}{2}$	{10$\bar{1}$1} Perfect	Vitreous, yellow to brown, white streak	$\omega = 1.875$ $\varepsilon = 1.635$ (−)	Chemical precipitation in sediments
$G = 4.4$ $H = 5$	{10$\bar{1}$1} Perfect	Vitreous, white or colored, white streak	$\omega = 1.849$ $\varepsilon = 1.621$ (−)	Weathering product
$G = 3.6$ $H = 3\frac{1}{2}–4$	{10$\bar{1}$1} Perfect	Vitreous, pink, red, or brown, white streak	$\omega = 1.816$ $\varepsilon = 1.600$ (−)	Limestones, marbles, carbonatites, veins

	Mineral, composition	System, class, space group	Unit cell dimension (Å)
59.	Dolomite $CaMg(CO_3)_2$	Trigonal $\bar{3}$ $R\bar{3}$	$a = 4.808$ $c = 16.010$
60.	ARAGONITE GROUP: Aragonite $CaCO_3$	Orthorhombic $\frac{2}{m}\frac{2}{m}\frac{2}{m}$ $Pmcn$	$a = 4.959$ $b = 7.968$ $c = 5.741$
61.	ARAGONITE GROUP: Strontiante $SrCO_3$	Orthorhombic $\frac{2}{m}\frac{2}{m}\frac{2}{m}$ $Pmcn$	$a = 5.107$ $b = 8.414$ $c = 6.029$
62.	ARAGONITE GROUP: Witherite $BaCO_3$	Orthorhombic $\frac{2}{m}\frac{2}{m}\frac{2}{m}$ $Pmcn$	$a = 5.314$ $b = 8.904$ $c = 6.430$
63.	ARAGONITE GROUP: Cerussite $PbCO_3$	Orthorhombic $\frac{2}{m}\frac{2}{m}\frac{2}{m}$ $Pmcn$	$a = 5.195$ $b = 8.436$ $c = 6.152$
64.	Azurite $Cu_3(OH)_2(CO_3)_2$	Monoclinic $2/m$ $P2_1/c$	$a = 5.008$ $b = 5.844$ $c = 10.336$ $\beta = 92.45°$
65.	Malachite $Cu_2(OH)_2CO_3$	Monoclinic $2/m$ $P2_1/a$	$a = 9.502$ $b = 11.974$ $c = 3.240$ $\beta = 98°45'$
BORATE			
66.	Borax $Na_2B_4O_7 \cdot 10H_2O$	Monoclinic $2/m$ $C2/c$	$a = 11.858$ $b = 10.674$ $c = 12.197$ $\beta = 106°41'$
SULFATES			
67.	Celestite $SrSO_4$	Orthorhombic $\frac{2}{m}\frac{2}{m}\frac{2}{m}$ $Pnma$	$a = 8.359$ $b = 5.352$ $c = 6.866$
68.	Barite $BaSO_4$	Orthorhombic $\frac{2}{m}\frac{2}{m}\frac{2}{m}$ $Pnma$	$a = 8.878$ $b = 5.540$ $c = 7.152$
69.	Anglesite $PbSO_4$	Orthorhombic $\frac{2}{m}\frac{2}{m}\frac{2}{m}$ $Pnma$	$a = 8.480$ $b = 5.398$ $c = 6.958$
70.	Anhydrite $CaSO_4$	Orthorhombic $\frac{2}{m}\frac{2}{m}\frac{2}{m}$ $Amma$ or $Ccmm$	$a = 6.991$ $b = 6.996$ $c = 6.238$
71.	Gypsum $CaSO_4 \cdot 2H_2O$	Monoclinic $2/m$ $C2/c$	$a = 5.68$ $b = 15.18$ $c = 6.29$ $\beta = 113°50'$

Specific gravity (G), hardness (H)	*Cleavage*	*Luster, color, streak*	*Refractive index, optic sign, 2V*	*Paragenesis*
$G = 2.86$ $H = 3\frac{1}{2}$–4	{10$\bar{1}$1} Perfect	Vitreous, white or pale gray, white streak	$\omega = 1.679$ $\varepsilon = 1.500\ (-)$	Carbonate sediments, veins
$G = 2.94$ $H = 3\frac{1}{2}$–4	{010} Fair {110} Poor	Vitreous, colorless or white, white streak	$\alpha = 1.530$ $\beta = 1.680$ $\gamma = 1.685$ 2V = 18° (−)	Carbonate sediments, cavities in volcanic rocks
$G \cong 3.72$ $H = 3\frac{1}{2}$	{110} Good {021}, {010} Poor	Vitreous, colorless, white, off-white, white streak	$\alpha = 1.518$ $\beta = 1.665$ $\gamma = 1.667$ 2V = 8° (−)	Hydrothermal veins
$G = 4.30$ $H = 3\frac{1}{2}$	{010} Distinct {110}, {012} Poor	Vitreous, colorless, white or off-white, white streak	$\alpha = 1.529$ $\beta = 1.676$ $\gamma = 1.677$ 2V = 16° (−)	Hydrothermal
$G = 6.57$ $H = 3$–$3\frac{1}{2}$	{110}, {021} Poor	Vitreous, colorless, white, off-white, white streak	$\alpha = 1.803$ $\beta = 2.074$ $\gamma = 2.076$ 2V = 8° (−)	Carbonation of galena in oxidation zone
$G = 3.80$ $H = 3\frac{1}{2}$–4	{100} Good {110} Poor	Vitreous, blue, blue streak	$\alpha = 1.730$ $\beta = 1.758$ $\gamma = 1.838$ 2V = 68° (+)	Oxidation zone of Cu deposits
$G = 4.0$ $H = 3\frac{1}{2}$–4	{001} Perfect {010} Fair	Vitreous, green, pale green streak	$\alpha = 1.655$ $\beta = 1.875$ $\gamma = 1.909$ 2V = 43° (−)	Oxidation zone of Cu deposits
$G = 1.7$ $H = 2$	{100} Excellent {110} Good	Vitreous to earthy white, grayish, bluish, white streak	$\alpha = 1.447$ $\beta = 1.469$ $\gamma = 1.472$ 2V = 39°36′ (−)	Saline-lake precipitate
$G = 3.97$ $H = 3$–$3\frac{1}{2}$	{001} Perfect {210} Good {010} Poor	Vitreous, colorless, white, off-white, white streak	$\alpha = 1.622$ $\beta = 1.624$ $\gamma = 1.631$ 2V = 51° (+)	Hydrothermal veins
$G = 4.5$ $H = 2\frac{1}{2}$–$3\frac{1}{2}$	{001} Perfect {210} Good {010} Poor	Vitreous, white, red, brown, gray, white streak	$\alpha = 1.636$ $\beta = 1.637$ $\gamma = 1.648$ 2V = 37° (+)	Hydrothermal veins
$G = 6.3$ $H = 3$	{001}, {210} Good	Adamantine to greasy, white and colored, white streak	$\alpha = 1.877$ $\beta = 1.882$ $\gamma = 1.894$ 2V = 36° (+)	Oxidation zone of Pb deposits
$G = 2.96$ $H = 3$–$3\frac{1}{2}$	{010} Perfect {100} Very good {001} Good	Pearly-vitreous white, grayish, reddish, off-white streak	$\alpha = 1.5700$ $\beta = 1.5757$ $\gamma = 1.6138$ 2V = 42° (−)	Evaporite deposits
$G = 2.30$–2.37 $H = 2$	{010} Perfect {100}, {011} Very good	Vitreous, colorless, white, gray, white streak	$\alpha = 1.5205$ $\beta = 1.5226$ $\gamma = 1.5296$ 2V = 58° (+)	Evaporite beds

	Mineral, composition	System, class, space group	Unit cell dimension (Å)
VANADATE			
72.	Carnotite $K_2(UO_2)_2(VO_4)_2(OH)_2$	Monoclinic $2/m$	$a = 10.47$ $b = 8.41$ $c = 6.59$ $\beta = 103°50'$
PHOSPHATES			
73.	Xenotime YPO_4	Tetragonal $\frac{4}{m}\frac{2}{m}\frac{2}{m}$ $I4_1/amd$	$a = 6.885$ $c = 5.982$
74.	Monazite $CePO_4$	Monoclinic $2/m$ $P2_1/n$	$a = 6.78$ $b = 7.00$ $c = 6.45$ $\beta = 104°$
75.	Lazulite $(Mg, Fe)Al_2(PO_4)_2(OH, F)_2$	Monoclinic $2/m$ $P2_1/c$	$a = 7.15$ $b = 7.28$ $c = 7.25$ $\beta = 102°35'$
76.	Apatite $Ca_5(PO_4)_3(OH, F, Cl)$	Hexagonal $6/m$ $P6_3/m$	$a = 9.43–9.63$ $c = 6.76–6.88$
77.	Turquois $CuAl_6(PO_4)_4(OH)_8 \cdot 5H_2O$	Triclinic $\bar{1}$	$a = 7.49$ $b = 9.95$ $c = 7.69$ $\alpha = 111°39'$ $\beta = 115°23'$ $\gamma = 69°26'$
ORTHOSILICATES			
78.	OLIVINE GROUP: Forsterite Mg_2SiO_4	Orthorhombic $\frac{2}{m}\frac{2}{m}\frac{2}{m}$ $Pbnm$	$a = 4.758$ $b = 10.214$ $c = 5.984$
79.	OLIVINE GROUP: Fayalite Fe_2SiO_4		$a = 4.817$ $b = 10.477$ $c = 6.105$
80.	Monticellite $CaMgSiO_4$	Orthorhombic $\frac{2}{m}\frac{2}{m}\frac{2}{m}$ $Pbnm$	$a = 4.827$ $b = 11.084$ $c = 6.376$
81.	Zircon $ZrSiO_4$	Tetragonal $\frac{4}{m}\frac{2}{m}\frac{2}{m}$ $I4/amd$	$a = 6.60$ $c = 5.98$
82.	GARNET GROUP: Almandine $Fe_3Al_2Si_3O_{12}$	Isometric $\frac{4}{m}\bar{3}\frac{2}{m}$ $Ia3d$	$a = 11.526$
83.	GARNET GROUP: Andradite $Ca_3Fe_2Si_3O_{12}$		$a = 12.048$
84.	GARNET GROUP: Grossular $Ca_3Al_2Si_3O_{12}$		$a = 11.851$
85.	GARNET GROUP: Pyrope $Mg_3Al_2Si_3O_{12}$		$a = 11.459$
86.	Sillimanite Al_2SiO_4	Orthorhombic $\frac{2}{m}\frac{2}{m}\frac{2}{m}$ $Pbnm$ or $Pnma$	$a = 7.484$ $b = 7.673$ $c = 5.771$

Specific gravity (G), hardness (H)	*Cleavage*	*Luster, color, streak*	*Refractive index, optic sign, 2V*	*Paragenesis*
$G = 4.5$ $H = 4$	{001} Perfect	Adamantine, yellow, pale yellow streak	$\alpha = 1.77$ $\beta = 2.01$ $\gamma = 2.09$ $2V = 52.5° (-)$	In sandstones
$G = 4.55$ $H = 4–5$	{110} Good	Adamantine to greasy, yellowish to reddish brown, pale reddish-brown streak	$\omega = 1.720$ $\varepsilon = 1.820 (+)$	Accessory in felsic plutonic rocks, also placer
$G = 5.0–5.3$ $H = 5–5\frac{1}{2}$	{001} Perfect {100} Fair	Adamantine to greasy, yellowish to reddish brown, pale reddish-brown streak	$\alpha = 1.786–1.800$ $\beta = 1.788–1.801$ $\gamma = 1.837–1.849$ $2V = 6–19° (+)$	Accessory in felsic plutonic rocks, also placer
$G = 3.08$ $H = 5–6$	{110}, {101} Poor	Adamantine, blue, white streak	$\alpha = 1.614$ $\beta = 1.633$ $\gamma = 1.644$ $2V = 69° (-)$	Hydrothermal veins
$G = 3.1–3.35$ $H = 5$	{0001} {10$\bar{1}$0} Poor	Adamantine-greasy, all colors, white streak	$\omega = 1.629–1.667$ $\varepsilon = 1.624–1.666 (-)$	Accessory in igneous rocks, hydrothermal veins, sedimentary
$G = 2.84$ $H = 5–6$		Vitreous, blue, green, white streak	$\alpha = 1.61$ $\beta = 1.62$ $\gamma = 1.65$ $2V = 40° (+)$	Hydrothermal veins
$G = 3.22$ $H = 7$	{010}, {100} Good	Vitreous, colorless, yellow, pale green, white streak	$\alpha = 1.635$ $\beta = 1.651$ $\gamma = 1.670$ $2V = 82° (+)$	Accessory to major mineral in mafic and ultramafic rocks
$G = 4.39$ $H = 6\frac{1}{2}$	{010} Fair	Vitreous, yellow green, dark green, black, white streak	$\alpha = 1.827$ $\beta = 1.869$ $\gamma = 1.879$ $2V = 46° (-)$	Accessory in felsic igneous rocks, some metamorphic
$G = 3.03$ $H = 5\frac{1}{2}$	{010} Poor	Vitreous, colorless, gray, white streak	$\alpha = 1.641$ $\beta = 1.649$ $\gamma = 1.655$ $2V = 80° (-)$	Metamorphosed siliceous dolomites
$G = 4.6–4.7$ $H = 7\frac{1}{2}$	{110} Imperfect	Adamantine, colorless, reddish, yellowish, gray	$\omega = 1.923–1.960$ $\varepsilon = 1.986–2.015 (+)$	Accessory in felsic rocks, placers
$G = 4.32$ $H = 6–7\frac{1}{2}$	—	Vitreous, red to brown, white streak	$n = 1.830$	Regional metamorphism of argillaceous sediments
$G = 3.76$ $H = 6–7\frac{1}{2}$	—	Vitreous, brown, green, black, white streak	$n = 1.887$	Skarns, contact metamorphism of calcareous sediments
$G = 3.59$ $H = 6–7\frac{1}{2}$	—	Vitreous, green, white streak	$n = 1.734$	Impure marbles
$G = 3.58$ $H = 6–7\frac{1}{2}$	—	Vitreous, bright red, white streak	$n = 1.714$	Ultramafic rocks, placers
$G = 3.23–3.27$ $H = 6\frac{1}{2}–7\frac{1}{2}$	{010} Good	Vitreous, colorless, white, gray, white streak	$\alpha = 1.654–1.661$ $\beta = 1.658–1.662$ $\gamma = 1.673–1.683$ $2V = 21–30°$	High-grade metamorphosed pelitic rocks

	Mineral, composition	*System, class, space group*	*Unit cell dimension* (Å)
87.	Andalusite Al_2SiO_5	Orthorhombic $\frac{2}{m}\frac{2}{m}\frac{2}{m}$ P*nnm*	$a = 7.796$ $b = 7.898$ $c = 5.558$
88.	Kyanite Al_2SiO_5	Triclinic $\bar{1}$ $P\bar{1}$	$a = 7.123$ $b = 7.848$ $c = 5.564$ $\alpha = 89°55'$ $\beta = 101°15'$ $\gamma = 105°58'$
89.	Topaz $Al_2SiO_4(OH, F)_2$	Orthorhombic $\frac{2}{m}\frac{2}{m}\frac{2}{m}$ P*mnb*	$a = 8.394$ $b = 8.792$ $c = 4.650$
90.	Staurolite $(Fe, Mg)_2(Al, Fe)_9O_6(SiO_4)_4(O, OH)_2$	Monoclinic 2/*m* C2/*m*	$a = 7.83–7.95$ $b = 16.50–16.82$ $c = 5.55–5.71$ $\beta = 90°$
91.	Sphene $CaTiSiO_4(O, OH, F)$	Monoclinic 2/*m* C2/*c*	$a = 7.44$ $b = 8.72$ $c = 6.56$ $\beta = 119°43'$
SOROSILICATES			
92.	Lawsonite $CaAl_2Si_2O_7(OH)_2H_2O$	Orthorhombic 222 C*cmm*	$a = 8.79$ $b = 5.84$ $c = 13.12$
93.	Clinozoisite $Ca_2Al_3(Si_2O_7)(SiO_4)O(OH)$	Monoclinic 2/*m* P2₁/*m*	$a = 8.887$ $b = 5.581$ $c = 10.14$ $\beta = 115°56'$
94.	Epidote $Ca_2(Al, Fe)Al_2(Si_2O_7)(SiO_4)O(OH)$	Monoclinic 2/*m* $P2_1/m$	$a = 8.88–8.98$ $b = 5.61–5.64$ $c = 10.17–10.30$ $\beta = 115°24'$
95.	Allanite *$(Ca, REE)_2(Fe^{2+}, Fe^{3+})(Al, Fe)_2(Si_2O_7)(SiO_4)O(OH)$	Monoclinic 2/*m* $P2_1/m$	$a = 8.98$ $b = 5.75$ $c = 10.33$ $\beta = 115°$
96.	Zoisite $Ca_2Al_3(Si_2O_7)(SiO_4)O(OH)$	Orthorhombic $\frac{2}{m}\frac{2}{m}\frac{2}{m}$ P*nma*	$a = 16.2 - 16.3$ $b = 5.45–5.63$ $c = 10.04–10.21$
CYCLOSILICATES			
97.	TOURMALINE GROUP: Dravite $NaMg_3Al_6Be_3Si_6O_{27}(OH, F)_4$		
98.	Schorl $Na(Fe, Mn)_3Al_6Be_3Si_6O_{27}(OH, F)_4$	Trigonal 3m R3*m*	$a = 15.84–16.03$ $c = 7.10–7.25$
99.	Elbaite $Na(Li, Al)_3Al_6Be_3Si_6O_{27}(OH, F)_4$		

*REE, rare-earth elements.

Specific gravity (G), hardness (H)	*Cleavage*	*Luster, color, streak*	*Refractive index, optic sign, 2V*	*Paragenesis*
$G = 3.13$–3.16 $H = 6\frac{1}{2}$–$7\frac{1}{2}$	{110} Good {100} Poor	Vitreous, pink, white gray, white streak	$\alpha = 1.629$–1.640 $\beta = 1.633$–1.653 $\gamma = 1.638$–1.660 2V = 73–86° (−)	Low- to medium-grade contact metamorphic argillaceous rocks
$G = 3.53$–3.65 $H = 5$ ∥ blade 7 ⊥ blade	{100} Perfect {010} Good	Vitreous, white, blue, reddish, gray, brown, white streak	$\alpha = 1.712$–1.718 $\beta = 1.721$–1.723 $\gamma = 1.727$–1.734 2V = 82–83° (−)	Regional metamorphism of pelitic rocks
$G = 3.49$–3.57 $H = 8$	{001} Perfect	Vitreous, colorless, pastel white streak	$\alpha = 1.606$–1.629 $\beta = 1.609$–1.631 $\gamma = 1.616$–1.638 2V = 48–68° (+)	Veins and cavities in felsic igneous rocks
$G = 3.74$–3.83 $H = 7\frac{1}{2}$	{010} Fair	Vitreous, reddish to yellowish brown, white streak	$\alpha = 1.739$–1.747 $\beta = 1.745$–1.753 $\gamma = 1.752$–1.761 2V = 82–90° (+)	Medium-grade regionally metamorphosed argillaceous sediments
$G = 3.45$–3.55 $H = 5$	{110} Good	Adamantine, colorless, green, brown, black, light gray streak	$\alpha = 1.843$–1.950 $\beta = 1.870$–2.034 $\gamma = 1.943$–2.110 2V = 17–40° (+)	Accessory in igneous and metamorphic rocks
$G = 3.05$–3.10 $H = 6$	{100}, {010} Perfect {101} Good	Vitreous, white, bluish, white streak	$\alpha = 1.665$ $\beta = 1.674$ $\gamma = 1.684$ 2V = 76–87° (+)	Low-temperature schists and retrograde mafic rocks
$G = 3.21$–3.38 $H = 6\frac{1}{2}$	{001} Perfect	Vitreous, white, pale yellow, green, white streak	$\alpha = 1.670$–1.715 $\beta = 1.674$–1.725 $\gamma = 1.690$–1.734 2V = 14–90° (+)	Low- to medium-grade regionally metamorphosed igneous and sedimentary rocks.
$G = 3.38$–3.49 $H = 6$	{001} Perfect	Vitreous, green, yellow-green, gray streak	$\alpha = 1.715$–1.751 $\beta = 1.725$–1.784 $\gamma = 1.734$–1.797 2V = 64–90° (−)	Low- to medium-grade regionally metamorphosed igneous and sedimentary rocks
$G = 3.4$–4.2 $H = 5$–$6\frac{1}{2}$	{001} Imperfect {100}, {110} Poor	Vitreous to resinous, brown to black	$\alpha = 1.690$–1.791 $\beta = 1.700$–1.815 $\gamma = 1.706$–1.828 2V = 40–123° (−)(+)	Accessory in felsic igneous rocks, commonly metamict
$G = 3.15$–3.27 $H = 6$	{100} Perfect {001} Imperfect	Vitreous, gray, pinkish, greenish, white streak	$\alpha = 1.685$–1.705 $\beta = 1.688$–1.710 $\gamma = 1.697$–1.725 2V = 0–60° (+)	Medium-grade regional metamorphism of argillaceous calcareous sandstones
$G = 3.03$–3.15 $H = 7$		Vitreous, brown to black, white streak	$\omega = 1.635$–1.661 $\varepsilon = 1.610$–1.632 (−)	
$G = 3.10$–3.25 $H = 7$	{11$\bar{2}$0} {10$\bar{1}$1} Poor	Vitreous, black, blue-black, white streak	$\omega = 1.655$–1.675 $\varepsilon = 1.625$–1.650 (−)	Accessory in granites, pegmatites, and veins, placers
$G = 3.03$–3.10 $H = 7$		Vitreous, colorless, red, blue, yellow, green, white streak	$\omega = 1.640$–1.655 $\varepsilon = 1.615$–1.630 (−)	

	Mineral, composition	System, class, space group	Unit cell dimension (Å)
INOSILICATES			
ORTHOPYROXENES			
100.	Enstatite $MgSiO_3$	Orthorhombic $\frac{2}{m}\frac{2}{m}\frac{2}{m}$ P*cab*	$a = 8.829$ $b = 18.228$ $c = 5.192$
101.	Bronzite *En 70–88%, Fs 12–30%		
102.	Hypersthene *En 50–70%, Fs 30–50%		
103.	Orthoferrosilite $FeSiO_3$	Orthorhombic $\frac{2}{m}\frac{2}{m}\frac{2}{m}$	$a = 18.433$ $b = 9.060$ $c = 5.258$
CLINOPYROXENES			
104.	Clinoenstatite $Mg_2Si_2O_6$	Monoclinic $2/m$ $P2_1/c$	$a = 9.62$ $b = 8.83$ $c = 5.19$ $\beta = 108°21'$
105.	Pigeonite $(Mg, Fe, Ca)(Mg, Fe)Si_2O_6$	Monoclinic $2/m$ $P2_1/c$	$a = 9.73$ $b = 8.95$ $c = 5.26$ $\beta = 108°33'$
106.	Diopside $CaMgSi_2O_6$	Monoclinic $2/m$ $C2/c$	$a = 9.743$ $b = 8.923$ $c = 5.251$ $\beta = 105°50'$
107.	Hedenbergite $CaFeSi_2O_6$	Monoclinic $2/m$ $C2/c$	$a = 9.85$ $b = 9.02$ $c = 5.26$ $\beta = 104°20'$
108.	Augite $(Ca, Mg, Fe, Ti, Al)_2(Si, Al)_2O_6$	Monoclinic $2/m$ $C2/c$	$a \cong 9.8$ $b \cong 9.0$ $c \cong 5.25$ $\beta \cong 105°$
109.	Spodumene $LiAlSi_2O_6$	Monoclinic $2/m$ $C2/c$	$a = 9.50$ $b = 8.30$ $c = 5.24$ $\beta = 110°20'$
110.	Jadeite $NaAlSi_2O_6$	Monoclinic $2/m$ $C2/c$	$a = 9.50$ $b = 8.61$ $c = 5.24$ $\beta = 107°26'$
111.	Aegirine $NaFeSi_2O_6$	Monoclinic $2/m$ $C2/c$	$a = 9.65$ $b = 8.79$ $c = 5.29$ $\beta = 107.4°$
ORTHO-AMPHIBOLE			
112.	Anthophyllite $(Mg, Fe)_7Si_8O_{22}(OH, F)_2$	Orthorhombic $\frac{2}{m}\frac{2}{m}\frac{2}{m}$ P*nma*	$a = 18.5–18.6$ $b = 17.7–18.1$ $c = 5.27–5.32$

*En, enstatite end member; Fs, ferrosilite end member.

Specific gravity (G), hardness (H)	*Cleavage*	*Luster, color, streak*	*Refractive index, optic sign, 2V*	*Paragenesis*
$G = 3.20$ $H = 5–6$	{210} Good	Vitreous, white, gray, green, brown, yellow, white streak	$\alpha = 1.650–1.662$ $\beta = 1.653–1.671$ $\gamma = 1.658–1.680$ 2V = 55–90° (+)	Mafic and ultramafic plutonic rocks, intermediate and mafic volcanic rocks, high-grade metamorphic rocks
$G = 3.96$ $H = 5–6$	{210} Good	Vitreous, green, dark brown, white streak	$\alpha = 1.755–1.762$ $\beta = 1.763–1.770$ $\gamma = 1.772–1.778$ 2V = 55–90° (+)	
$G = 3.19$ $H = 6$	{110} Good	Vitreous, colorless, yellowish, greenish, white streak	$\alpha = 1.651$ $\beta = 1.654$ $\gamma = 1.660$ 2V = 53½° (+)	Mafic igneous rocks, stony meteorites
$G = 3.30–3.46$ $H = 6$	{110} Good	Vitreous, brown, dark green, black, white streak	$\alpha = 1.682–1.722$ $\beta = 1.684–1.722$ $\gamma = 1.705–1.751$ 2V = 0–30° (+)	Extrusive igneous rocks
$G = 3.22–3.38$ $H = 5½–6½$	{110} Good	Vitreous, white, green, white streak	$\alpha = 1.664$ $\beta = 1.672$ $\gamma = 1.694$ 2V = 59° (+)	Skarns and metamorphosed siliceous limestones and dolomites
$G = 3.50–3.56$ $H = 6$	{110} Good	Vitreous, brown, green, dark green, black, white streak	$\alpha = 1.716–1.726$ $\beta = 1.723–1.730$ $\gamma = 1.741–1.751$ 2V = 52–62° (+)	Skarns, metamorphosed iron-rich sediments, also granophyres
$G = 2.96–3.50$ $H = 5½–6$	{110} Good	Vitreous, brown, dark green, black, white streak	$\alpha = 1.662–1.735$ $\beta = 1.672–1.741$ $\gamma = 1.703–1.761$ 2V = 25–50° (+)	Characteristic pyroxene of mafic igneous rocks, also ultramafics and intermediates
$G = 3.03–3.22$ $H = 6½–7$	{110} Good	Vitreous, gray, pale blue, pale green, yellowish-white streak	$\alpha = 1.648–1.663$ $\beta = 1.655–1.669$ $\gamma = 1.662–1.679$ 2V = 58–68° (+)	Li-bearing granite pegmatites
$G = 3.24–3.43$ $H = 6$	{110} Good	Vitreous, white, bluish green, green, white streak	$\alpha = 1.640–1.658$ $\beta = 1.645–1.663$ $\gamma = 1.652–1.673$ 2V = 67–70° (+)	High- and low-grade metamorphic rocks
$G = 3.55–3.60$ $H = 6$	{110} Good	Vitreous, dark green to greenish black, white streak	$\alpha = 1.750–1.776$ $\beta = 1.780–1.820$ $\gamma = 1.800–1.836$ 2V = 60–70° (−)	Alkaline plutonic rocks
$G = 2.85–3.57$ $H = 5½–6$	{210} Perfect {010}, {100} Fair	Vitreous, white, gray, green, brown, white streak	$\alpha = 1.596–1.694$ $\beta = 1.605–1.710$ $\gamma = 1.615–1.722$ 2V = 78–110° (−)(+)	Medium- to high-grade metamorphic rocks

	Mineral, composition	System, class, space group	Unit cell dimension (Å)
CLINO-AMPHIBOLES			
113.	Tremolite $Ca_2Mg_5Si_8O_{22}(OH, F)_2$	Monoclinic 2/*m* C2/*m*	**a** ≅ 9.85 **b** ≅ 18.1 **c** ≅ 5.3 β ≅ 104°50′
114.	Actinolite $Ca_2(Mg, Fe)_5Si_8O_{22}(OH, F)_2$		
115.	Hornblende (common) $(Ca, Na, K)_{2-3}(Mg, Fe, Al)_5Si_6(Al, Si)_2O_{22}(OH, F)_2$	Monoclinic 2/*m* C2/*m*	**a** ≅ 9.9 **b** ≅ 18.0 **c** ≅ 5.3 β ≅ 105°
116.	Hornblende (basaltic) $Ca_2(Na, K)_{\frac{1}{2}-1}(Mg, Fe)_{3-4}(Fe, Al)_{2-1}(Si_6Al_2)O_{22}(O, OH, F)_2$	Monoclinic 2/*m* C2/*m*	**a** ≅ 10.0 **b** ≅ 18.1 **c** ≅ 5.35 β ≅ 106°
117.	Glaucophane $Na_2Mg_3Al_2Si_8O_{22}(OH)_2$	Monoclinic 2/*m* I2/*m*	**a** = 9.99 **b** = 17.92 **c** = 5.27 β = 108°22′
118.	Riebeckite $Na_2Fe^{2+}_3Fe^{3+}_2Si_8O_{22}(OH, F)_2$	Monoclinic 2/*m* C2/*m*	**a** ≅ 9.75 **b** ≅ 18.0 **c** ≅ 5.3 β ≅ 103°
OTHER INOSILICATE			
119.	Wollastonite $CaSiO_3$	Triclinic $\bar{1}$ P$\bar{1}$	**a** = 7.94 **b** = 7.32 **c** = 7.07 α = 90°3′ β = 95°22′ γ = 103°26′
PHYLLOSILICATES			
120.	Kaolinite $Al_4(Si_2O_5)_2(OH)_8$	Triclinic $\bar{1}$ P$\bar{1}$	**a** = 5.155 **b** = 8.959 **c** = 7.407 α = 91°41′ β = 104°52′ γ = 89°56′
121.	Antigorite $Mg_6(Si_2O_5)_2(OH)_8$	Monoclinic 2/*m* C*m*	**a** ≅ 5.3 **b** ≅ 9.2 **c** ≅ 7.3 β ≅ 90°
122.	Chrysotile $Mg_6(Si_2O_5)_2(OH)_8$	Monoclinic 2/*m*	**a** ≅ 5.3 **b** ≅ 9.2 **c** ≅ 7.3 β ≅ 90°
123.	Talc $Mg_3(Si_2O_5)_4(OH)_2$	Monoclinic 2/*m* C2/*c*	**a** = 5.28 **b** = 9.15 **c** = 18.9 β = 99°30′
124.	Muscovite $K_2Al_4(Si_3AlO_{10})_2(OH, F)_4$	Monoclinic 2/*m* C2/*c*	**a** = 5.203 **b** = 8.995 **c** = 20.03 β = 94°28′
125.	Phlogopite $K_2Mg_6(AlSi_3O_{10})_2(OH, F)_4$	Monoclinic 2/*m* C*m*	**a** = 5.314 **b** = 9.204 **c** = 10.314 β = 99°54′

Specific gravity (G), hardness (H)	*Cleavage*	*Luster, color, streak*	*Refractive index, optic sign, 2V*	*Paragenesis*
$G = 3.02–3.44$ $H = 5–6$	{110} Good	Vitreous, white or gray, white streak Vitreous, light to dark green, white streak	$\alpha = 1.599–1.688$ $\beta = 1.612–1.697$ $\gamma = 1.622–1.705$ 2V = 86–65° (−)	Typical of medium-grade metamorphic rocks
$= 3.02–3.45G$ $H = 5–6$	{110} Good	Vitreous, green, dark green, black, white streak	$\alpha = 1.615–1.705$ $\beta = 1.618–1.714$ $\gamma = 1.632–1.730$ 2V = 95–27° (−)(+)	Felsic to intermediate plutonic rocks, schists, retrograde alteration of augite
$G = 3.19–3.30$ $H = 5–6$	{110} Perfect	Vitreous, dark brown to black, white streak	$\alpha = 1.662–1.690$ $\beta = 1.672–1.730$ $\gamma = 1.680–1.760$ 2V = 60–82° (−)	Mafic volcanic rocks, also known as *oxyhornblende*
$G = 3.08–3.30$ $H = 6$	{110} Good	Vitreous, gray, pale blue, white streak	$\alpha = 1.606–1.661$ $\beta = 1.622–1.667$ $\gamma = 1.627–1.670$ 2V = 50–0° (−)	Metamorphic rocks
$G = 3.08–3.30$ $H = 5$	{110} Good	Vitreous, dark blue, black, white streak	$\alpha = 1.654–1.701$ $\beta = 1.662–1.711$ $\gamma = 1.668–1.717$ 2V = 80–90° (−)(+)	Iron-rich felsic igneous rocks
$G = 2.87–3.09$ $H = 4\frac{1}{2}–5$	{100} Perfect {001}, {$\bar{1}$02} Good	Vitreous, white, grayish green, white streak	$\alpha = 1.616–1.640$ $\beta = 1.628–1.650$ $\gamma = 1.631–1.653$ 2V = 38–60° (−)	Metamorphosed impure limestones, neither pyroxene nor amphibole
$G = 2.61–2.68$ $H = 2–2\frac{1}{2}$	{001} Perfect	Earthy, white, gray, white streak	$\alpha = 1.553–1.565$ $\beta = 1.559–1.569$ $\gamma = 1.560–1.570$ 2V = 24–50° (−)	Weathering or hydrothermal alteration of alumino-silicates
$G = 2.6$ $H = 2\frac{1}{2}–3\frac{1}{2}$	{001} Perfect	Resinous to waxy, white, green, white streak	$\alpha = 1.588–1.567$ $\beta \cong 1.565$ $\gamma = 1.562–1.574$ 2V = 37–61° (−)	Hydrothermal alteration of ultramafic rocks, also low-grade metamorphism, hydrothermal veins
$G \cong 2.55$ $H = 2\frac{1}{2}$	Fibrous	Resinous, white, gray, green, shining white streak	$\alpha = 1.532–1.549$ $\beta = −$ $\gamma = 1.545–1.556$ 2V = − (−)	Hydrothermal veins
$G = 2.58–2.83$ $H = 1$	{001} Perfect	Pearly to dull, white, greenish, brownish-white streak	$\alpha = 1.539–1.550$ $\beta = 1.589–1.594$ $\gamma = 1.589–1.600$ 2V = 0–30° (−)	Metamorphic as soapstone and steatite
$G = 2.77–2.88$ $H = 2\frac{1}{2}–3$	{001} Perfect	Vitreous to pearly, colorless or pale green, red, brown, white streak	$\alpha = 1.552–1.574$ $\beta = 1.582–1.610$ $\gamma = 1.587–1.616$ 2V = 30–47° (−)	Most common mica, accessory in granites, abundant in schists and some sediments
$G = 2.76–2.90$ $H = 2–2\frac{1}{2}$	{001} Perfect	Pearly to submetallic, yellowish, brown, green, reddish brown, dark brown, coppery	$\alpha = 1.530–1.590$ $\beta = 1.557–1.637$ $\gamma = 1.558–1.637$ 2V = 0–15° (−)	In metamorphosed limestones and in ultramafic rocks

	Mineral, composition	System, class, space group	Unit cell dimension (Å)
126.	Biotite $K_2(Mg, Fe)_{6-4}(Fe, Al, Ti)_{0-2}(Si_{6-5}Al_{2-3}O_{20})O_{0-2}(OH, F)_{4-2}$	Monoclinic $2/m$ Cm	$a \cong 5.3$ $b \cong 9.2$ $c \cong 10.2$ $\beta \cong 100°$
127.	Lepidolite $K_2(Li, Al)_{5-6}(Si_{6-7}Al_{2-1}O_{20})(OH, F)_4$	Monoclinic (or Trigonal) $2/m$ (or 32) Cm (or $P3_1/12$)	$a \cong 5.3$ $b \cong 9.2$ $c \cong 10.2$ $\beta \cong 100°$
128.	Glauconite $(K, Na, Ca)_{1.2-2}(Fe, Al, Mg)_4(Si_{7-7.6}Al_{1-.4}O_{20})(OH)_4$	Monoclinic $2/m$ Cm or $C2/m$	$a = 5.25$ $b = 9.09$ $c = 10.03$ $\beta = 95°$
129.	Montmorillonite $(\frac{1}{2}Ca, Na)_{0.7}(Al, Mg, Fe)_4(Si, Al)_8O_{20}(OH)_4 \cdot nH_2O$	Monoclinic $2/m$	$a = 5.23$ $b = 9.06$ $c =$ variable $\beta = 90°$
130.	Vermiculite $(Mg, Fe, Al)_6(Al, Si)_8O_{20}(OH)_4 \cdot 8H_2O$	Monoclinic m Cc	$a \cong 5.3$ $b \cong 9.2$ $c \cong 28.9$ $\beta = 97°$
131.	Margarite $Ca_2Al_4(SiAlO_5)_4(OH)_4$	Monoclinic $2/m$ $C2/c$	$a = 5.13$ $b = 8.92$ $c = 19.5$ $\beta = 100°$
132.	Chlorite $(Mg, Al, Fe)_{12}(Si, Al)_8O_{20}(OH)_{16}$	Monoclinic $2/m$ $C2/m$ or $C2/c$	$a \cong 5.3$ $b \cong 9.2$ $c \cong 14.3$ $\beta \cong 97°$
133.	Apophyllite $KFCa_4(Si_2O_5)_4 \cdot 8(H_2O)$	Tetragonal $\frac{4}{m}\frac{2}{m}\frac{2}{m}$ $P4/mnc$	$a = 9.00$ $c = 15.84$
TEKTOSILICATES			
FELDSPATHOIDS			
134.	Nepheline $Na_3K(AlSiO_4)_4$	Hexagonal 6 $C6_3$	$a = 9.99$ $c = 8.33$
135.	Analcime $NaAlSi_2O_6 \cdot H_2O$	Isometric $\frac{4}{m}\bar{3}\frac{2}{m}$ $Ia3d$	$a = 13.7$
136.	Leucite $KAlSi_2O_6$	Tetragonal $4/m$ $I4_1/a$	$a = 13.0$ $c = 13.8$
FELDSPARS			
137.	Sanidine $(K, Na)AlSi_3O_8$	Monoclinic $2/m$ $C2/m$	$a = 8.617$ $b = 13.030$ $c = 7.175$ $\beta = 116°5'$
138.	Orthoclase $KAlSi_3O_8$	Monoclinic $2/m$ $C2/m$	$a = 8.562$ $b = 12.996$ $c = 7.139$ $\beta = 116°1'$

Specific gravity (G), hardness (H)	*Cleavage*	*Luster, color, streak*	*Refractive index, optic sign, 2V*	*Paragenesis*
G = 2.7–3.3 H = $2\frac{1}{2}$–3	{001} Perfect	Vitreous to pearly, black, dark brown, dark green, white streak	α = 1.565–1.625 β = 1.605–1.696 γ = 1.605–1.696 2V = 0–25° (−)	Most common dark mica, accessory in felsic igneous rocks, common in schists, also kimberlites
G = 2.80–2.90 H = $2\frac{1}{2}$–4	{001} Perfect	Pearly, shades of pink and purple, white streak	α = 1.525–1.548 β = 1.551–1.585 γ = 1.554–1.587 2V = 0–58° (−)	In Li-rich granite pegmatites
G = 2.4–2.95 H = 2	{001} Perfect	Pearly to earthy, colorless, yellow, green, blue green, white streak	α = 1.592–1.610 β = 1.614–1.641 γ = 1.614–1.641 2V = 0–20° (−)	Marine sediments, weathering product as *celadonite*
G = 2–3 H = 1–2	{001} Perfect	Earthy, white, pale yellow or green, white streak	α = 1.48–1.61 β = 1.50–1.64 γ = 1.50–1.64 2V = small (−)	Weathering product in clays and soils
G = 2.3 $H \simeq 1\frac{1}{2}$	{001} Perfect	Pearly to bronze-like, yellow, green, brown, white streak	α = 1.525–1.564 β = 1.545–1.585 γ = 1.545–1.582 2V = 0–8° (−)	In soils, as alteration product of biotite, in mafic pegmatites
G = 3.0–3.1 H = $3\frac{1}{2}$–$4\frac{1}{2}$	{001} Perfect	Pearly, grayish pink, yellowish, greenish, white streak	α = 1.630–1.638 β = 1.642–1.648 γ = 1.644–1.650 2V = 40–67° (−)	In metamorphic emery deposits and some schists
G = 2.6–3.3 H = 2–3	{001} Perfect	Vitreous to pearly, green, brown, yellow, white, white streak	α = 1.57–1.67 β = 1.57–1.69 γ = 1.57–1.69 2V = 0–60° (+) (−)	Low-grade schists, hydrothermal alteration of mafic minerals, sediments
G = 2.33–2.37 H = $4\frac{1}{2}$–5	{001} Perfect	Vitreous, colorless, pastel, white streak	ω = 1.534–1.535 ε = 1.535–1.537 (+)	Hydrothermal in cavities in basalts
G = 2.56–2.66 H = $5\frac{1}{2}$–6	{10$\bar{1}$0} {0001} Poor	Vitreous to greasy, colorless, white, gray, white streak	ω = 1.529–1.546 ε = 1.526–1.542 (−)	Silica-undersaturated igneous and metamorphic rocks
G = 2.24–2.29 H = $5\frac{1}{2}$	{001} Very poor	Vitreous, white, off-white, white streak	n = 1.479–1.493	In cavities in igneous rocks
G = 2.47–2.50 H = $5\frac{1}{2}$–6	{110} Very poor	Vitreous, white, gray, white streak	$\omega \simeq \varepsilon$ = 1.508–1.511	K-rich, Si-poor volcanic rocks
G = 2.56–2.62 H = 6	{001}, {010} Perfect	Vitreous, white, off-white, white streak	α = 1.518–1.527 β = 1.522–1.532 γ = 1.523–1.534 2V = 18–54° (−)	Felsic volcanic rocks
G = 2.55–2.63 H = 6	{001}, {010} Perfect	Vitreous, white, yellow, red, green, white streak	α = 1.518–1.529 β = 1.522–1.533 γ = 1.522–1.539 2V = 33–84° (−)	Felsic igneous rocks, schists and gneisses

		Mineral, composition	System, class, space group	Unit cell dimension (Å)
139.		Microcline $KAlSi_3O_8$	Triclinic $\bar{1}$ $C\bar{1}$	$a = 8.57$ $b = 12.98$ $c = 7.22$ $\alpha = 90°41'$ $\beta = 115°59'$ $\gamma = 87°30'$
140.	PLAGIOCLASE GROUP	Albite $NaAlSi_3O_8$	Triclinic $\bar{1}$ $C\bar{1}$	$a = 8.14$ $b = 12.79$ $c = 7.16$ $\alpha = 94°34'$ $\beta = 116°31'$ $\gamma = 87°42'$
141.		Oligoclase *An 10–30		
142.		Andesine *An 30–50		
143.		Labradorite *An 50–70		
144.		Bytownite *An 70–90		
145.		Anorthite $CaAl_2Si_2O_8$	Triclinic $\bar{1}$ $P\bar{1}$	$a = 8.18$ $b = 12.88$ $c = 7.08$ $\alpha = 93°10'$ $\beta = 115°51'$ $\gamma = 91°13'$
ZEOLITES				
146.		Natrolite $Na_2Al_2Si_3O_{10}\cdot 2H_2O$	Orthorhombic $2mm$ $Fdd2$	$a = 18.30$ $b = 18.63$ $c = 6.60$
147.		Phillipsite $(\frac{1}{2}Ca, Na, K)_3Al_3Si_5O_{16}\cdot 6H_2O$	Monoclinic $2/m$ $P2_1/m$ or $P2_1$	$a = 10.02$ $b = 14.28$ $c = 8.64$ $\beta = 125°40'$
OTHER TEKTOSILICATES				
148.		Beryl $Al_2Be_3Si_6O_{18}$	Hexagonal $\frac{6}{m}\frac{2}{m}\frac{2}{m}$ $C6/mcc$	$a = 9.215$ $c = 9.192$
149.		Cordierite $Mg_2Al_3(AlSi_5O_{18})$	Orthorhombic $\frac{2}{m}\frac{2}{m}\frac{2}{m}$ $Cccm$	$a = 9.739$ $b = 17.08$ $c = 9.345$
150.		Lazurite $(Na, Ca)_8(AlSiO_4)_6(S, SO_4, Cl)_2$	Isometric $\bar{4}3m$ $P\bar{4}3m$	$a = 9.13$

*Percentage of anorthite (An) in plagioclase.

Specific gravity (*G*), *hardness* (*H*)	*Cleavage*	*Lustes, color, streak*	*Refractive index, optic sign, 2V*	*Paragenesis*
$G = 2.56–2.63$ $H = 6$	{001}, {010} Perfect	Vitreous, white, yellow, red, green, white streak	$\alpha = 1.514–1.529$ $\beta = 1.518–1.533$ $\gamma = 1.521–1.539$ $2V = 66–84°\ (-)$	Felsic igneous rocks, hydrothermal veins, schists and gneisses
$G = 2.63$ $H = 6$	{001} Perfect {010} Good	Vitreous, colorless, white, gray, white streak	$\alpha = 1.527$ $\beta = 1.531$ $\gamma = 1.538$ $2V = 77°\ (+)$	Plagioclases are commonest rock-forming minerals, although Ca-rich varieties are relatively rare; albite characterizes felsic igneous rocks, pegmatites, veins, and low-grade metamorphic rocks, also sediments; oligoclase and andesine characterize intermediate igneous and medium-grade metamorphics; calcic plagioclase is found in mafic and ultramafic rocks and in meteorites
$G = 2.76$ $H = 6$	{001} Perfect {010} Good	Vitreous, white, gray, white streak	$\alpha = 1.577$ $\beta = 1.585$ $\gamma = 1.590$ $2V = 78°\ (-)$	
$G = 2.24$ $H = 5$	{110}, {1$\bar{1}$0} Very good	Vitreous to pearly, colorless, white, gray, white streak	$\alpha = 1.473–1.489$ $\beta = 1.476–1.491$ $\gamma = 1.485–1.501$ $2V = 58–64°\ (+)$	Hydrothermal veins, vesicles, and fissures in basalts and syenites
$G = 2.2$ $H = 4–4\frac{1}{2}$	{010}, {100} Good	Vitreous, white, pink, white streak	$\alpha = 1.483–1.504$ $\beta = 1.484–1.509$ $\gamma = 1.486–1.514$ $2V = 60–80°\ (+)$	Cavities in basalts and phonolites
$G = 2.66–2.92$ $H = 7\frac{1}{2}–8$	{0001} Fair	Vitreous, colorless, white, bluish green, yellow, rose, white streak	$\omega = 1.569–1.598$ $\varepsilon = 1.565–1.590\ (-)$	Cavities in granites and pegmatites, also in schists
$G = 2.53–2.78$ $H = 7$	{010} Fair {001}, {100} Poor	Vitreous, shades of blue, white streak	$\alpha = 1.522–1.558$ $\beta = 1.524–1.574$ $\gamma = 1.527–1.578$ $2V = 65–104°\ (-)(+)$	High-grade metamorphism of argillaceous sediments
$G = 2.30–2.50$ $H = 5\frac{1}{2}$	{110} Poor	Vitreous, blue, greenish blue, white streak	$n \cong 1.50$	Contact metamorphism of limestone

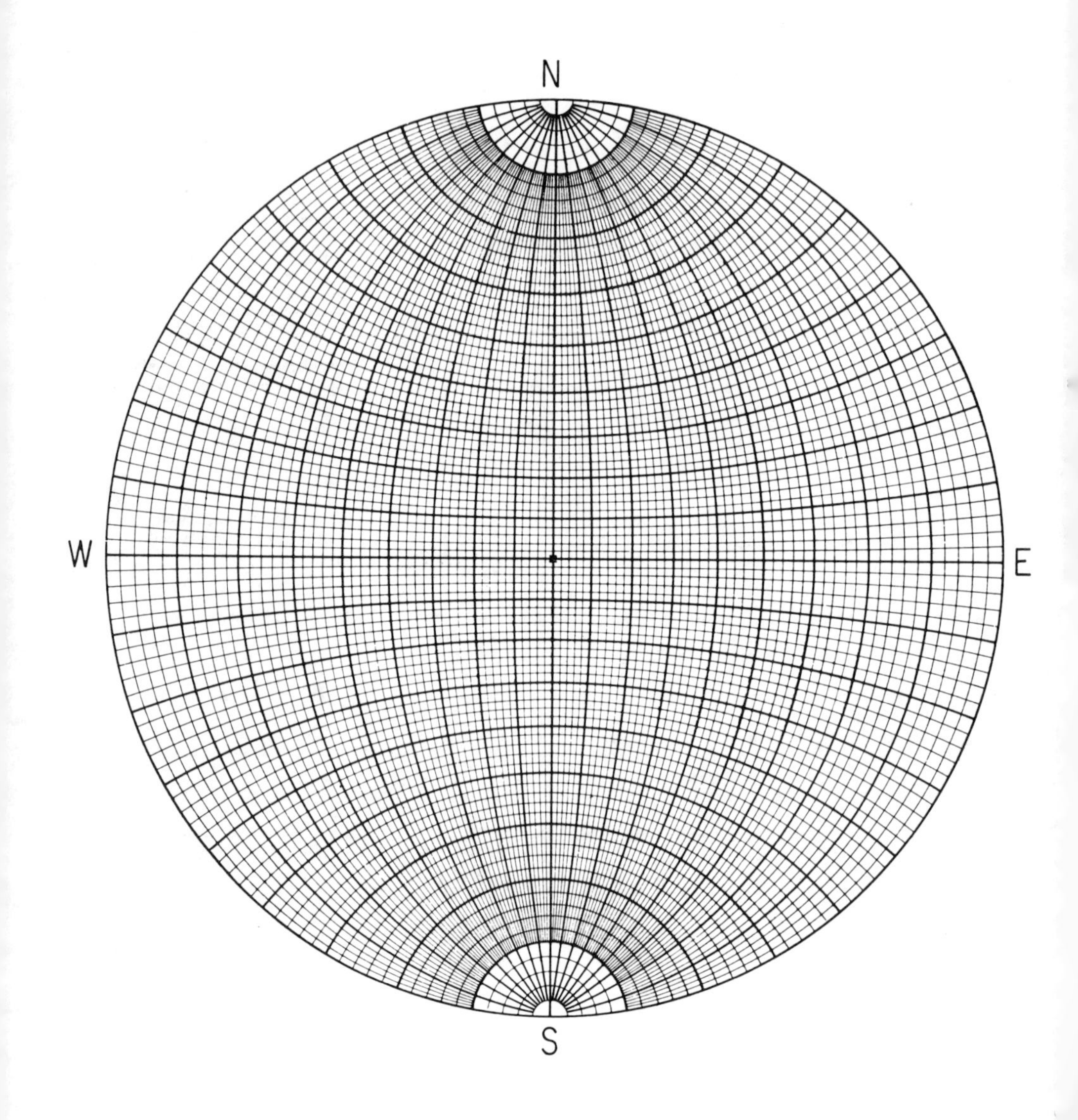
N
W
E
S

Index of Names

Y

Z

Index of Mineral Names

Italic numbers indicate figures.

D

E

F

G

H

General Index